Springer Proceedings in Materials

Volume 39

Springer Proceedings in Materials publishes the latest research in Materials Science and Engineering presented at high standard academic conferences and scientific meetings. It provides a platform for researchers, professionals and students to present their scientific findings and stay up-to-date with the development in Materials Science and Engineering. The scope is multidisciplinary and ranges from fundamental to applied research, including, but not limited to:

- Structural Materials
- Metallic Materials
- Magnetic, Optical and Electronic Materials
- Ceramics, Glass, Composites, Natural Materials
- Biomaterials
- Nanotechnology
- Characterization and Evaluation of Materials
- Energy Materials
- Materials Processing

To submit a proposal or request further information, please contact one of our Springer Publishing Editors according to your affiliation:

European countries: **Mayra Castro** (mayra.castro@springer.com)
India, South Asia and Middle East: **Priya Vyas** (priya.vyas@springer.com)
South Korea: **Smith Chae** (smith.chae@springer.com)
Southeast Asia, Australia and New Zealand: **Ramesh Nath Premnath** (ramesh.premnath@springer.com)
The Americas: **Michael Luby** (michael.luby@springer.com)
China and all the other countries or regions: **Mengchu Huang** (mengchu.huang@springer.com)

This book series is indexed in **SCOPUS** and **EI Compendex** database.

Mulakaluri Rama Manohara Babu ·
Manoj Kumar Buragohain · Srinivas Kuchipudi
Editors

Recent Advances in Composites Engineering and Research

Proceedings of 17th ISAMPE National Conference on Composites (INCCOM-17)

Springer

Editors
Mulakaluri Rama Manohara Babu
Advanced Systems Laboratory
Defence Research and Development Organisation
Hyderabad, India

Manoj Kumar Buragohain
Advanced Systems Laboratory
Defence Research and Development Organisation
Hyderabad, Telangana, India

Srinivas Kuchipudi
Advanced Systems Laboratory
Defence Research and Development Organisation
Hyderabad, India

ISSN 2662-3161 ISSN 2662-317X (electronic)
Springer Proceedings in Materials
ISBN 978-981-99-8806-8 ISBN 978-981-99-8807-5 (eBook)
https://doi.org/10.1007/978-981-99-8807-5

This Springer imprint is published by the registered company Springer Nature Singapore Pte Ltd.
The registered company address is: 152 Beach Road, #21-01/04 Gateway East, Singapore 189721, Singapore

Organization

Program Committee Chairs

Dr. Manoj Kumar Buragohain
Dr. Srinivas Kuchipudi
Dr. Anil Kumar

Program Committee Members

Dr. Rajarajan Arumugham
Dr. Manoj Kumar Buragohain
Chandraprakash C.
Sharada Prabhaker Chavali
Dr. Rohini Devi G.
Ramarao Golla
Prof. Padmanabhan K.
Dr. Basant Kumar
Dr. Srinivas Kuchipudi
Anil Kumar
Prakash Mangalgiri
Sudarshan Nakka
Dr. Padmanabha Raju
Dr. Rama Subba Reddy

Reviewers

Dr. Rajarajan Arumugham
Dr. Manoj Kumar Buragohain
Chandraprakash C.
Dr. Sri Chaitanya Ch
Sharada Prabhaker Chavali
Dr. Rohini Devi G.
Ramarao Golla
Dr. Ajit Raymond James
Prof. Padmanabhan K.
Dr. Basant Kumar
Dr. Srinivas Kuchipudi
Anil Kumar
Prakash Mangalgiri
Sudarshan Nakka
Dr. Padmanabha Raju
Dr. Rama Subba Reddy
Prof. Suryanarayana S.V.
Dr. Sanjaya Kumar Sahoo
Dr. Sourav Sarkar
Kishora Shetty
Prof. Narendra Babu Simhachalam
Dr. Shylaja Srihari
Dr. Karthikeyan T. V.

Preface

It is with great pleasure that I present the conference proceedings of the 17th ISAMPE National Conference on Composites, held in Hyderabad, hosted by the ISAMPE Hyderabad chapter. The conference, with its theme of *Recent Advances in Composite Engineering and Research* provided a platform for researchers, engineers, academicians, and industry professionals to come together and discuss the latest developments in the field of composites. Composites have revolutionized numerous industries, offering lightweight, high-strength materials with exceptional properties. This conference aimed to explore various aspects of composite engineering, including design, manufacturing, testing, and quality control. The event covered a wide range of topics, from fundamental research to practical applications, highlighting advancements in high-temperature ceramics, elastomers, and carbon-carbon composites.

Over the course of the conference, a total of 72 papers were presented and deliberated upon, showcasing the significant research and innovation happening in the field. These papers covered a broad spectrum of topics, providing valuable insights into the latest trends and breakthroughs in composites engineering. The diverse range of presentations ensured that attendees gained a comprehensive understanding of the advancements in the field and their potential impact on various industries.

In addition to the paper presentations, the conference also featured an exhibition, creating a platform for manufacturers to showcase their products and services. This exhibition provided a unique opportunity for participants to explore cutting-edge technologies, materials, and manufacturing processes related to composites. The presence of industry representatives and researchers facilitated valuable interactions, fostering collaborations and knowledge exchange between academia and industry.

I extend our heartfelt gratitude to the organizing committee members, the ISAMPE Hyderabad chapter, and all the sponsors for their invaluable support and dedication in making this conference a resounding success. We also express our sincere appreciation to all the authors and presenters who contributed their research, knowledge, and expertise to enrich the conference proceedings.

Finally, I would like to thank all the participants who attended the conference, actively engaging in discussions, networking, and sharing their insights. I hope that the conference proceedings will serve as a valuable resource for researchers, engineers, and professionals working in the field of composites. May the knowledge shared and the discussions held during this conference inspire further advancements and innovations, propelling the field of composites engineering to new heights.

Hyderabad, India

Dr. Mulakaluri Rama Manohara Babu
Conference Chair and Chief Editor

Introduction

With ever increasing use, composites have now become an integral aspect of technological advancement in almost all industrial sectors—Defence and Aerospace, Automotive, Building and Infrastructure, Marine, Chemicals, Conventional and Non-conventional Energy, and so on. The national scenario, as reflected by tremendous interest shown by industry in and diversification into composites, has been very positive in recent times. Composites technology has witnessed tremendous progress in respect of engineering application that has been supported by commensurate basic and application-oriented research. R&D institutes, academia and industry have been partners in this progression towards technological growth. ISAMPE National Conference on Composites (INCCOM-17) was conducted by the Hyderabad Chapter of ISAMPE during November 2–4, 2022 in Hyderabad. One of the primary objectives of the conference was to facilitate presentations on the developments in composites and exchange of knowledge by professionals from the three sectors mentioned above. And, it was a huge success. There was overwhelming participation on all the three days of the conference by authors, delegates and exhibitors of innovative products and technologies. A large number of abstracts were received. They were reviewed by a panel of experts, and total 61 contributory and 14 invited papers were presented on various aspects ranging from design and analysis, manufacturing and product development, and material characterization and product testing.

This book *Proceedings of the ISAMPE National Conference on Composites (INCCOM-17)* contains thirty-four papers selected from the papers presented during the conference. Contents of the proceedings are divided into three parts—*Mechanics and Analysis of Composites*, *Applications and Product Development*, and *Characterization of Composites.*

The first part is related broadly to the study of performance in terms of stress and strain, buckling behavior, and natural frequency. Different material forms, viz., laminated composites, rubberized textile composites and sandwich have been considered by the authors. Design issues such as joints, delaminations and cut out in different environments have also been addressed.

Mechanics of composites involves study in a building-block approach that starts with the constituents and moves through lamina, laminate, joints and ends finally at a complete product. Stress-strain behavior of composites is studied with experimental data on constituents in **Paper 1**. Process-induced defects have always been a subject of keen interest. Such defects are known to influence stress distribution in the material, which in turn adversely influences performance of the structure. **Paper 2** presents a work on stress distribution with broken fibers of unidirectional composites. Sandwich structures are a unique class of composites; effect of variability in fiber orientation on performance of a conical sandwich structure is discussed in **Paper 3**. Configuration of a joint for high temperature application and study of delamination in a pressure vessel are presented respectively in **Papers 4** and **5**.

The second part is devoted to applications of composites and product development. **Papers 6** and **7** give general description of design and development of composite products. First, an account of integrated design including design specification, material selection, configuration, analysis, manufacturing, and testing of satellite bus structure, and second, design and analysis of a composite tube for under water application are presented. Defects in composite products are not uncommon, and sometimes repair of such defective products becomes necessary. **Paper 8** is a description of repair carried out on a composite pressure vessel. Composite product development has several integral aspects—development of materials, development of processing technologies, and product testing. Process development and associated aspects are addressed in **Papers 9** to **13**. Additive manufacturing, automation in vacuum assisted resin transfer molding (VARTM), microwave assisted curing, process modelling of chemical vapor infusion and mandrel making for rocket motor application are the subjects in these papers. Non-destructive testing (NDT) plays a crucial role in the qualification and acceptance of a composite product. **Paper 14** presents contributions of NDT in the development of dimensionally stable space composite camera structures. Product testing for evaluation of static strength and thermo-structural performance respectively are given in **Papers 15** and **16**. Structural health monitoring of an aircraft wing using fiber brag grating sensors is presented in **Paper 17**.

Material characterization is an integral aspect of composites technology, in general, and composite product development, in particular. Papers on characterization of composites and related topics are presented in the third part. One of the primary objectives of material testing is to generate data for design and analysis. Evaluation of physical, mechanical, thermal, and electro-magnetic properties of Quartz/Cyanate Ester composites is given in **Paper 18**. **Papers 19** to **23** address various issues associated with characterization of composites with unique features and specialty composites. They include: shock response and low-velocity impact characterization of E-glass/epoxy composites, uncertainty of measurement in the evaluation of longitudinal modulus, failure behavior of honeycomb sandwich composites with inserts, and characteristic behavior of composite panel with aluminium facings and polyurethane core under dynamic loading. Work on improvement in characterization techniques has been a continuous process as reflected by **Papers 24** to **26**. Novel techniques presented in these papers are in respect of evaluation of co-efficient of thermal expansion, non-destructive techniques in characterization of interfacial bond and joints, and prediction of propellant characteristics. Progress in materials research has been a continuous process, and development of new and more efficient materials is a major objective of material testing. **Papers 27** to **34** belong broadly to this category. These topics include: basalt fabric composites and metal-intermetallic laminates for impact resistance, synthesis of polymer and polymer-ceramic nanocomposites, and automation in extraction of natural fibers for largescale production. Influence of various factors on performance of composite materials has been a research subject for many. Some of the topics addressed in this respect are: effect of coagent concentration on performance of hydrogenated acrylonitrile butadiene rubber (HNBR), effect of different liquid media on interlaminar shear strength of polymer matrix composites, environmental effect on glass transition temperature and resin matrices on thermal, and effect of resin matrices on thermal, mechanical and ablation properties of carbon fiber reinforced composites.

Mulakaluri Rama Manohara Babu
Manoj Kumar Buragohain
Srinivas Kuchipudi

Contents

Characterisation of Composite Materials

About the Editors

Dr. Mulakaluri Rama Manohara Babu is Distinguished Scientist, Director ASL obtained M.S. in Mechanical Engineering from JNTU, Hyderabad in 1994. He was conferred with Ph.D. in Marine Engineering on Acoustic Emission Testing of Carbon Epoxy Filament Wound Pressure Vessels from College of Engineering, Andhra University in 2014. He became the director of Advanced Systems Laboratory and spearheaded the design and development of various Advanced versions of Long Range weapon systems. He is the recipient of numerous Awards. Later, he took over as the Programme Director—Missile Programme consisting of eight projects across different DRDO labs covering various configurations of missile systems, ground support systems and sites. He is also responsible for the design, development, production and deployment of long range missile systems of various missile variants.

Dr. Manoj Kumar Buragohain is an outstanding scientist at Advanced Systems Laboratory, Defence Research and Development Organization (DRDO), India. He received his B.Sc. in Engineering from National Institute of Technology (NIT) Rourkela, M.Tech. and Ph.D. from Indian Institute of Technology (IIT) Madras, and PGDFA from The Institute of Chartered Financial Analysts of India. Dr. Buragohain has over three decades of hands-on experience in design and development of composite products. His primary research interests are in the fields of geodesic and non-geodesic filament winding, tape winding, rosette lay-up and contact lay-up. Some of his major contributions have been in large size composite pressure vessels, grid-stiffened composite structures, tubular structures, ablative liners, and composite rotor blade. He has several journal and conference publications to his credit and is a life member of Indian Society for Advancement of Materials and Process Engineering (ISAMPE) and Aeronautical Society of India (AeSI). His contributions to the field of composites have been well recognized and he was awarded Laboratory Scientist of the Year Award, National Science Day Commendation Certificate and Silicon Medal, DRDO Award for Performance Excellence, and Agni Award for Excellence in Self-reliance, and DRDo Scientist of the year. His areas of interest includes AL in composites.

Dr. Srinivas Kuchipudi is a senior scientist at Advanced Systems Laboratory, Defence Research and Development Organization (DRDO), India. He received his Ph.D. in Physics from Osmaina University, India and conducted his postdoctoral research at National Tsing-hua University, Taiwan. He has over 20 years experience in non-destructive evaluation of composite structures using various advanced techniques including acousto-ultrasonics, thermography, microwave and nuclear magnetic resonance. He is recipient of several awards during his career. He has over 60 publications in peer reviewed journals of national and international repute.

Mechanics and Analysis of Composites

Study on the Stress–Strain Behaviour of Rubberized Textile Composite Used in Aerospace Applications

S. B. M. Guruvayurappan, S. P. Sivapirakasam, R. Suresh, Adarsh Kumar Sahu, and S. Ratheesh

1 Introduction

Rubberized textile composites (RTCs) offer high strength-to-weight ratio, which enables them to be suitable for realizing inflatable structures in aerospace applications. In RTC the fibre reinforcement phase is used in the form of fabrics made by the textile manufacturing processes. The nylon fabric provides requisite strength to composite, whereas rubber matrix provides necessary impermeability to enable them to achieve required shape. In woven fabric, two sets of yarns pass over and under in a specific pattern. Generally, warp yarns run along the length and weft yarn along the width. Two sets of yarns are at right angles to each other in the plane of fabric.

The RTCs essentially find applications in the launch vehicle and spacecraft in the form of collar sets, thermal screen, floats for recovery, inflatable aerodynamic decelerator, and bellows for structural testing of hardware and inflatable antennas. The stress–strain behaviour of RTC is essential not only to predict structural response of the articles but also it is essential to achieve correct shape and size upon inflation. The classical laminate, lamina, has very good elastic properties only in the direction of fibre orientation and poor properties in the transverse and out-of-plane direction. The effective stiffness and strength properties under transverse tension are derived from an analytical and experimental test campaign. In the present paper, the modulus of the constituents is evaluated experimentally, and these data are used to obtain the response of the RTC under mechanical load using macro-mechanics and hyperelastic approaches. The limitations of the approaches are brought out by correlating the

S. B. M. Guruvayurappan (✉) · R. Suresh · A. K. Sahu · S. Ratheesh
Vikram Sarabhai Space Centre, ISRO, Thiruvananthapuram 695022, India
e-mail: guru_gayu@hotmail.com

S. B. M. Guruvayurappan · S. P. Sivapirakasam
Department of Mechanical Engineering, National Institute of Technology, Tiruchirappalli 620015, India

M. R. Manohara Babu et al. (eds.), *Recent Advances in Composites Engineering and Research*, Springer Proceedings in Materials 39,
https://doi.org/10.1007/978-981-99-8807-5_1

obtained results with composite test results. The numerical modelling of RTC using finite element method is also compared with test results.

2 Rubberized Textile Composites

The technical requirements for deployable/inflatable structures are defined as: good foldability, low fabric weight, resilience, heat resistance, low gas permeability, high fabric strength, good abrasion resistance, and stability upon ageing. To satisfy these requirements, an inflatable structure is usually constructed of woven textile membranes that are mostly fabricated from multi-filament nylon yarns. The woven fabric structure studied is a plain weave, where the warp and fill yarns are interlaced in a regular sequence of interchange as shown in Fig. 1.

Woven fabrics exhibit high flexibility, which allows them to fold easily and conform to different shapes. Fundamental parameters contributing to the performance of an inflatable space structures are the mechanical properties of the fabric layer. The fraction of load carried by the fibre in RTC is function of the fibre modulus, matrix modulus, and fibre volume fraction. The response of the constituents, namely nylon fabric and nitrile butadiene rubber–chloroprene rubber (NBR-CR), is evaluated through experimental campaign. The failure strain of nylon fabric in the transverse direction is 0.28%, whereas failure strain of rubber is 8%. Though failure strain of NBR-CR is high, as a composite its failure strain is limited by fabric. It can be noted that for the present nylon reinforced rubber composite, the fibre-to-matrix modulus ratio is more than ten. Similar to polymeric composites, the RTC also exhibits more than 70% of load capacity by fibre even with less than 20% fibre volume fraction. The mechanical behaviour of a RTC is studied using micro-mechanics approach. It can also be noted that behaviour of rubber is that hyperelastic lends itself to both geometric deformations and material nonlinearity of the RTC. The orthotropic nature of fabric and hyperelastic nature of matrix make the micro-mechanics study complex.

Fig. 1 Typical woven fabric structure

The structural response of the composite is captured through mathematical model explained in the subsequent section.

3 Mathematical Model

A plain-woven nylon fibre fabric embedded in a rubber matrix results in plain-woven textile composite (PWTC) or plain-woven lamina. The PWTC-type polymeric composites were idealized by Ishikawa [1] assuming that as a chain of two-layered cross-ply laminates placed alternatively. The concept of classical laminate theory (CLT) is adopted to compute the lower and upper bounds of the averaged stiffness of the woven composite by applying uniform stress and uniform strain conditions, respectively [2]. The numerical modelling methodology using hyperelastic model [3] is also proposed in this section.

3.1 Models for Predicting Transverse (Weft) Directions

The constitutive material behaviour [4] of RTC under transverse tension is derived from inverse rule of mixture, the iso-stress formulation, or series model (S model) for the effective transverse modulus (E2) studied by Jones [5]. These models assumed that both the fibres and matrix are having same cross-sectional area. The fibre-packing geometry is not considered in those models. The experimentally evaluated modulus nylon fabric and NBR-CR are used to obtain the response of the RTC under mechanical load using following models [6].

Series–Parallel (Sp) Model

One of the models used to predict weft properties of RTC is called series–parallel (SP) model, which has been formulated using the representative element as shown in Fig. 2. It is formulated considering square fibre-packing geometry. It is obtained by the conversion of circular cross-sectional area of the fibre into an equivalent square cross-sectional area. The element is divided in to three subregions parallel to the loading direction as shown in Fig. 2. The subregion B is subjected to iso-strain condition as Gibson [7].

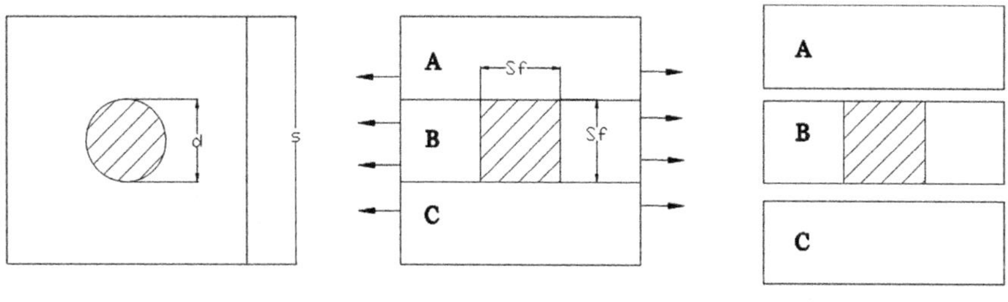

Weft direction modulus in series model is given by,

$$E_2 = \frac{E_m}{1 - \sqrt{V_f}\left(1 - E_m/E_{f2}\right)}.$$

Similarly, using series–parallel model modulus is computed as,

$$E_2 = E_m\left[\left(1 - \sqrt{V_f}\right) + \frac{\sqrt{V_f}}{1 - \sqrt{V_f}\left(1 - E_m/E_{f2}\right)}\right].$$

Parallel-Series (PS) Model

Another model used to predict weft properties of RTC is called parallel-series (PS) model, which has been formulated using the representative element as shown in Fig. 2. The element is divided in to three subregions perpendicular to the loading direction as shown in Fig. 2. The subregion B is subjected to iso-strain condition as Gibson [7].

The equation for transverse modulus is given by,

$$E_2 = E_m\left[\frac{\sqrt{V_f}E_{f2} + \left(1 - \sqrt{V_f}\right)E_m}{V_f E_m + \left(1 - \sqrt{V_f}\right)\left[E_m + \sqrt{V_f}E_{f2}\right]}\right].$$

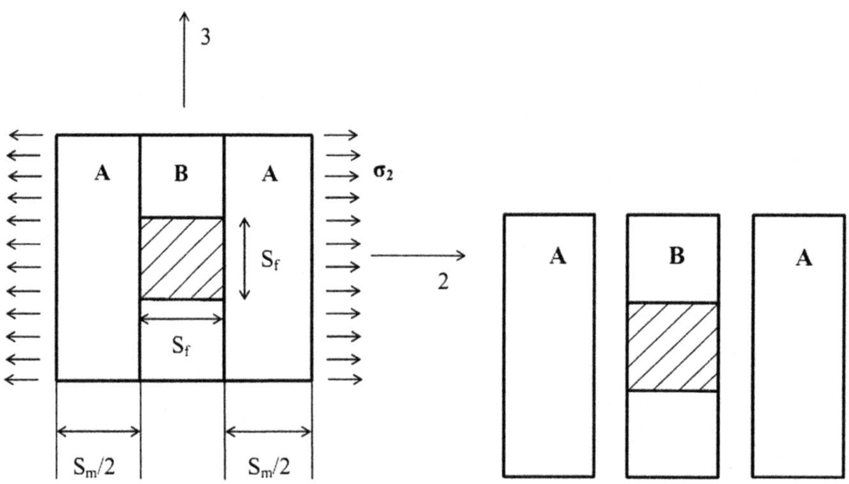

Fig. 2 Parallel-series model

3.2 "Yeoh" Hyperelastic Model

The big challenge in nonlinear elasticity theory is to come up with a reasonable and applicable elastic law (strain energy function), which is the crucial step to the development of dependable analysis tools. A considerable number of attempts have been made to develop a theoretical stress–strain relation that can match the experimental results for hyperelastic materials. Yeoh's third-order constitutive model was one of them, in which the elastic properties of rubber are represented by strain energy functions involving only the first strain invariant, $I1$, and expressed as,

$$W = \sum_{i=1}^{3} C_i (I_1 - 3)^i \tag{1}$$

where C_i is the material constants, and the quantity $2C_1$ can be interpreted as the initial shear modulus. For consistency with linear elasticity, the Yeoh model has to satisfy the following condition:

For third-order Yeoh model, the constitutive model for hyperelastic material can be expressed as

$$W = C_1(I_1 - 3) + C_2(I_1 - 3)^2 + C_3(I_1 - 3)^3 \tag{2}$$

The model is solved for weft direction, and coefficients were derived.

4 Experimental Study

In order to evaluate the transverse modulus of the RTC, characteristics of nylon fabric and NBR-CR are obtained under mechanical load using UTM. The experimental study consists of preparation and testing of nylon fabric, rubber matrix, and RTC composite. Uniaxial tests were performed, and stress–strain curves were generated from the obtained data [8].

4.1 Sample Preparation

Test coupons of size 200 mm × 25 mm are cut from a bigger size roll to carry out the uniaxial test. The sample preparation and experiment for the fabric and RTC are carried out according to ASTM-D751, whereas ASTM-D412 standard is employed to prepare sample to evaluate the properties of NBR/CR.

Fig. 3 Experimental setup

4.2 *Testing Procedure*

Instron 5969 UTM is used to conduct the uniaxial testing of the samples as shown in Fig. 3. The test was carried out by fixing the two ends of the sample in the machine. The gauge length was set as 75 mm. A constant displacement of 100 mm/min is considered for the test specimen, and corresponding load and displacement data were recorded [9].

4.3 *Test Results*

Details of test results of constituents and RTC are discussed in this section.

Nylon Fabric

Three samples are tested for stress versus strain which is shown in Fig. 4. Summary of results is provided in Table 1. It can be observed that the average failure stress and strain of 97.3 MPa and 0.28% respectively are obtained for nylon fabric. The average modulus is 511.5 MPa.

NBR-CR

The stress vrsus strain of five samples is shown in Fig. 5. It can be seen from Table 2 that the average failure stress of 11.45 MPa and strain of 8.43% are obtained for rubber. It can be noted that failure stress is very low compared to reinforcement. The shore hardness of this rubber is in the range of 63.2–65.8.

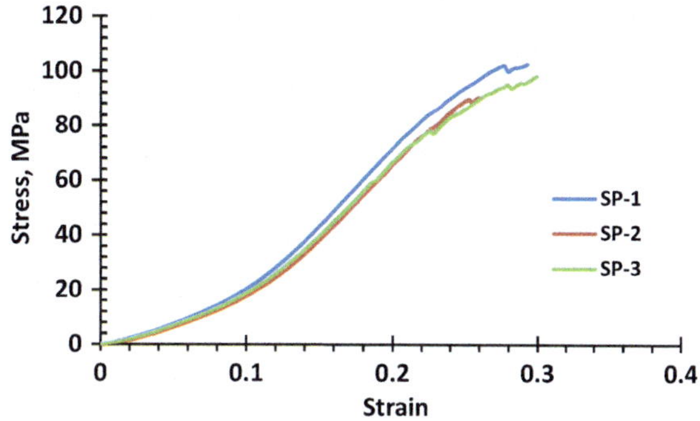

Fig. 4 Stress versus strain plot for nylon fabric

Table 1 Mechanical *properties of nylon fabric*

S. No.	Tensile strength, MPa	Strain	Modulus, MPa
1	102.79	0.29	536.25
2	90.69	0.26	495.57
3	98.49	0.30	502.88
Avg.	97.30	0.28	511.57

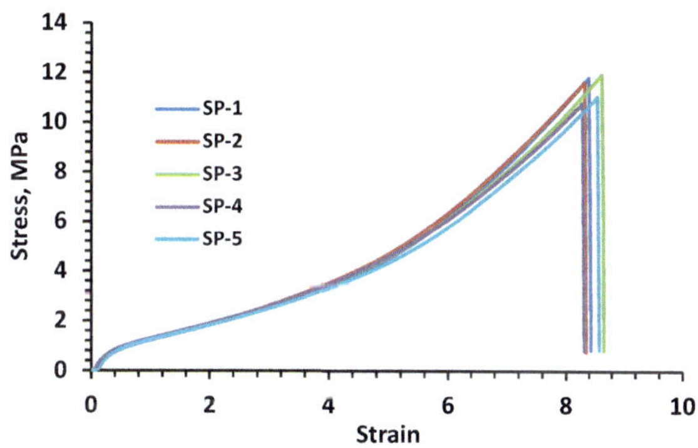

Fig. 5 Stress versus strain plot for NBR-CR

Table 2 Mechanical *properties of NBR-CR*

S. No.	Tensile strength, MPa	Strain	Stress @ 100% strain, MPa	Hardness, shore A
1	11.82	8.39	1.25	63.2–65.8
2	11.65	8.32	1.30	
3	11.95	8.62	1.32	
4	10.80	8.28	1.32	
5	11.05	8.54	1.28	
Avg.	11.45	8.43	1.29	

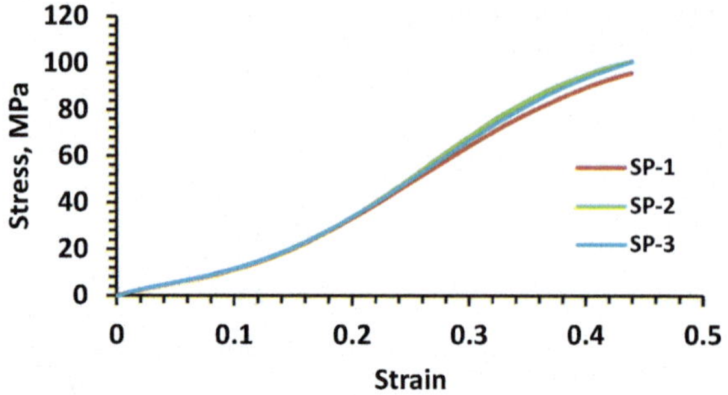

Fig. 6 Stress versus strain plot for RFR RTC in weft direction

Table 3 Mechanical *properties of RFR RTC in weft direction*

S. No.	Tensile strength, MPa	Strain	Modulus, MPa
1	96.06	0.438	300.36
2	101.13	0.439	333.40
3	100.74	0.440	314.86
Avg.	99.31	0.439	316.21

Rubberized Textile Composite
The stress versus strain of three samples is shown in Fig. 6. It can be seen from Table 3 that the average failure stress of 99.31 MPa and strain of 0.43% are obtained for RTC which is very close to nylon fabric. The average modulus is 311.2 MPa.

5 Numerical Results of Mathematical Model

The constituent properties are used to obtain the RTC properties, and obtained transverse modulus are compared with test results obtained through transverse coupon test of RTC samples.

5.1 Series, Series–Parallel (SP), Parallel-Series (PS) Models

Weft direction properties of RTC were predicted using mechanics models, viz., series, series–parallel, and parallel-series. The comparison between the moduli obtained through the models is compared with RTC sample results in the following Fig. 6. It can be seen that nonlinear behaviour could be captured by mathematical model [10], but predicted modulus is high. This may be due to the assumption in the load sharing of the constituent material of RTC (Fig. 7).

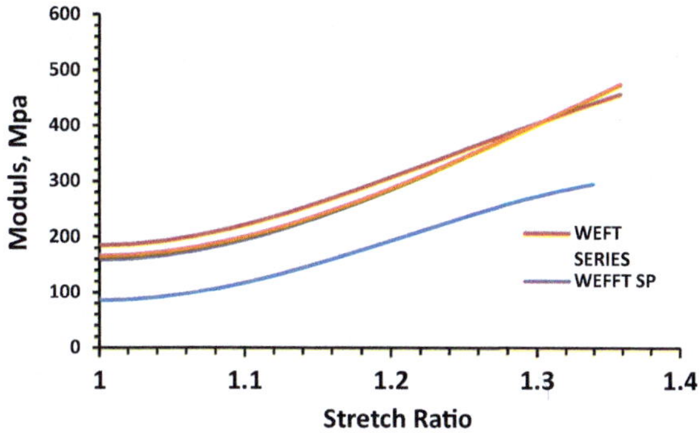

Fig. 7 Comparison of mathematical model and test results of RTC in weft direction

5.2 Yeoh Hyperelastic Models

The model [11] is solved for weft direction using Generalized Reduced Gradient—nonlinear solver using MS Excel—and arrived at convergence after 7 and 9 iterations for RTC. The coefficients of Yeoh material model are derived and provided in Table 4. It can be seen in Fig. 8 that the experimental data are compared well with numerical model RTC in weft direction.

Table 4 Third-order Yeoh model material constants C_1, C_2, and C_3

Volume fraction (v_f)	C_1 (N/mm^2)	C_2 (N/mm^2)	C_3 (N/mm^2)
0.099	15.22	103.58	-91.91

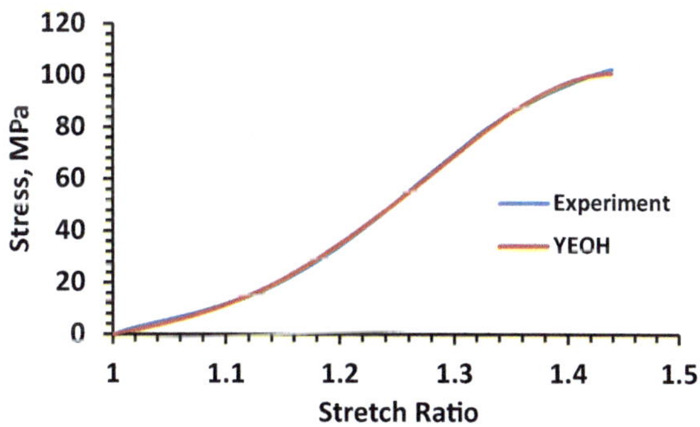

Fig. 8 Experimental data versus numerical model RTC in weft

Fig. 9 FE model details

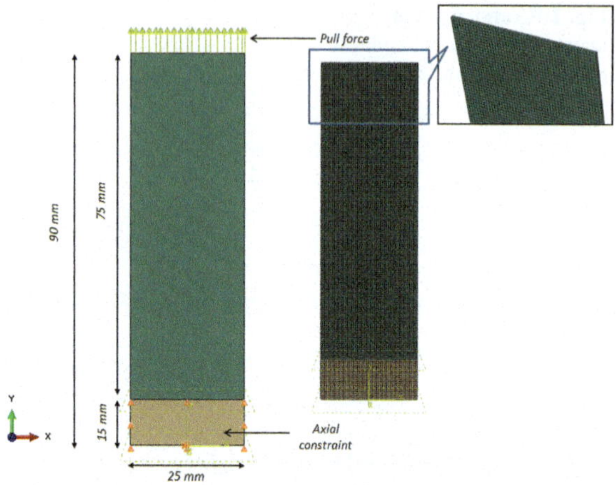

6 Numerical Simulation and Validation Using ABAQUS

The validation of the proposed mathematical model was carried out through FEM [12]-based software ABAQUS to simulate the uniaxial tensile test of fibre reinforced composites. An equivalent shell model was generated to simulate the tensile test coupon for the fibre reinforced composite. Coupon length and width are 90 mm and 25 mm, respectively. Thickness of the RTC is 0.5 mm. The gauge length of 75 mm is simulated in the model. It can be noted that these dimensions are consistent with the dimensions of the tested coupons w.r.t. ASTM standards.

6.1 Loads and Boundary Conditions

The composite was stretched by applying a tensile force at the one edge. This tensile force is equal to the maximum load taken by the composite. Nonlinear simulation is invoked due to hyperelastic material model. At the other end, axial constraint is provided over a region of axial height of 15 mm so as to obtain the gauge length of 75 mm as shown in Fig. 9.

6.2 Results and Discussion

Figure 10 shows the deformed configuration of the RTC-weft composite simulated under uniaxial tensile load. The maximum axial deformation obtained, i.e. 24.4 mm shows a very good match obtained from experiment, i.e. 24.3 mm at the corresponding load of 990 N. Figure 11 shows the comparison between the stress–strain curve obtained from FE analysis and the experimental data for RTC-weft composite. The

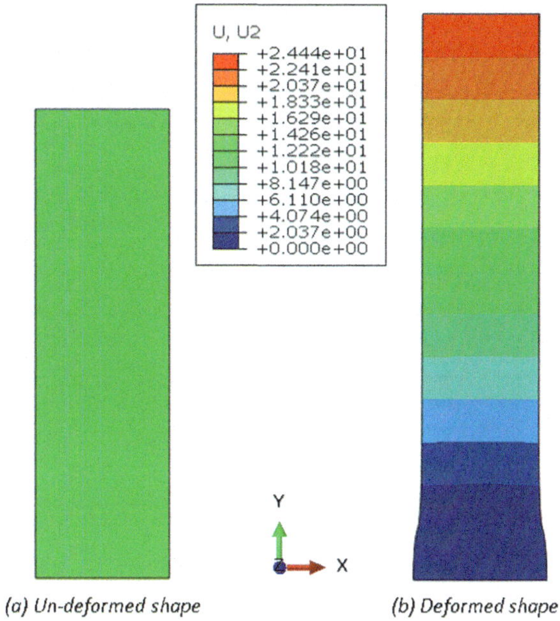

Fig. 10 Axial deformation (mm) of RFR-weft composite

stress–strain curve obtained from FE analysis represents the proposed mathematical model wherein the material constants were derived from the experiments. It can be seen that an excellent match is obtained between the data obtained from FE analysis and the experimental data.

Hence, it can be concluded that the proposed mathematical model accurately captures the behaviour of rubber reinforced nylon fabric.

7 Conclusions

The effective stiffness properties under transverse tension of RTC are derived from an analytical and experimental test campaign. The modulus of the constituents evaluated experimentally is used to obtain the response of the RTC under mechanical load using macro-mechanics, and hyperelastic approaches are discussed in detail. The mathematical model obtained through different combinations of parallel-series models is over-predicting the modulus. The third-order Yeoh material model predicts the RTC characteristic up to the failure strain level. The coefficient of obtained Yeoh material model is used in the finite element simulation, and the numerical and test results are showing good agreement.

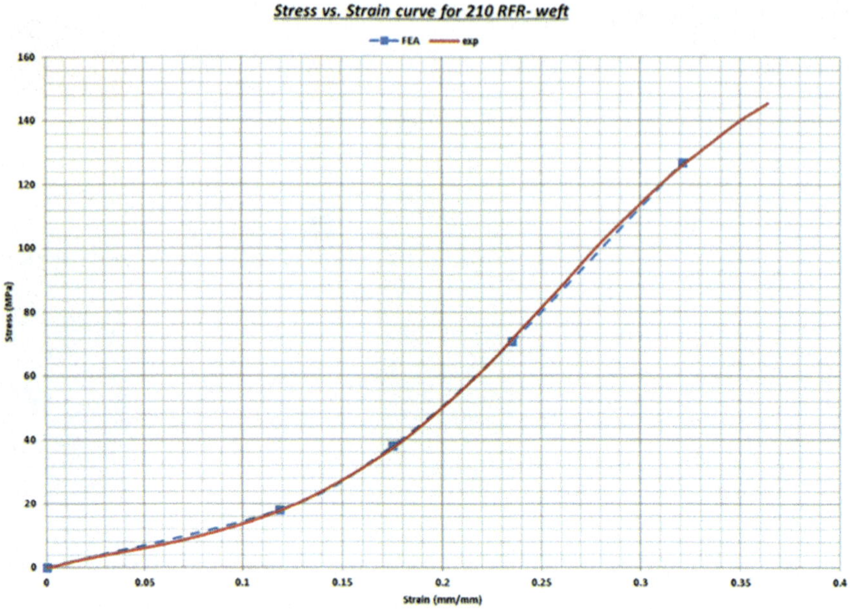

Fig. 11 Stress–strain curve comparison for RTC-weft composite

References

1. Ishikawa (1981) Antisymmetric elastic properties of composite plates of satin weave cloth. Fibre Sci Technol 15:127–145
2. Udhayaraman R, Shantanu S, Mulay (2017). Multi-scale approach based constitutive modelling of plain woven textile composites. Mech Mater 117:172–192
3. Ogden RW (1972) Large deformation isotropic elasticity—on the correlation of theory and experiment for the compressible rubberlike solids. Proc R Soc Lond A 328(1575):565–584
4. Boyce MC, Arruda EM (2000) Constitutive models of rubber elasticity: a review. Rubber Chem Technol 73(3):504–523
5. Jones R (1998) Mechanics of composite materials. CRC Press
6. Whitcomb J, Srirengan K, Chapman C (1995) Evaluation of homogenization for global/local stress analysis of textile composites. Compos Struct 31:137–149
7. Gibson R (2012) Principles of composite material mechanics. CRC Press
8. Chaphalkar P, Kelkar A (1999) Analytical and experimental elastic behaviour of twill woven laminate. In: Proceedings of the 12th international conference on composite materials, Paris, France
9. Selvadurai A (2006) Deflections of a rubber membrane. J Mech Phys Solids 54(6):1093–1119
10. Dixit A, Mali HS (2013) Modelling techniques for predicting the mechanical properties of woven-fabric textile composites: A review. Mech Compos Mater 49:1–20
11. Yeoh OH (1993) Some forms of the strain energy function for rubber. Rubber Chem Technol 66(5):754–771
12. Scida D, Aboura Z (1997) Prediction of the elastic behavior of hybrid and non-hybrid woven composites. Compos Sci Technol 59:1727–1740

Prediction of Stress Distribution Around Broken Fiber in a Unidirectional Composite

Abhay Kaushik Nudurupati and Ashish Mishra

1 Introduction

Vastness of composite applications in the field of material sciences showed an exponential progression to the contemporary times. Unique characteristics of immense strength with the least weight and numerous resistive properties hover through large-scale applications over extensive fields. Production and manufacturing of composites are however a notable challenge with no ideal progress in material planning along the woven fibers in the matrix. Scores of material tests are performed primarily including the non-destructive testing for internal manufacturing defects across the material. Not with any unique defects, the study lines up to an effective and efficient characterization of the effects and the results as a cause of broken fibers in the composite lamina. The three-dimensional model in consideration for the analysis quantifies the study across the fibers with multiple considerations and assumptions. A research gap along the three-dimensional quantitative analysis of material behavior that includes broken fibers is fulfilled. Stress concentration analysis along with deformation plots are presumed for the study to include the behavior of broken as well as the intact fibers. A result-oriented analysis to provide effects of the broken fiber lamina persisting a composite structure in evaluation to the performed study shows forth the composite material performance holding any similar reinforcement fractures [1–3].

A. K. Nudurupati (✉) · A. Mishra
University of Petroleum and Energy Studies, Dehradun 248007, India
e-mail: nabhaykaushik@gmail.com

M. R. Manohara Babu et al. (eds.), *Recent Advances in Composites Engineering and Research*, Springer Proceedings in Materials 39,
https://doi.org/10.1007/978-981-99-8807-5_2

2 Methodology

Composite material manufacturing, including the later stages of applications, proves to be responsible for non-ideal structures with a range of defects, primarily companying broken fibers across the composite material. Three-dimensional analysis of the same is performed with computer-based modeling. A computer-aided design model of the composite lamina is constructed on Autodesk Fusion 360, with further analysis structured on finite element analysis tool, COMSOL Multiphysics. Carbon fibers along with the epoxy matrix pertain to certain elemental properties are opted to form an epoxy-carbon fiber unidirectional composite material. Analysis performed with the role of fiber in the center to be freely suspended on one end, a relative study over results obtained for the model with both broken and intact fiber shows the effect of the broken fiber on the lamina and on the composite as a whole in greater scale. The perfectly bonded fibers to the matrix in the unidirectional composite structure distribute the applied load over the intact fibers to the broken fiber in a proportionately decreasing manner, saturating with successive progression [4–6].

Finite section of composite lamina supports computerized analysis in the company with the material modeling and simulation environment. Intact fibers to the broken fibers represent the nomenclature (see Fig. 1) across the periodic boundaries counting fibers to 'F_n' in either directions (Fig. 2).

Composite lamina dimensions along with parameters that include inter-fiber spacing, fiber length, and thickness show follow-up across the computerized modeling of the composite material.

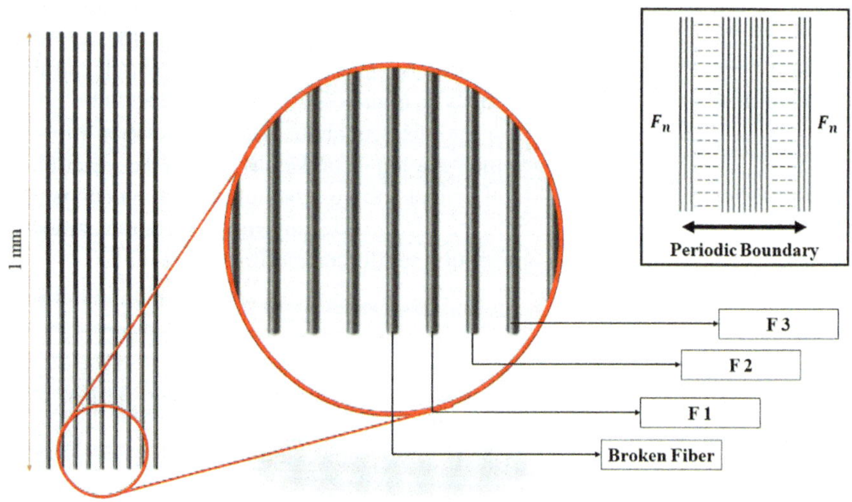

Fig. 1 Front view of unidirectional composite lamina

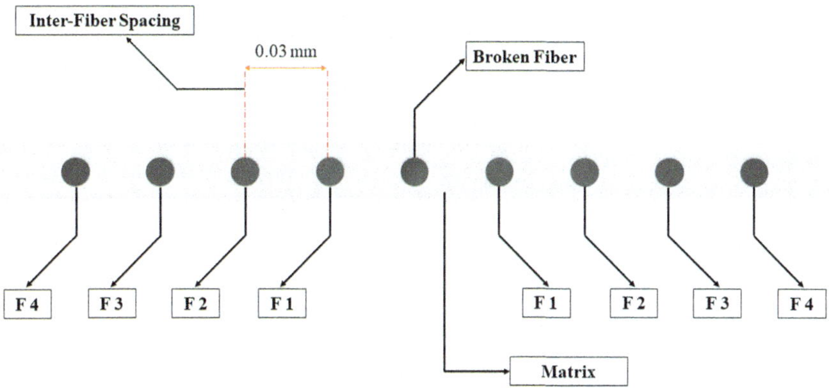

Fig. 2 Top view of composite lamina

Table 1 Material properties of reinforcement and matrix

Material	Property	Value
Carbon fiber	Young's modulus	270 GPa
	Poisson's ratio	0.27
	Density	1700 kg/m^3
Epoxy matrix	Young's modulus	3 GPa
	Poisson's ratio	0.30
	Density	1100 kg/m^3

2.1 *Computational Modeling*

The lamina designed using the software Autodesk Fusion 360 including nine fibers enclosed with matrix is considered to use periodic boundary condition along the cut sections. Matrix material along with reinforcement material represents epoxy matrix with carbon fibers respectively holding the material properties (see Table 1).

The epoxy matrix, reinforced with carbon fiber is modeled with dimensions to resemble actual materials. The composite material in analysis uses dimensions (see Table 2) with defined boundary conditions focusing on the internal physics.

Modeled geometry of composite is analyzed using the software COMSOL Multiphysics with structural mechanics.

2.2 *Simulation Environment*

Computational simulation performed on COMSOL Multiphysics, the designed model of composite material is imported to assign material properties and boundary

Table 2 Composite material overview

Parameter	Specifications
Fibers	Carbon fiber
Matrix	Epoxy resin
Number of fibers	09 fibers
Fiber length	1.00 mm
Fiber thickness	0.01 mm
Inter fiber spacing	0.03 mm
Lamina height	1.00 mm
Lamina width	0.30 mm (periodic boundary)

conditions with structural physics. A cross section of lamina defines boundary conditions of geometry (see Fig. 3). The simulation model uses prescribed displacement across surface, acting load on (pulling).

Composite material geometry with nine fibers across the periodic matrix impregnates the center fiber 'F_O' as the broken fiber at the bottom boundary. The physical broken condition of the fiber along with the enclosed matrix defaces the free boundary condition on the broken boundary while the other section is fixed to one end with the fixed boundary condition (see Fig. 3). A prescribed displacement of 0.05 mm acting in the positive direction of y-axis is defined along the upper boundary companying all fibers and the matrix. The sectional model of the composite material geometry uses extremely fine triangular grid elements across both the matrix and fibers, distributed evenly.

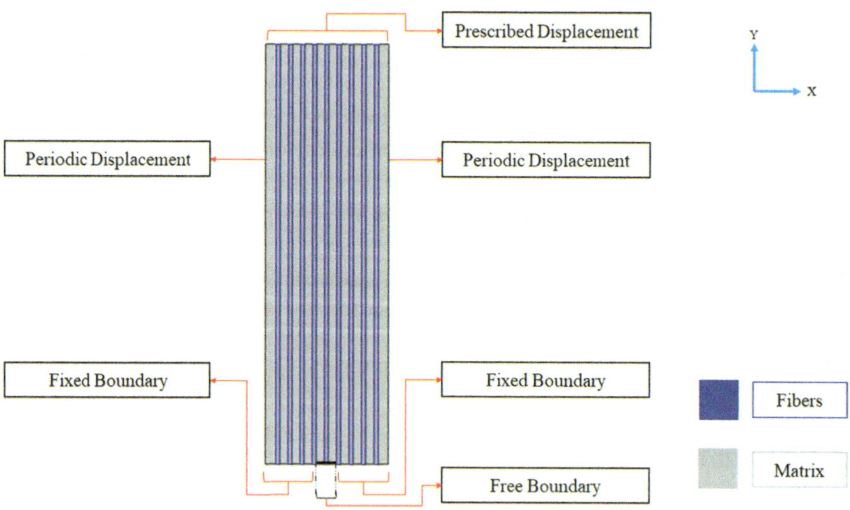

Fig. 3 Simulation physics and boundary conditions

2.3 Mesh Analysis

The geometric mesh for the composite material uses extremely fine triangular mesh elements with finer resolution (see Table 3) in accordance with the parameters analyzed using mesh approximation.

The resultant mesh (see Fig. 4) with appropriate mesh convergence analysis holds in suitable statistics (see Table 4).

Table 3 Mesh elements and parameters for composite material geometry

Parameter	Specifications
Mesh elements	Triangular
Maximum element size	0.0010 mm
Minimum element size	0.0009 mm
Quality	Extremely fine

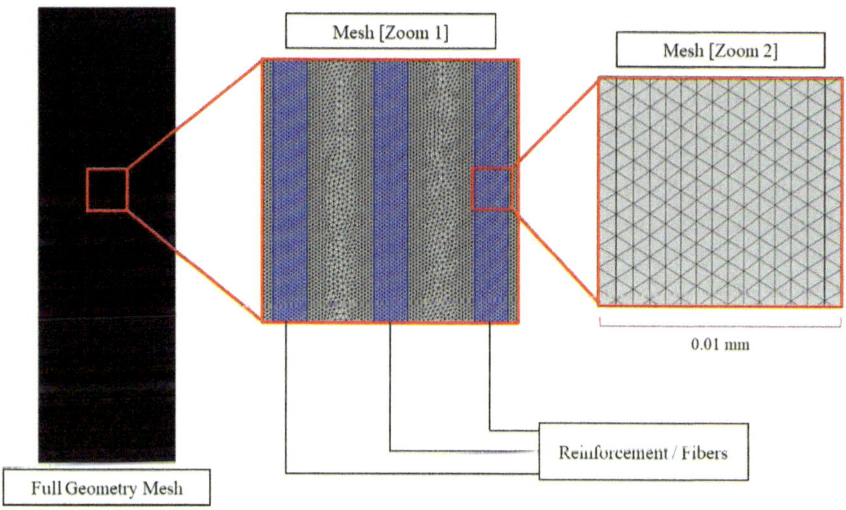

Fig. 4 Magnified representation of mesh for composite material geometry

Table 4 Complete mesh statistics

Parameter	Specifications
Number of mesh elements	5,71,880
Number of mesh vertices	2,86,455
Minimum element quality	0.5656
Average element quality	0.9073
Mesh area	3.0 E−7 m^2
Element area ratio	0.0234

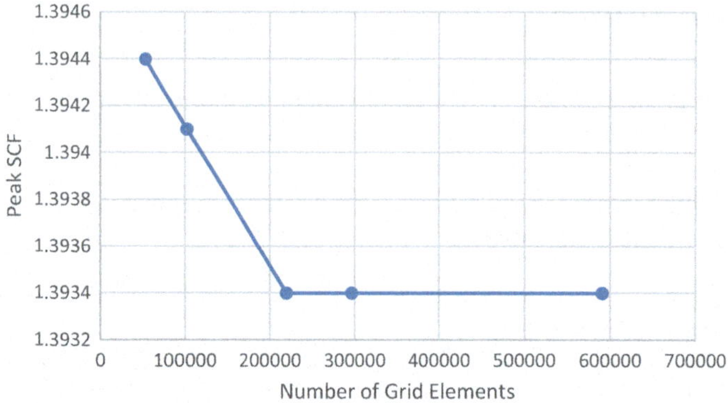

Fig. 5 Mesh convergence

The mesh convergence study for varying quality of geometry grids with different number of elements is studied to saturate the gradient in results with appropriate mesh quality (see Fig. 5). The plot focuses variation of peak stress concentration factor of the composite material over the change in number of mesh elements. A saturation of result tends to show up while the mesh elements go past 2,00,000. However, with more certainty for precision, a mesh with the number of mesh elements close to around 5,00,000 is generated in the final geometry grid (see Table 4).

Computerized analysis using COMSOL Multiphysics is performed in the domain of solid mechanics over the linearly elastic matrix and reinforcements in unidirectional characteristics. The simulation results provide the stress distribution as a function of prescribed displacement along with the deformation of the entire carbon fiber—epoxy composite lamina. The studies follow results from the solution drawn into two-dimensional plots.

3 Results and Discussion

Modeling of computerized simulation over the composite lamina with carbon fiber and epoxy matrix, impregnated with nine fibers in the representative volume element can be extended to a complete lamina or a laminate following a series of RVEs. The center fiber 'F_O' holds a free end with the enclosed matrix, considered as physically broken end. A prescribed displacement assigned to the top surface of unidirectional lamina is a result of shearing and stress distribution across the fibers and matrix in the composite. Analysis of composite lamina behavior with respect to deformation and stress are plotted.

Result of equalized displacement for a lamina with perfect fibers shows equal stress distributions across the reinforcements along with the matrix. The presence of a broken fiber infers in stress concentration gradients across fibers 'F_0' to 'F_n' as a result of disturbed displacement across the lamina. With free on the broken plane of the fiber 'F_0', a maximum displacement of the broken fiber is noted, while the intact fibers are least affected by the displacement on the broken plane (see Fig. 6).

Prescribed displacement of 0.05 mm assigned over the upper surface of composite lamina, the stress experienced by the broken fiber 'F_0' at the broken end shows negligible to zero. However, saturates to equalize the stress evenly along the lamina while gradually reaching a saturation limit at the upper surface, surface of action for displacement and hence, load (see Fig. 7). The intact fibers 'F_1's' in the RVE take in added stresses across the broken plane of fibers and tend to saturate along the length equalizing the saturation limit of fiber 'F_0'. Representation of the same in account with stress concentration factor positions the same curve with saturation value at 1 (see Fig. 8).

$$\mathrm{SCF} = \mathrm{Stress}_{(\mathrm{Local})}/\mathrm{Stress}_{(\mathrm{Infinity})} \tag{1}$$

Fig. 6 Displacement plots of 'F_0' with 'F_1' along the fiber length

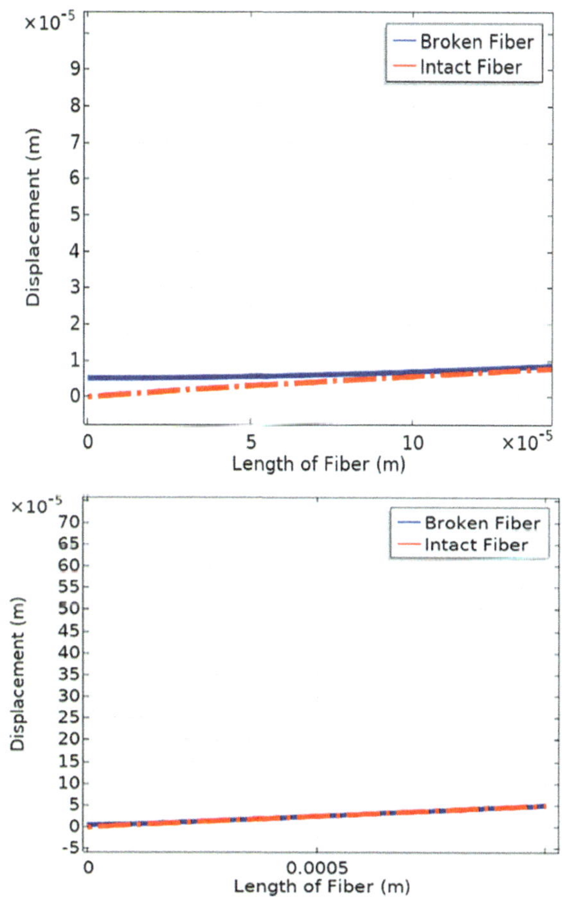

The stress concentration factor of 'F_1' tends to reach a maximum value of 1.4 while the broken fiber initiates from the zero value as relative length presumed from the broken plane of the lamina. The guided stresses show the tendency of saturation toward the load-acting surface prescribed with displacement. A gradient in stress concentration across the fibers 'F_0', 'F_1', 'F_2', 'F_3', 'F_4'.... 'F_n' tends to saturate reaching zero, proportional to the distance of intact fibers along the length of the representative volume element taken from the composite lamina (see Fig. 9).

A gradual decrease in the stress concentration factor with increasing distance of the intact fibers is observed, saturating, and nullifying at the fiber 'F_n'. The peak stress concentration factor of the fibers on the broken plane resembles a similar tendency (see Fig. 10).

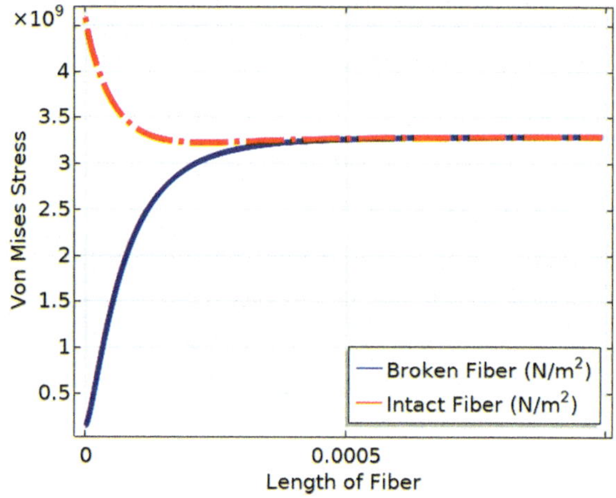

Fig. 7 Stress distribution along the length of fibers 'F_0' & 'F_1'

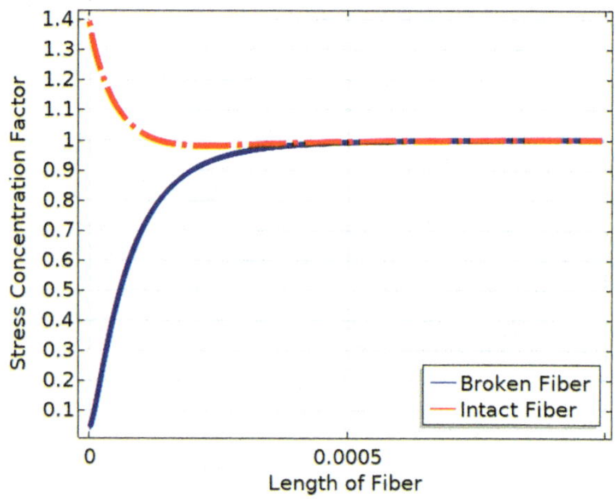

Fig. 8 Stress concentration factor along the length of fibers 'F_0' & 'F_1'

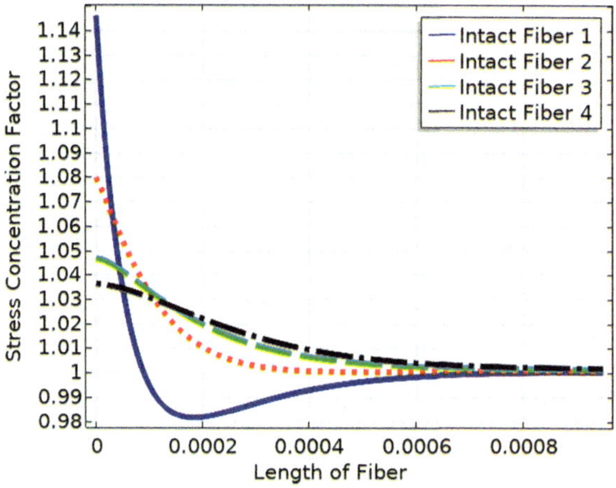

Fig. 9 Stress concentration factor across the length of intact fibers

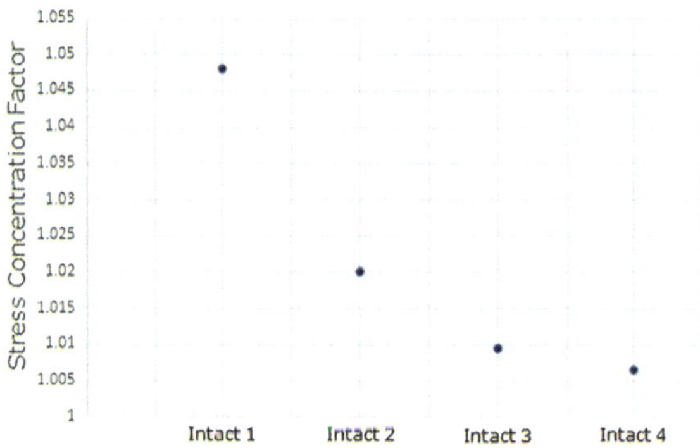

Fig. 10 Stress concentration factor across intact fibers in the broken plane

4 Conclusion

The numerical study predicts the stress distribution in a unidirectional composite lamina around the broken fiber. The analysis shows forth maximum stress concentrations over intact fibers 'F_1', with a gradual decrease to saturation at fibers 'F_n'. However, a zero displacement is observed in the intact fibers with the broken fiber acting over displaced. The present study is performed and validated to contest with the analytical model of the broken fiber stress distribution by Guerrero [7].

4.1 Future Scope

The study elaborates on prerequisites for advanced future works to be performed on the crack propagation in a unidirectional composite along with the prediction of stress distribution across a unidirectional or bidirectional composite laminate over varying orientation of reinforcements.

Acknowledgements Authors wish to sincerely thank their family, friends, and colleagues for all the support provided in successfully completing the study. A note of special gratitude is expressed to the University of Petroleum and Energy Studies, Dehradun, India, for immense support with facilities to perform the work. Authors would like to greet and thank all the researchers responsible for providing the availability of reliable public data in support of the present study.

References

1. Landis CM, McMeeking RM (1999) A shear-lag model for a broken fiber embedded in a composite with a ductile matrix. Compos Sci Technol 59(3):447–457
2. Goree JG, Gross RS (1980) Analysis of a unidirectional composite containing broken fibers and matrix damage. Eng Fract Mech 13(3):563–578
3. Lagoudas DC, Chung-Yuen H, Leigh Phoenix S (1989) Time evolution of overstress profiles near broken fibers in a composite with a viscoelastic matrix. Int J Solids Struct 25(1):45–66
4. Ochiai S, Schulte K, Peters PWM (1991) Strain concentration factors for fibers and matrix in unidirectional composites. Compos Sci Technol 41(3):237–256
5. Ma Y et al (2016) A study on the failure behavior and mechanical properties of unidirectional fiber reinforced thermosetting and thermoplastic composites. Compos Part B: Eng 99:162–172
6. Mishnaevsky Jr L, Brøndsted P (2009) Micromechanical modeling of damage and fracture of unidirectional fiber reinforced composites: a review. Comput Mater Sci 44(4):1351–1359
7. Guerrero JM et al (2019) An analytical model to predict stress fields around broken fibres and their effect on the longitudinal failure of hybrid composites. Compos Struct 211:564–576

Effect of Variation of Fibre Orientation on the Natural Frequency of a CFRP Conical Sandwich Structure

Prashant Kumar, Vivek Raj, Abhishek Parida, Sonali Gyan, K. Senthil, and B. Santhosh

1 Introduction

There is extensive use of truncated conical shells made of CFRP in launch vehicle structures as it provides an interface to components of lesser to bigger diameter. Fabrication of such conical CFRP sandwich structure possesses an inherent issue of fibre angle deviation as the radius of cross sections changes along the meridian. Angle deviation is low for cones of small semi-vertex angles as structures are close to the cylinder [1, 2]. While for cones of large semi-vertex angle, the deviation can be significant and result in a change of the stiffness from the designed configuration. Khakimova et al. [3] reported difficulty in maintaining the fibre angle variation for the large semi-vertex angles of the sandwich cone realised through the prepreg/autoclave method and suggested a ply topology optimisation method to minimise the fibre angle variation and thickness imperfection. The study also shows larger cone angle results in larger angle deviation and with narrow ply strips, the deviation could be minimised. However, increasing the number of ply strips increases the time to cut and lay prepreg, also increases the possibilities of deviations associated with these processes.

This paper deals with a truncated sandwich cone made of CFRP skin and aluminium honeycomb core. The structure is realised through the autoclave/prepreg method which is widely used in the aerospace industry to render high-quality components. The structure is being used in the upper stage of the launch vehicle which

P. Kumar · V. Raj · A. Parida (✉) · S. Gyan · K. Senthil · B. Santhosh
Composites Entity, Vikram Sarabhai Space Centre, Trivandrum, India
e-mail: abhishek.parida1995@gmail.com

V. Raj
e-mail: vivek_raj@vssc.gov.in

M. R. Manohara Babu et al. (eds.), *Recent Advances in Composites Engineering and Research*, Springer Proceedings in Materials 39,
https://doi.org/10.1007/978-981-99-8807-5_3

has very stringent frequency requirements. Hence the effect of inherent fibre angle deviation on the natural frequency was studied. The study also highlights the impact on the overall thickness due to the overlapping of strips and equivalent properties at various angles of a sector.

2 Fabrication Complexities

While designing a conical structure, the meridian of the cone is taken as the reference axis and fibre orientations are defined with respect to it. Fibres in a given ply diverge at the larger diameter end and converge at the smaller diameter end without any overlapping and make a constant angle with the meridian. This idealisation cannot be met through the prepreg/autoclave method. To avoid overlapping at the smaller diameter end, prepregs are cut into trapezoidal-shaped pieces, herein referred to as sectors and laid on the mould. In each sector, fibres run parallel to each other. Thus, only one of the fibres in each sector is oriented as per the desired fibre orientation as shown in Fig. 1. Fibre angle deviation increases as we move away from the centre of a sector.

3 Structural Configuration

The structure is a single-shell sandwich construction with high modulus carbon fibre and a medium pcf aluminium honeycomb core. The diameter of the cone varies from 4 m at the aft-end to 2 m at the fore-end over the height of 1 m as shown in Fig. 2. The conical CFRP sandwich shell has a semi-cone angle of 45°. Aluminium rings are provided at both ends of the shell to provide the interface for attachment. Four cut-outs are provided to meet the functional requirements.

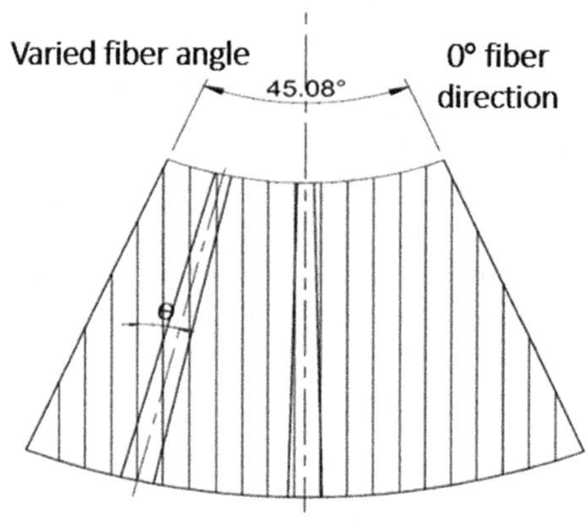

Fig. 1 Variation of fibre angle w.r.t. generator

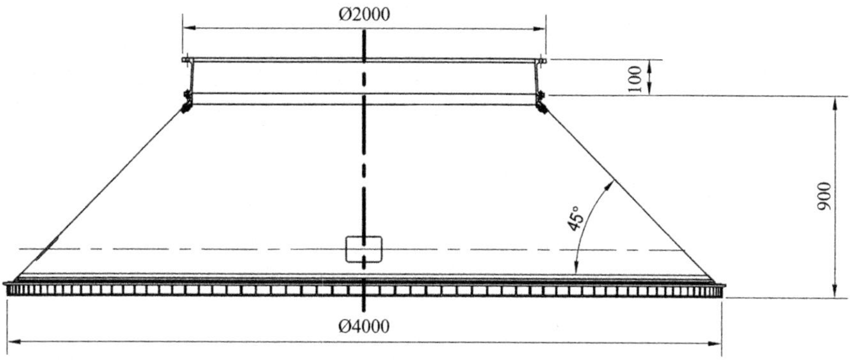

Fig. 2 Geometric configuration of the structure

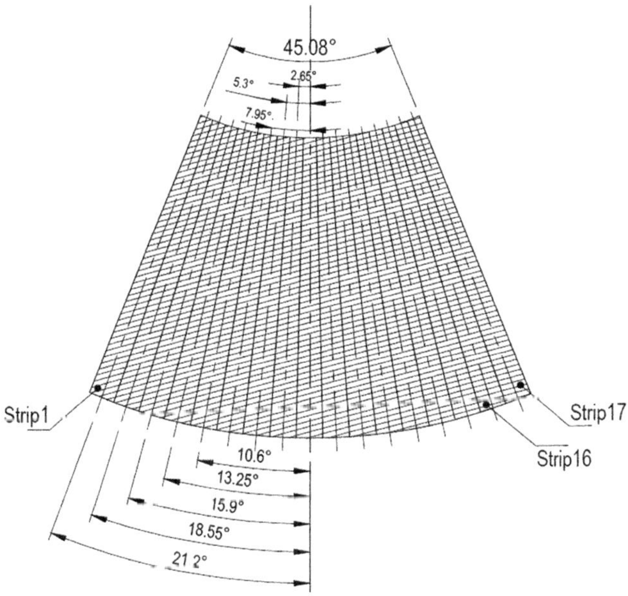

Fig. 3 Fibre angle variation in a typical prepreg sector shown in the developed view

In the basic region, the shell has 16 layers in the inner and outer skin of the sandwich. Additional reinforcement layers are provided near the fore-end and cut-outs. Shapes of the rings and layup sequence in the shell are optimised to meet frequency requirements.

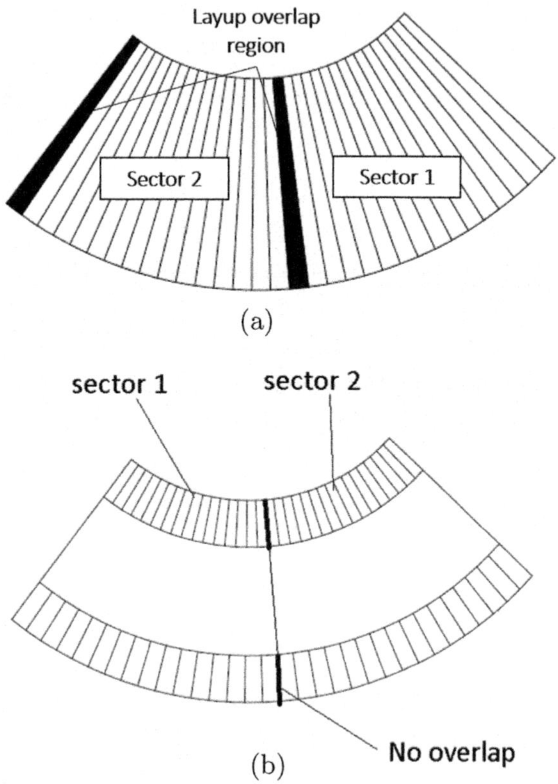

Fig. 4 Placement of layups in **a** basic and **b** reinforcement region

The conical shell is fabricated in a full shell configuration. Where each layer is having six sectors (60° segments). Increasing number of sector will reduce the fibre angle variation but lead to fabrication complexities with regard to cutting and laying of prepreg. So number of sectors are restricted to six. The minimum 60 mm overlap is provided between adjacent sectors so that at the end of each skin layup, one uniform extra layer exist. The developed view of a typical sector is shown in Fig. 3. The overlap region is provided only for a basic layup, reinforcement layers at fore-end and aft-end are provided with the butt joint as shown in Fig. 4.

4 FE Modelling of Fibre Angle Variation

The sector angle and length of overlap region used for fabrication are replicated in the FE model. Full shell is divided into six sectors and each sector is further divided into 16 strips, thus forming a total of 96 strips. Basic region, potting region, reinforcement

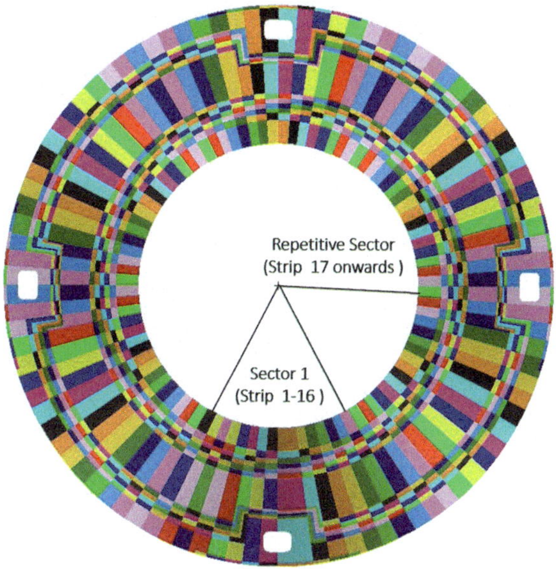

Fig. 5 FE model of CFRP conical shell having angular variation

Table 1 Natural frequency comparison between designed and realised with fibre angle deviation

S. No.	Mode shape	Designed configuration with ideal fibre orientation (Hz)	Realised configuration with fibre angle deviation (Hz)
1	Lateral	8.40	8.38
2	Longitudinal	42.2	44.6

region, and staggering as in the fabricated hardware are modelled in detail to capture the difference from the ideal condition. FE model of the CFRP shell with 96 strips is shown in Fig. 5.

For basic skin layup, when the first layer of prepreg is placed on the first sector, last strip (17th) of the sector overlaps with the first strip of the subsequent sector. Overlapping takes place between each consecutive sector for every layer of prepreg. At the end of the prepreg layup for each skin, an additional layer is formed across each sector. Thus, the realised shell has 17 basic layers instead of the designed 16 layers. Accounting for fibre angle deviation in each strip across all sectors, a total of 32 different layup sequences are formed. The first sector will have 16 different property strips and the other five sectors will be identical 16 sets of layup sequences which will be repetitive in each of the five sectors. Tables 2, and 3 show the fibre angle deviation in each strip of the first and other five sectors, respectively. Numbering of strips have

Table 2 Variation of fibre angle in each layer of the strips w.r.t generator in the first sector for the inner basic skin with reinforcement

	Strips No.															
Designed lay-up sequence	1	2	3	4	5	6	7	8	9	10	11	12	13	14	15	16
0	21	19	16	13	11	8	5	3	0	− 3	− 5	− 8	− 11	− 13	− 16	− 19
0	− 21	21	19	16	13	11	8	5	3	0	− 3	− 5	− 8	− 11	− 13	− 16
30	− 19	− 21	51	49	46	43	41	38	35	33	30	27	25	22	19	17
− 30	14	11	9	− 9	− 11	− 14	− 17	− 19	− 22	− 25	− 27	− 30	− 33	− 35	− 38	− 41
0	− 43	− 46	− 49	− 51	21	19	16	13	11	8	5	3	0	− 3	− 5	− 8
0	− 11	− 13	− 16	− 19	− 21	21	19	16	13	11	8	5	3	0	− 3	− 5
− 30	− 8	− 11	− 13	− 16	− 19	− 21	− 9	− 11	− 14	− 17	− 19	− 22	− 25	− 27	− 30	− 33
30	− 35	− 38	− 41	− 43	− 46	− 49	− 51	51	49	46	43	41	38	35	33	30
0	27	25	22	19	17	14	11	9	21	19	16	13	11	8	5	3
0	0	− 3	− 5	− 8	− 11	− 13	− 16	− 19	− 21	21	19	16	13	11	8	5
60	3	0	− 3	− 5	− 8	− 11	− 13	− 16	− 19	− 21	81	79	76	73	71	68
− 60	65	63	60	57	55	52	49	47	44	41	39	− 39	− 41	− 44	− 47	− 49
0	− 52	− 55	− 57	− 60	− 63	− 65	− 68	− 71	− 73	− 76	− 79	− 81	21	19	16	13
0	11	8	5	3	0	− 3	− 5	− 8	− 11	− 13	− 16	− 19	− 21	21	19	16
− 60	13	11	8	5	3	0	− 3	− 5	− 8	− 11	− 13	− 16	− 19	− 21	− 39	− 41
60	− 44	− 47	− 49	− 52	− 55	− 57	− 60	− 63	− 65	− 68	− 71	− 73	− 76	− 79	− 81	81
	79	76	73	71	68	65	63	60	57	55	52	49	47	44	41	39
0	19	16	13	11	8	5	3	0	0	− 3	− 5	− 8	− 11	− 13	− 16	− 19
0	19	16	13	11	8	5	3	0	0	− 3	− 5	− 8	− 11	− 13	− 16	− 19
60	79	76	73	71	68	65	63	60	60	57	55	52	49	47	44	41
− 60	− 41	− 44	− 47	− 49	− 52	− 55	− 57	− 60	− 60	− 63	− 65	− 68	− 71	− 73	− 76	− 79
0	19	16	13	11	8	5	3	0	0	− 3	− 5	− 8	− 11	− 13	− 16	− 19
0	19	16	13	11	8	5	3	0	0	− 3	− 5	− 8	− 11	− 13	− 16	− 19
− 60	− 41	− 44	− 47	− 49	− 52	− 55	− 57	− 60	− 60	− 63	− 65	− 68	− 71	− 73	− 76	− 79
60	79	76	73	71	68	65	63	60	60	57	55	52	49	47	44	41
0	19	16	13	11	8	5	3	0	0	− 3	− 5	− 8	− 11	− 13	− 16	− 19
0	19	16	13	11	8	5	3	0	0	− 3	− 5	− 8	− 11	− 13	− 16	− 19

Table 3 Variation of fibre angle in each layer of the strips w.r.t generator in sectors other than first sector for the inner basic skin with reinforcement

	Strips No. (similar for all 5 sectors)															
Designed lay-up sequence	17	18	19	20	21	22	23	24	25	26	27	28	29	30	31	32
0	− 21	19	16	13	11	8	5	3	0	− 3	− 5	− 8	− 11	− 13	− 16	− 19
0	21	− 21	19	16	13	11	8	5	3	0	− 3	− 5	− 8	− 11	− 13	− 16
30	− 19	21	9	49	46	43	41	38	35	33	30	27	25	22	19	17
− 30	14	11	51	− 51	− 11	− 14	− 17	− 19	− 22	− 25	− 27	− 30	− 33	− 35	− 38	− 41
0	− 43	− 46	− 49	− 9	− 21	19	16	13	11	8	5	3	0	− 3	− 5	− 8
0	− 11	− 13	− 16	− 19	21	− 21	19	16	13	11	8	5	3	0	− 3	− 5
− 30	− 8	− 11	− 13	− 16	− 19	21	− 51	− 11	− 14	− 17	− 19	− 22	− 25	− 27	− 30	− 33
30	− 35	− 38	− 41	− 43	− 46	− 49	− 9	9	49	46	43	41	38	35	33	30
0	27	25	22	19	17	14	11	51	− 21	19	16	13	11	8	5	3
0	0	− 3	− 5	− 8	− 11	− 13	− 16	− 19	21	− 21	19	16	13	11	8	5
60	3	0	− 3	− 5	− 8	− 11	− 13	− 16	− 19	21	39	79	76	73	71	68
− 60	65	63	60	57	55	52	49	47	44	41	81	− 81	− 41	− 44	− 47	− 49
0	− 52	− 55	− 57	− 60	− 63	− 65	− 68	− 71	− 73	− 76	− 79	− 39	− 21	19	16	13
0	11	8	5	3	0	− 3	− 5	− 8	− 11	− 13	− 16	− 19	21	− 21	19	16
− 60	13	11	8	5	3	0	− 3	− 5	− 8	− 11	− 13	− 16	− 19	21	− 81	− 41
60	− 44	− 47	− 49	− 52	− 55	− 57	− 60	− 63	− 65	− 68	− 71	− 73	− 76	− 79	− 39	39
	79	76	73	71	68	65	63	60	57	55	52	49	47	44	41	81
0	19	16	13	11	8	5	3	0	0	− 3	− 5	− 8	− 11	− 13	− 16	− 19
0	19	16	13	11	8	5	3	0	0	− 3	− 5	− 8	− 11	− 13	− 16	− 19
60	79	76	73	71	68	65	63	60	60	57	55	52	49	47	44	41
− 60	− 41	− 44	− 47	− 49	− 52	− 55	− 57	− 60	− 60	− 63	− 65	− 68	− 71	− 73	− 76	− 79
0	19	16	13	11	8	5	3	0	0	− 3	− 5	− 8	− 11	− 13	− 16	− 19
0	19	16	13	11	8	5	3	0	0	− 3	− 5	− 8	− 11	− 13	− 16	− 19
− 60	− 41	− 44	− 47	− 49	− 52	− 55	− 57	− 60	− 60	− 63	− 65	− 68	− 71	− 73	− 76	− 79
60	79	76	73	71	68	65	63	60	60	57	55	52	49	47	44	41
0	19	16	13	11	8	5	3	0	0	− 3	− 5	− 8	− 11	− 13	− 16	− 19
0	19	16	13	11	8	5	3	0	0	− 3	− 5	− 8	− 11	− 13	− 16	− 19

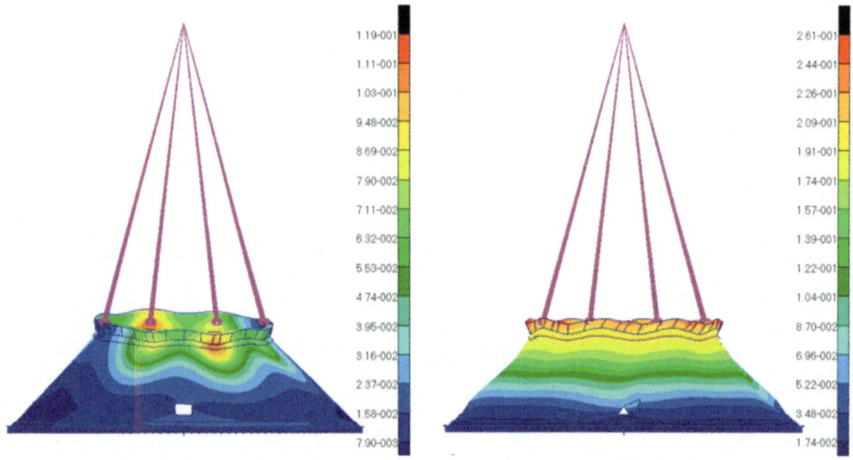

(a) (b)

Fig. 6 First natural mode shape in **a** lateral and **b** axial direction with designed fibre layup

started from the left. Strip 9 locates at the centre of sector and fibre in the strips run parallel to the generator without any angle deviation for first layer of prepreg.

The strips are at different angles from the generator of each sector of the cone. As sign convention, the angle deviation for the strips on the left side of the generator is taken as positive and the angle deviation for the strips on the right side of the generator is taken as negative.

5 Results

Frequency analysis is carried out for both the designed and the varied fibre angles and the first natural mode shape in lateral and axial directions for the designed and realised configuration with fibre angle deviation is shown in Figs. 6 and 7, respectively. Table 1 summarises the frequency analysis results.

The variation in equivalent skin properties along fibre (Ex) and hoop direction (Ey) and corresponding inplane rigidity (Ex*t and Ey*t) variations due to an increase in skin thickness from 1.6 to 1.7 mm are also compared for each side of the skin of the sandwich. Ex*t and Ey*t varies between + 11% to − 7%, and 0.5% to + 39% from nominal layup respectively (Figs. 8 and 9).

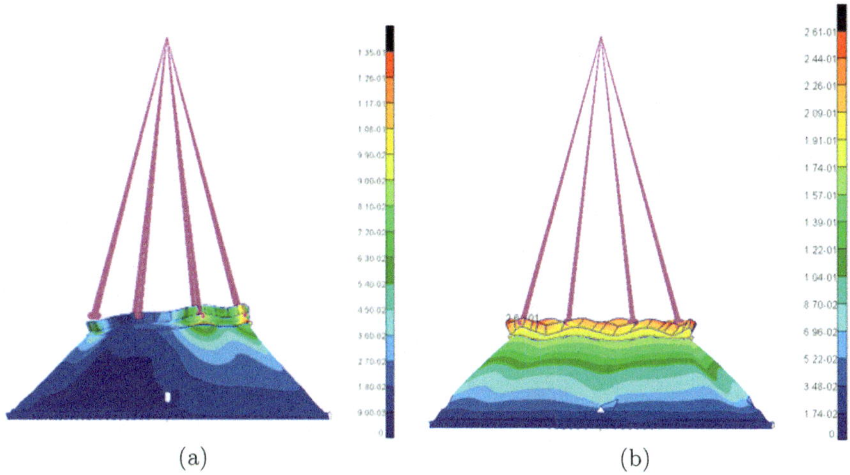

Fig. 7 First natural mode shape in **a** lateral and **b** axial direction with realised fibre layup

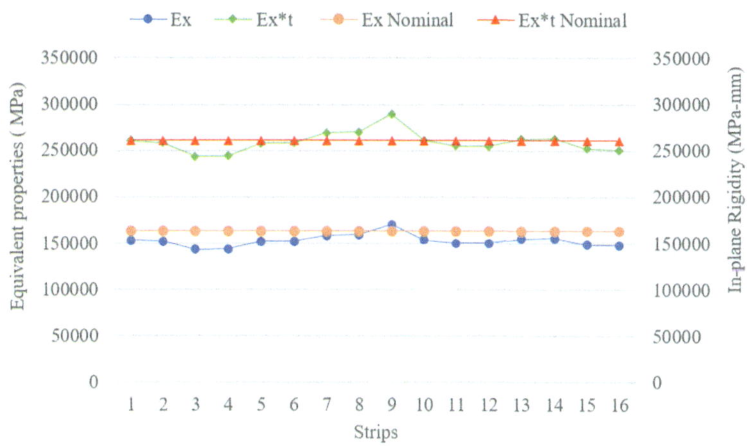

Fig. 8 Variation of equivalent properties in the fibre direction

6 Discussion and Conclusion

A detailed study is carried out to examine the effect of fibre angle variation in a CFRP conical structure which is inherent to the fabrication process. No significant variation in frequency is observed between the design and realised configuration of the structure. This is because the natural frequency of the specific structure is largely governed by geometrical stiffness of the CFRP cone and insensitive to the change in the stiffness of the truncated conical shell due to variation of fibre angle. It is noticed from classical laminate theory calculation [4, 5] that there is a minor reduction in axial modulus and there is improvement in circumferential modulus.

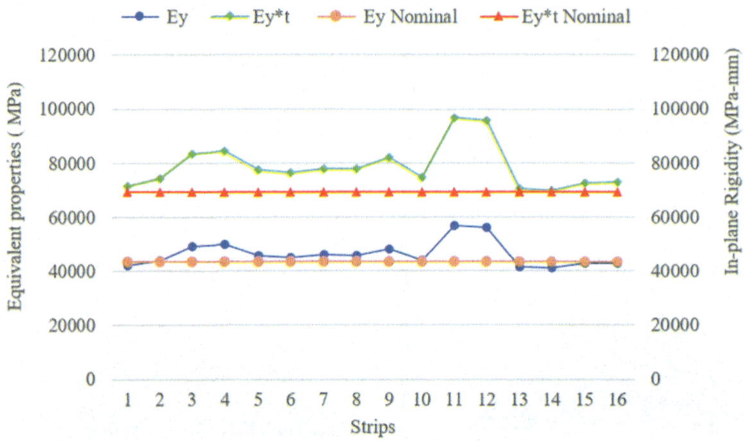

Fig. 9 Variation of equivalent properties in the transverse direction

The processing of a truncated conical structure can be challenging to meet the design requirement. However, the discussed approach can be used to account for the fibre angle variation and predict the actual behaviour of hardware. It can also be used to select optimum sector angle and overlap lengths to be used during fabrication to meet the requirement.

References

1. Bert C, Crisman W, Nordby G (1968) Fabrication and full-scale structural evaluation of glass-fabric reinforced plastic shells. J Aircr 5(1):27–34
2. Bert C, Crisman W, Nordby G (1969) Buckling of cylindrical and conical sandwich shells with orthotropic facings. AIAA J 7(2):250–257
3. Khakimova R, Burau F, Degenhardt R, Siebert M, Castro SG (2016) Design and manufacture of conical shell structures using prepreg laminates. Appl Compos Mater 23(3):289–312
4. Jones RM (2018) Mechanics of composite materials. CRC Press
5. Kaw AK (2005) Mechanics of composite materials. CRC Press

Joint Configuration of Composites with Metals for High-Temperature Applications Through Thermal Stress-Free Fastener

Suresh Lanka and Jhon Paul

Nomenclature

b	Constant of integration
E	Modulus of elasticity
L	Length
m, n	Parameter that determines characteristic shape in x–y plane
x, y	Rectangular Cartesian coordinates
r, θ, z	Cylindrical coordinates
α	Coefficient of thermal expansion (CTE)
ν	Poisson's ratio
T	Temperature
R	Radius of fastener shank
W	Washer thickness
P	Point on material
$\overline{\alpha}$	Average CTE over a given temperature range
Δ	Difference or change

Subscripts

0	Initial or reference condition
1	Material 1
2	Material 2

S. Lanka (✉) · J. Paul
Azista Composites Pvt. Ltd., Hyderabad, Telangana 501401, India
e-mail: sureshlanka.anits@gmail.com

M. R. Manohara Babu et al. (eds.), *Recent Advances in Composites Engineering and Research*, Springer Proceedings in Materials 39,
https://doi.org/10.1007/978-981-99-8807-5_4

1 Introduction

The re-entry and hypersonic vehicles demand operation at very high Mach number at higher altitudes to meet mission requirements. During the operation, these vehicles are subjected to higher aero-thermal loads. High-temperature materials or hot structures are ingenious design options for re-entry and hypersonic vehicles to reduce weight and increase vehicle efficiency [1]. Low density, low thermal expansion coefficient, and retained strength at elevated temperature are ideal characteristics of materials used in re-entry and hypersonic vehicles. Ceramic matrix composite (CMC's) possess high specific strength and stiffness, improved damage tolerance, low thermal expansion coefficient, and excellent oxidation resistance at high temperatures [2, 3]. High-temperature composites of carbon or ceramic-based materials are regarded a material of choice due to these properties[4].

Conventionally most of these high-temperature structures are joined to the substructure with metal fasteners. The coefficients of linear thermal expansion (CTE) of high-temperature composites are much lower than those of metal fastener materials. This thermal expansion mismatch can cause carbon–carbon material failure around standard cylindrical fasteners.

CMC and superalloy bolted joint systems are subjected to combined mechanical and thermal loads in high-temperature service. The difference in thermal expansion between CMC and superalloy plates would result in complicated thermal stress and strain distributions at the hole-edge area, as well as variations in joint assembly parameters such as preload loosening, and bolt-hole clearance variation as presented by Zhao et al. [5].

Figure 1 depicts the issues encountered when using conventional cylindrical fasteners in high-temperature joints [6]. Various connection conditions for conventional fasteners are summarized in Table 1. When the temperature rises, a fastener that is radially tight at room temperature will face substantial thermal stresses. It can be too loose at room temperature if there is a large enough radial gap all the way around the fastener to simply tighten as the temperature rises. Similar to this, a fastener that is axially tight at room temperature can either get looser as the temperature increases or needs to be severely preloaded at room temperature to remain tight at high temperatures. Premature joint failure can be brought on by the axial preload or radial thermal stress needed to maintain a satisfactory tight joint throughout the operating temperature range.

The problem has been extensively studied, and various approaches have been proposed as solution. In one approach, the radial gap and preload are predicted at ambient temperature which induces reasonable thermal stresses and snug joint at high temperatures for cylindrical fastener joint. In another approach, fastener material is chosen presenting CTE close to structural members CTE which reduces the thermal stresses at joint. However, the CTE of metals widely varies from composite materials. New joining methods are therefore required for high-temperature composite materials.

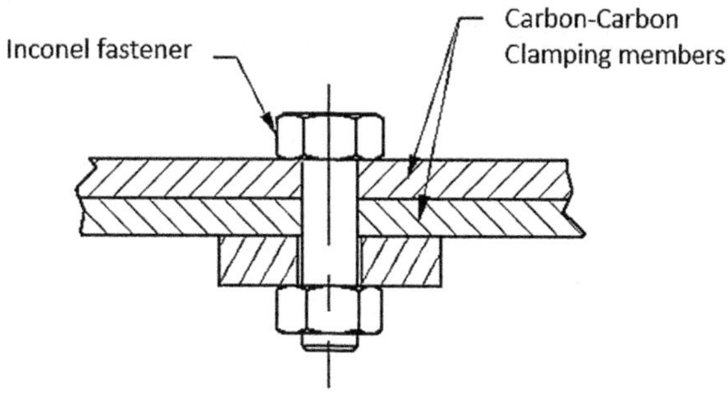

Fig. 1 Major issues for standard cylindrical joints at elevated temperatures applications [7]

Table 1 Different joint conditions for conventional fasteners [7]

	Joint condition	
	Room temperature	Elevated temperature
Radial direction	Snug	High thermal stresses
	Loose	Snug
Axial direction	Snug	Loose
	Pre-stressed	Snug

The state-of-the-art technology employed for joining high-temperature composite materials is known in literature as "thermal stress-free concept". This concept reduces the thermal stresses at the interface to acceptable levels by properly configuring the fasteners and mating hole. The concept involves a technique that allows fasteners and surrounding materials to expand freely. Even if the two materials are in constant contact, the movement of one material does not constrain the movement of the adjacent material.

The primary objective of this work is to propose a solution for a two-dimensional thermally stress-free boundary between two orthotropic materials and to use finite-element analysis to analyze the behavior of these fasteners. The answer to the problem of a two-dimensional, thermal stress-free boundary between two dissimilar materials at high temperature is presented in this study. To solve this problem, a mathematical formulation has been developed with following assumptions:

(1) For both materials, the CTE remains constant with respect to location and temperature.
(2) In both materials, the temperature rises uniformly.
(3) No friction along the boundary.

The results of a finite-element analysis of thermal stresses in conventional joints and thermally stress-free joints of isotropic materials are reported.

2 Mathematical Analysis

2.1 Development of Interface Equation

The governing principle of the approach lies in generating an interface equation between two homogeneous materials having orthotropic CTE through which the two materials are connected without interference or separation as temperature rises [1, 8].

The following equation expresses the essential relationship between increase in temperature dT, thermal expansion, and increase in length dL:

$$\frac{dL}{L} = \alpha dT, \tag{1}$$

where

L = length and α = linear coefficient of thermal expansion (CTE).

From Eq. (1), through integration within two temperature limits T_0 and T

$$L = L_0 e^{\overline{\alpha}(T - T_0)}, \tag{2}$$

where $\overline{\alpha}$ is the average CTE value within the range of T_0 and T.

At an initial temperature of T_0, a point P is considered on the interface, whereas adjacent particles in the two materials are indicated by the numbers P_1 and P_2. The neighboring two points separate yet continue to lie on a similar boundary following an infinitesimal temperature shift, as depicted in Fig. 2. At time $T_0 + T$, each point (x, y) on the boundary at temperature T_0 shifts to a new position on the new boundary.

The movement of particles is represented by the new coordinates:

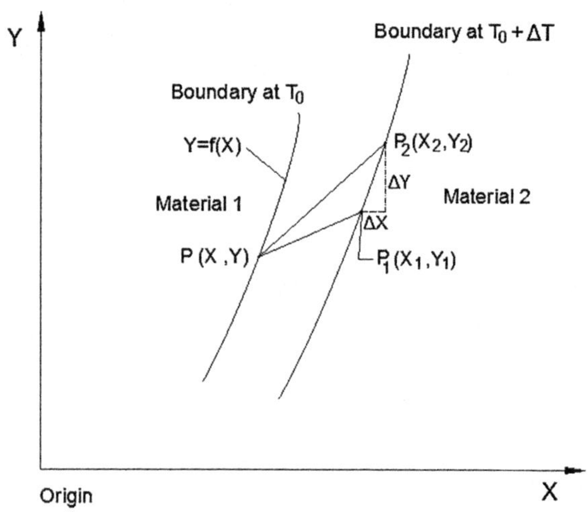

Fig. 2 Thermal stress-free boundary between materials with different CTE's at T_0 and $T_0 + \Delta T$ [7]

From Eq. (2),
For material 1:

$$\left.\begin{aligned} x_1 &= x e^{\alpha_{x_1}(\Delta T)} \\ y_1 &= y e^{\alpha_{y_1}(\Delta T)} \end{aligned}\right\} \tag{3}$$

For material 2:

$$\left.\begin{aligned} x_2 &= x e^{\alpha_{x_2}(\Delta T)} \\ y_2 &= y e^{\alpha_{y_2}(\Delta T)} \end{aligned}\right\} \tag{4}$$

where

α_{x_1} CTE of material 1 in x-direction.
α_{y_1} CTE of material 1 in y-direction.
α_{x_2} CTE of material 2 in x-direction.
α_{y_2} CTE of material 2 in y-direction.

The original shape of the boundary, $y = f(x)$, be such that (x_1, y_1) and (x_2, y_2) are positioned on coincident boundaries after a temperature increase of ΔT. The slope of this boundary at $T = T_0 + \Delta T$ can be approximated by

$$\frac{\Delta y}{\Delta x} = \frac{y_2 - y_1}{x_2 - x_1} = \frac{(y)\left(\mathrm{e}^{\alpha_{y_2}(\Delta T)} - \mathrm{e}^{\alpha_{y_1}(\Delta T)}\right)}{(x)\left(\mathrm{e}^{\alpha_{x_2}(\Delta T)} - \mathrm{e}^{\alpha_{y_2}(\Delta T)}\right)} \tag{5}$$

Now as $\Delta T \to 0$, the boundary approaches $y = f(x)$, and the approximate slope of the boundary at $T = T_0 + \Delta T$ becomes the exact slope of the initial boundary at $T = T_0$.

As $\Delta T \to 0$; $\Delta y \to 0$; and $\Delta x \to 0$; therefore,

$$\lim_{\Delta T \to 0} \frac{\Delta y}{\Delta x} = \frac{dy}{dx} \tag{6}$$

For infinitesimal $\Delta T \to 0$ yields

$$\frac{dy}{dx} = \frac{\left(\alpha_{y_2} - \alpha_{y_1}\right)y}{\left(\alpha_{x_2} - \alpha_{x_1}\right)x} \tag{7}$$

Integration of Eq. (7) gives the initial boundary shape

$$y = bx^m, \tag{8}$$

where

$$m = \frac{\alpha_{y_2} - \alpha_{y_1}}{\alpha_{x_2} - \alpha_{x_1}}$$

and b is constant of integration which is obtained by the fastener diameter and washer thickness.

$$b = \frac{W}{R^m}$$

At any other temperature, $T = T_0 + \Delta T$, and the boundary of each material undergoes thermal expansion governed by Eqs. (3) and (4) to the shape given by

$$y = be^{n(T-T_0)}x^m, \tag{9}$$

where

$$n = \frac{\alpha_{y_2}\alpha_{x_1} - \alpha_{y_1}\alpha_{x_2}}{\alpha_{x_1} - \alpha_{x_2}}$$

and where Eq. (9) is obtained by direct substitution of either Eq. (3) or (4) into Eq. (8). Thus, any point on the boundary given by Eq. (8) at $T = T_0$, for material 1 or material 2, will be located on the boundary given by Eq. (9) after a temperature change.

3 Characteristic Shapes

The characteristic shapes give an idea about thermal stress-free interface for different values of m, where m is the ratio of the difference in CTE for the two dissimilar materials in the through-the-thickness and in-plane directions, respectively. The values of m are used to create a family of thermal stress-free boundary in Eq. (8). When determining a specific fastener shape, the value of b is calculated from the fastener diameter and washer thickness. The correlation between the CTE of the two materials, as defined by m, determines the different boundaries. The thermal stress-free boundaries that are displayed are all derived in two dimensions, but if the in-plane CTE of each material is directionally independent, the boundary can be revolved to create axisymmetric thermal stress-free fasteners. The thermal stress-free boundary for $m = 1$, which would occur in joints involving various isotropic materials, is a straight line, resulting in a conical fastener. The basic conical fastener concept as depicted in Fig. 3 is therefore a subset of the thermal stress-free limits.

Using the calculated thermal stress-free equation, typical thermal stress-free fastener configurations are depicted in Fig. 4. For creating typical joints, it is feasible to use the fastener configurations described for $m > 0$. The fastener has a concave shape when $0 < m < 1$, but a convex shape when $m > 1$. As was previously mentioned, a conical fastener is produced when $m = 1$. However, the washer thickness varies depending on the value of m if the diameter of the fastener's cylindrical part is kept fixed as illustrated.

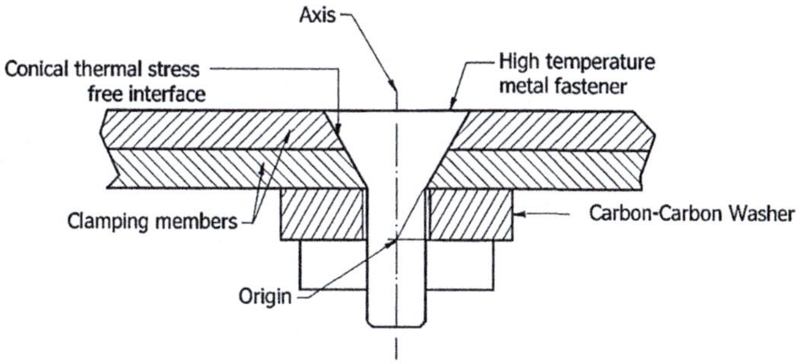

Fig. 3 Thermal stress-free interface for isotropic materials

Fig. 4 Solution to equation $y = bx^m$ [7]

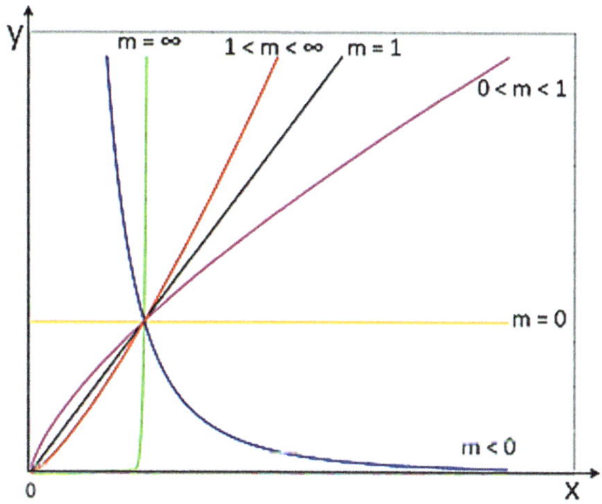

To prevent thermal stress, a space between the washer and the fastener's shank must always be provided. The thermal stress-free limits do not lead to a typical joint configuration for values of $m < 0$. The diagram depicts a potential thermal stress-free arrangement for the scenario in which $m < 0$. The setup displayed is fairly inefficient because the fastener's bending would have to support the joint shear load. Positive values of m result in more realistic thermal stress-free arrangement in most of the high-temperature composite material and metal fastener combinations (Fig. 5).

4 Finite-Element Analysis

The stresses in joints are quantitatively analyzed using the finite-element method. As a result, finite-element analysis outputs were obtained for comparison with the theoretically derived solutions provided in this paper. In this study, a conical fastener

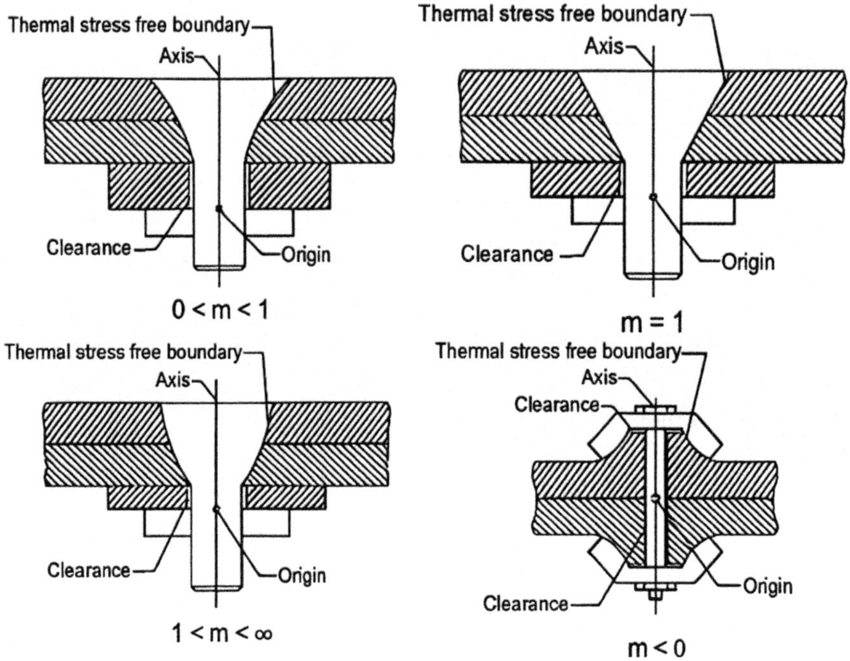

Fig. 5 Thermal stress-free profiles for various m values [7]

was used to connect a material with transversely isotropic CTEs, which caused thermal stresses that were analyzed using finite-element analysis (thermal stress-free only with isotropic materials). The analyses were performed using Ansys Workbench, an available general-purpose finite-element tool.

4.1 Modeling

Three-dimensional finite-element models were used to depict a number of joint configurations using isotropic and transversely isotropic materials. The axisymmetric fasteners were modeled as a 90° wedge with the surrounding material. Figures 1 and 6 depict the two joints that were modeled. Figure 1 depicts a finite-element model of a cylindrical fastener and its surroundings, while Fig. 6 depicts a model of a conical fastener with a washer and its surroundings.

For each of the two joints modeled, arbitrary dimensions were chosen. Figure 6 depicts these dimensions. The washer thickness depicted in Fig. 6 was selected such that the vertex of the conical surface was situated on the model's bottom plane. The conical fastener's half-angle was 30°. Equation 9 was used to create the fastener in Fig. 4. In Figs. 1 and 6, the washer radii were selected to be marginally bigger than the maximum radii of the respective fasteners (Fig. 7).

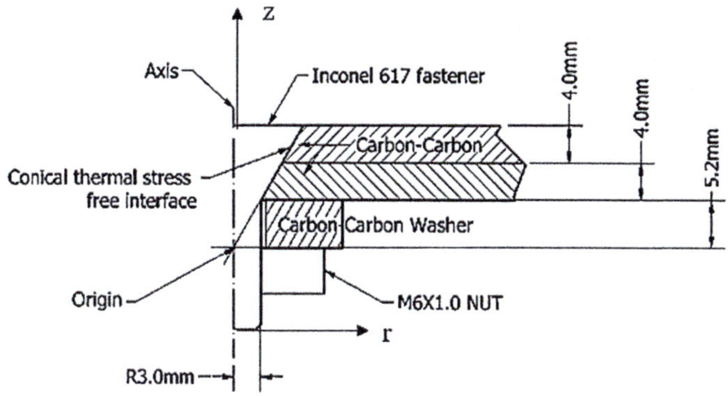

Fig. 6 Finite-element model of conical fastener and surrounding material

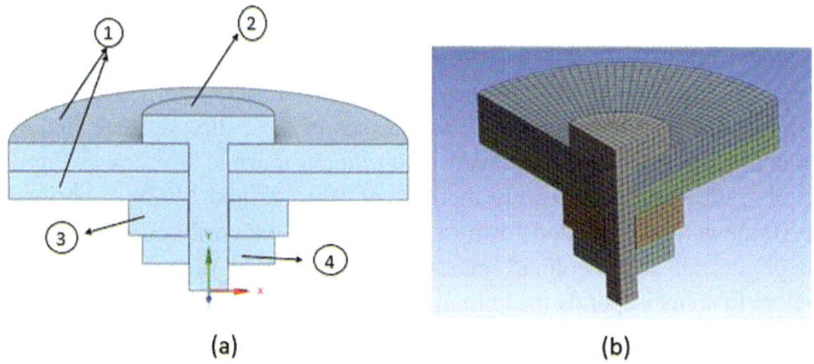

Fig. 7 Details of cylindrical fastener assembly **a** 3D model (1: carbon–carbon composite materials, 2: Inconel 617 bolt, 3: carbon–carbon composite washer, and 4: Inconel 617 nut) and **b** mesh model

A crucial component of the finite-element study was modeling the interaction between the fastener and the surrounding materials along the interface. Between the fastener and the carbon–carbon materials, as well as between the plates, frictionless contact is used. The displacement constraint is applied in such a way that it allows relative motion along the boundary between neighboring nodes that are not perpendicular to it, i.e., the displacement between the plates is zero along the axis. Table 2 shows the mechanical properties of materials used in models.

5 Numerical Results

Thermal stresses were calculated using three finite-element models by applying both thermal and structural loads of a uniform temperature increase of 1000 °C and a preload of 330 N. Axisymmetric analysis is performed to calculate the thermal stresses since geometry and loading conditions are symmetric. Therefore, the r-θ

Table 2 Mechanical properties of materials

	Inconel 617	Isotropic C–C composite	2D C–C composite	
Density (kg/m^3)	8360	1700	1700	
Young's modulus (GPa)	211	30	E_x, E_y	30
			E_z	8
Poisson ratio	0.3	0.3	0.2	
Thermal expansion coefficient (1 × 10^{-6} m/m/k)	13.9	0.2	α_x, α_y	0.2
			α_z	3.9

shear and the θ-z shear stresses are insignificant. The thermal stresses for an isotropic material combination in a cylindrical fastener joint are shown in Fig. 8a–d. The fastener material has a CTE of 13.9 × 10^{-6} m/m/°C, while the material surrounding the fastener has a CTE of 0.2 × 10^{-6} m/m/°C in all directions. The materials are in radial compression of – 274 MPa and tensile hoop stress of approximately the equal magnitude, as shown in Fig. 8a, b. The stresses in the z direction are insignificant. The conical fastener model displayed insignificant thermal stresses in Fig. 9 when subjected to the same temperature increase of 1000 °C and 330 N of preload (a–d). Figure 10 shows the extended form of an isotropic material conical fastener joint. The two materials at the conical interface do not separate or interfere even though the conical fastener expands more than the clamping members.

In the same way, the thermal stresses for transversely isotropic materials of m = 0.73 are calculated. Based on m value, need to generate a thermal stress-free interface and incorporate and do the simulation but due to difficulty in the manufacturing point of view, a conical interface is used instead of an actual thermal stress boundary for transversely isotropic materials joining using same loading conditions. The expanded shape of the conical fastener joint in isotropic materials and transversely isotropic materials is illustrated in Fig. 10a, b.

Both materials are allowed to expand without restriction during thermal expansion by sliding along the conical contact. However, the real difference in vertical expansion is just 0.19 mm, which is negligible in comparison to fastener size, despite the enhanced thermal expansion appearing to be significant, as illustrated in Fig. 10.

The initial preload given for all three models at room temperature is 330 N. After the temperature changes to 1000 °C, the preload changes in cylindrical, conical isotropic, and conical transverse isotropic models are 0 N, 300 N, and 1080 N, respectively. The conical interface is best feasible for isotropic materials since there is not much preload loss/gain as temperature changes and all the stresses induced are very low. The conical interface can be used for transversely isotropic materials if the induced stresses are within the acceptable range at all working temperatures (Fig. 11).

To determine the preload relaxation in fasteners, an experiment was carried out. The models utilized in the experiment are depicted in Fig. 12. Using a digital torque

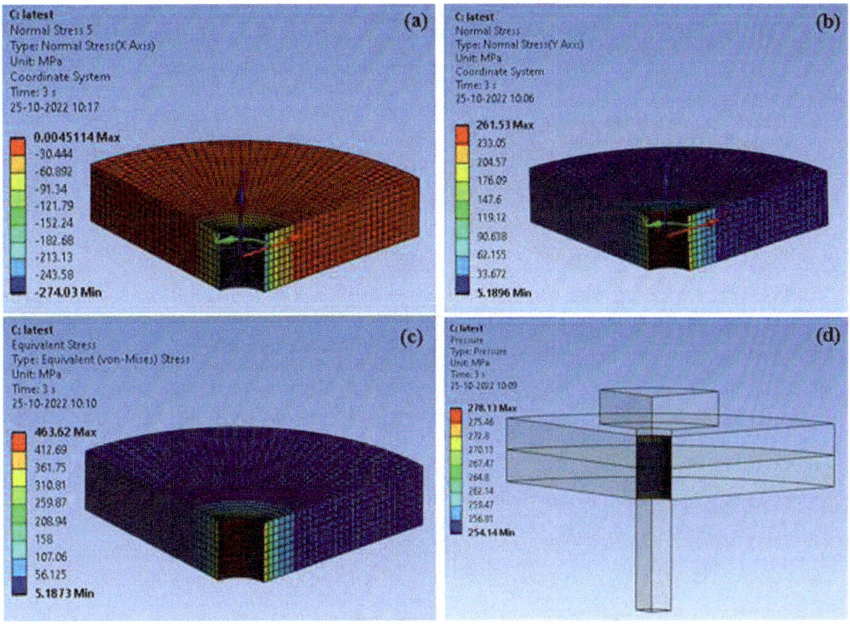

Fig. 8 **a** Radial compressive stress in the material surrounding the cylindrical fastener, **b** tensile hoop stress in the material surrounding the cylindrical fastener, **c** equivalent stress in the material surrounding the cylindrical fastener, and **d** contact stress along the interface

wrench, these models are initially preloaded to 330 N when they are at room temperature. These models are then heated uniformly to a high temperature of 1000 °C in the furnace. These models are then cooled to room temperature. Then, using a dial torque wrench, measure the fastener's preload loss/gain. Table 3 compares the experimental results with the simulated results.

Table 3 demonstrates that the cylindrical fastener model experiences relaxation. The conical fastener model has maintained the preload around 330 N, making it the best model for joining and maintaining snug joints at all working temperatures for isotropic materials. In contrast, the preload in the cylindrical model has decreased from 330 to 260 N.

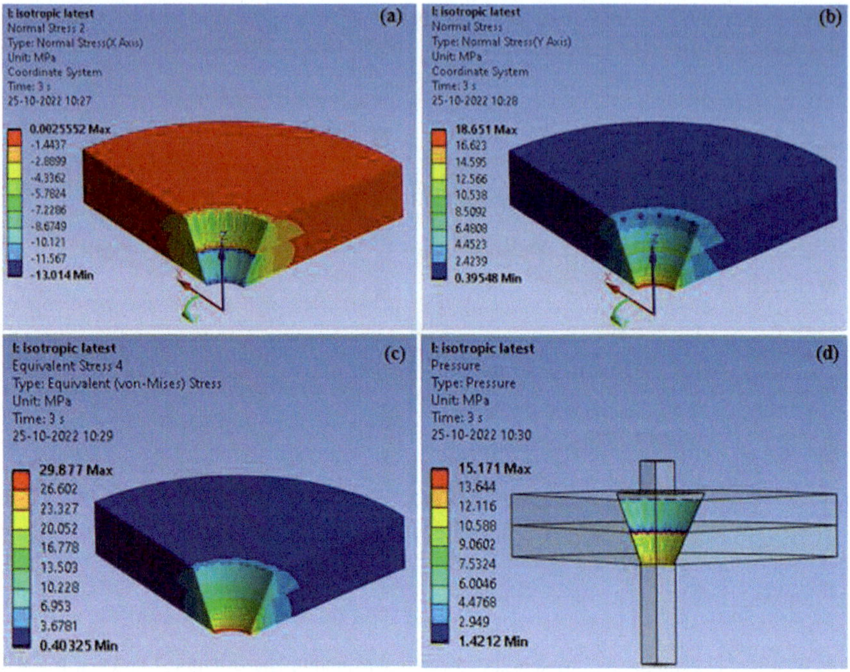

Fig. 9 **a** Radial compressive stress in the material surrounding the conical fastener, **b** tensile hoop stress in the material surrounding the conical fastener, **c** equivalent stress in the material surrounding the conical fastener, and **d** contact stress along the conical interface

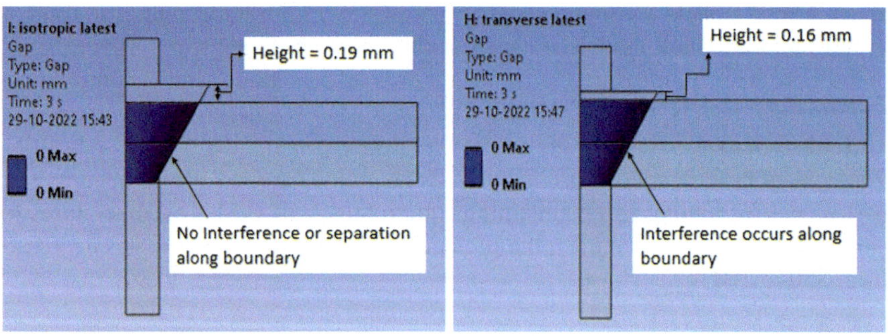

Fig. 10 Thermally expanded shape of finite-element model of conical fastener for **a** isotropic material and **b** transversely isotropic material (thermal expansion enhanced)

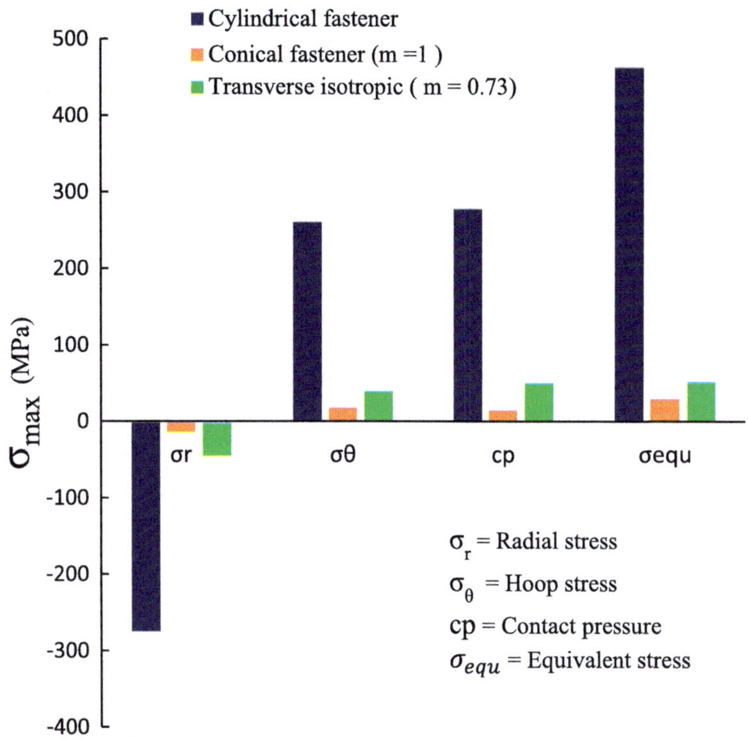

Fig. 11 Comparison of stresses for all three models

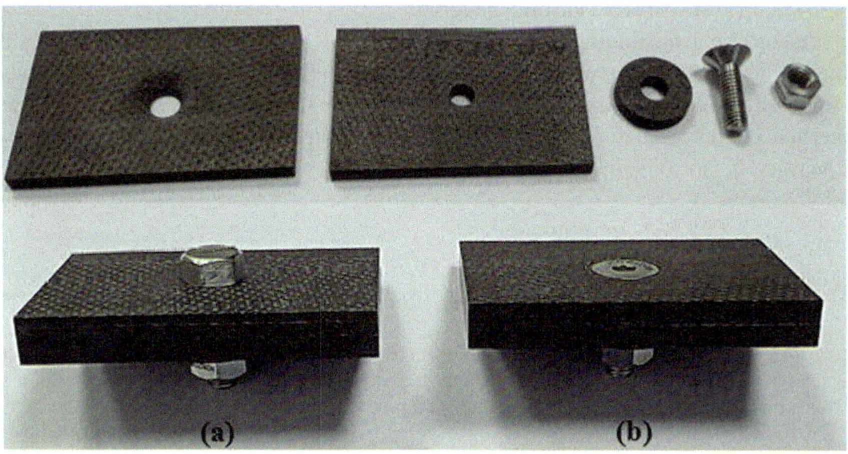

Fig. 12 **a** Cylindrical fastener along with 2D C–C composite plates' assembly and **b** conical fastener along with 2D C–C composite plates' assembly

Table 3 Comparison of preload values after cooling between experiment and simulation

Joint condition	Experimental preload (N)	Simulation preload (N)	% Error between experimental and simulation values	%Error between preload 330 N and achieved preload
Cylindrical model	260	275	5.7	21.21
Conical model	320	344	7.5	3.03

6 Conclusion

High-temperature composite materials have very low linear coefficients of thermal expansion (CTE's) than metallic fasteners. An ingenious method has been developed to reduce thermal stresses at interface by configuring the fastener shape and a mating hole. The mathematical equations have been developed to create stress-free interface for fastener and surrounding orthotropic materials. The developed algebraic equations result conical interface is the thermal stress-free interface for isotropic materials.

The stresses in joints are numerically analyzed using the finite-element method. In this study, the use of finite-element analysis served two purposes: first, to confirm that a typical joint with a fastener designed to reduce thermal stresses maintains contact while remaining thermally stress-free at high temperatures; and second, to examine the thermal stresses that arise when using a conical fastener to join a material with transversely isotropic CTEs. According to finite-elements' study, conical fasteners can eliminate thermal stresses for isotropic materials and reduce the contact thermal stresses for transversely isotropic materials.

According to the findings in this study, fasteners with a cylindrical shape can result in significant high stresses at interface for orthotropic materials joining at elevated temperatures. When compared to a cylindrical fastener, conical fasteners, which are thermally stress-free for isotropic materials, can lower thermal stresses for some transversely isotropic materials.

References

1. Blosser ML, McWithey RR (1983) Theoretical basis for design of thermal-stress-free fasteners
2. Naslain R (2004) Design, preparation and properties of non-oxide CMCs for application in engines and nuclear reactors: an overview. Compos Sci Technol 64:155–170. https://doi.org/10.1016/S0266-3538(03)00230-6
3. Böhrk H, Beyermann U (2010) Secure tightening of a CMC fastener for the heat shield of re-entry vehicles. Compos Struct 92:107–112. https://doi.org/10.1016/j.compstruct.2009.07.002
4. Blosser ML (1989) Thermal-stress-free fasteners for joining orthotropic materials. AIAA J 27:472–478
5. Zhao SY, Li ZY, Pu ZL, Sun XY, Zhang WJ (2021) Thermal stress and assembling parameters evolution of CMC bolted joint under transient thermal loads. J Harbin Inst Technol (New Ser) 28:1–8. https://doi.org/10.11916/j.issn.1005-9113.2019038

6. Blosser ML (1988) Thermal stress in high temperature cylindrical fasteners
7. Sawyer JW, Blosser ML, McWithey RR (1985) Derivation and test of elevated temperature thermal-stress-free fastener concept. Welding Bond Fasten 1984
8. Wulz H (1999) Joining, fastening and sealing of hot CMC structures. In: 33rd Thermophysics conference, p 3552

Multiple Reinforced Composites, the Possible Solutions to Avoid Delamination in Composite Rocket Motor Casings

Rakesh Kumar and Basant Kumar

1 Introduction

Composite products can have many types of defects like delamination, de-bond, air pockets, blisters, etc., and these defects may cause damage to the composite products. The damage caused by separation of layers due to decomposition of material between the layers of composite laminate is called delamination. Delamination may occur during the manufacturing as well as load application. Delamination can reduce the load-carrying capacity which can lead to failure of the product during the load application.

Practically, it is very important to identify the onset of delamination and its progression resulting the failure. Maimi et al. [4] have determined the onset of matrix cracking for in-plane stress states through his proposed plastic damage model. The prediction for the onset of delamination is very important aspect of the product design for its reliability and reproducibility. Short et al. [5] have studied the effect of the delamination geometry and its location for the compressive behaviour of laminated glass reinforced composite flat plate. Through experiments, it was concluded that the buckling load of composite plates is decreasing in the presence of delamination. The buckling loads depend on delamination size, geometry, and its location inside the plate. Kutlu and Chang [2] have investigated analytically and experimentally and reported the studies on the effects of multiple delamination on the compression behaviour of laminates. Li et al. [3] used an extended layer-wise method to model composite laminated beam with multiple delamination and matrix cracks. Suematsu et al. [6] analyse the composite plate with multiple circular delaminations and investigate the failure mechanism by considering the delamination growth using cohesive interface element. It is concluded that the load capacity of a composite structure

R. Kumar (✉) · B. Kumar
Advanced Systems Laboratory (DRDO), Kanchanbagh- Post, Hyderabad 500058, India
e-mail: rakanitp@gmail.com

M. R. Manohara Babu et al. (eds.), *Recent Advances in Composites Engineering and Research*, Springer Proceedings in Materials 39,
https://doi.org/10.1007/978-981-99-8807-5_5

with several delaminations depends on the growth of delamination. Jayatilake et al. [1] have developed a three-dimensional finite element model and investigate the free vibration behaviour of composite sandwich panel with delamination by using the first-order shear deformation theory-based plate elements. It is concluded that the size of delamination and boundary conditions are mainly responsible for the reduction in stiffness in composite laminate.

Composite rocket motor casing (CRMC) is widely used in various aerospace vehicles. Rocket motor case's functional requirement is as a Stage Rocket Motor (SRM) to provide interfaces to other components, contain propellant, withstand inner pressure to provide required thrust.

The general configuration of CRMC is shown in Fig. 1. CRMC is filament-wound composite pressure vessel which consists of cylindrical geometry at the central portion and adjoined with domes at both the ends. These end domes are called nozzle-end (NE) and igniter-end (IE) domes. The metallic end fittings are attached to domes through rubber shear ply at the both end domes. The interface to the launch vehicles is obtained by the forward and aft skirts. The skirts are connected to the vessel by means of a rubber shear ply. They are laid up of carbon/epoxy layup interspersed with hoop winding. The skirts terminate with high strength aluminium alloy bulkheads (BH) to interface the Launch vehicle interstage.

The detail B provided in Fig. 1 is termed as Y-joint. This is providing a smooth load transfer mechanism from the bulkhead to cylindrical section of the CRMC through

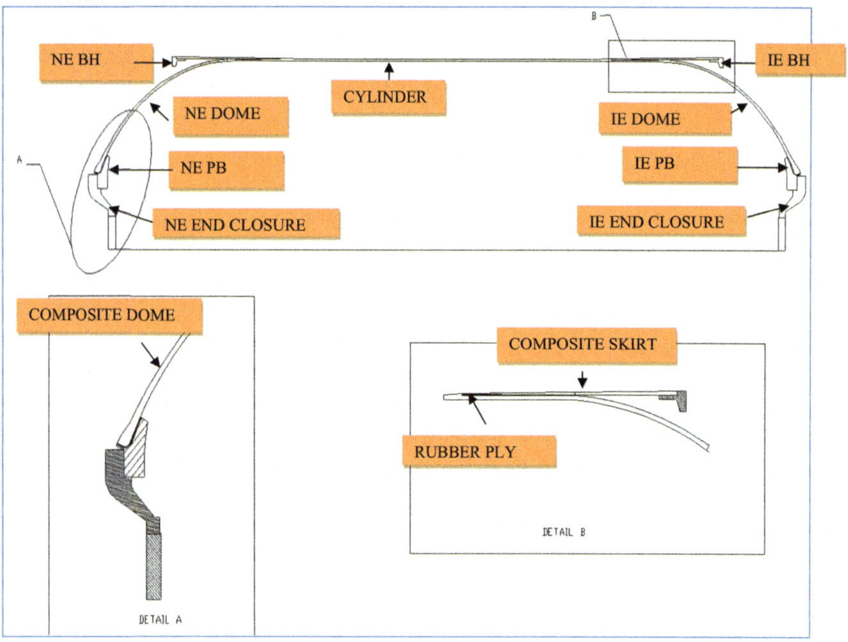

Fig. 1 Composite case architecture

skirt or vice versa. Any delamination at this zone will reduce the load-carrying capacity of CRMC and can lead to its failure during loading operations.

In this paper, the detail study of occurrence of delamination at Y-joint during the PPT of a typical CRMC is described systematically.

2 Problem Definition

The Y-joint configuration of CRMC is shown in Fig. 2. The gap between the casing and the shear ply is filled by the carbon hoop layer of variable thickness known as the skirt seating. Shear ply interfacing the skirt seating and the skirt.

After the proof pressure test of CRMC, the crack zone is identified inside the skirt seating by the NDT methods as shown in Fig. 3.

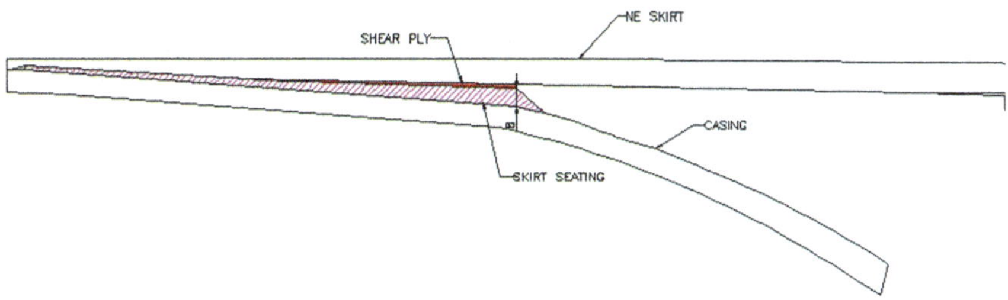

Fig. 2 Y-joint configuration

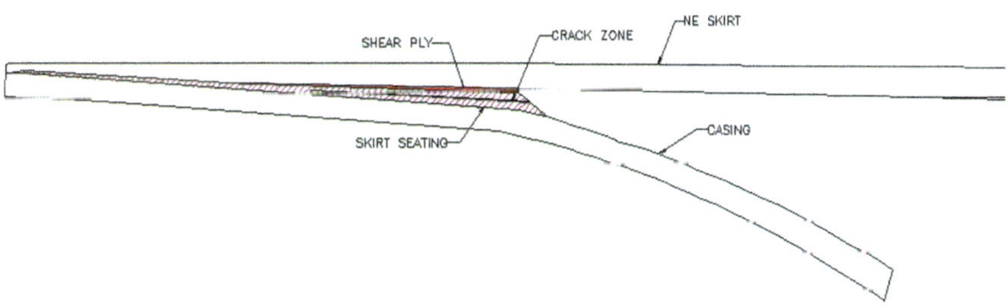

Fig. 3 Crack position at y-joint after PPT

3 Theory

There are many theories for the design and analysis of laminated composites. These theories can be classified as equivalent single-layer (ESL), layer-wise, and zig-zag theories. The ESL theory is broadly accepted for the design and analysis of laminated composites' theory which can be divided in to three main categories: classical laminated plate theory (CLPT), first-order shear deformation theory (FSDT), and higher order shear deformation theory (HSDT).

The CLPT ignores shear deformation effects and provides reasonable results for thin plates. CLPT theories hold good for the design and analysis of CRMC.

If we consider a three-dimensional elastic body subjected to external loads is under static equilibrium stages. The orthogonal Cartesian system as shown in Fig. 4 as reference material system of the body is considered. The differential parallelepiped element of the sides (dx, dy, dz) parallel to the coordinate plane is assumed. Stresses on the faces of the element will be produced due to external forces. These stresses are normal stress and shear stress. If body forces are negligible and assumed to be zero and body is in equilibrium, then the following differential equation of equilibrium must be satisfied. Theory of reciprocity is used in deriving these equations.

$$\frac{\partial \sigma_x}{\partial x} + \frac{\partial \tau_{xy}}{\partial y} + \frac{\partial \tau_{xz}}{\partial z} = 0,$$
$$\frac{\partial \sigma_y}{\partial y} + \frac{\partial \tau_{yx}}{\partial x} + \frac{\partial \tau_{yz}}{\partial z} = 0,$$
$$\frac{\partial \sigma_z}{\partial z} + \frac{\partial \tau_{zx}}{\partial x} + \frac{\partial \tau_{zy}}{\partial y} = 0.$$

From these above relations, it is obvious that interlaminar stress is always present inside the externally loaded plate even if plate is assumed to be classical plate. For out of plane loaded plate, in-plane shear stress are diminished at the edges, and to maintain the equilibrium condition, interlaminar stresses have to be increased

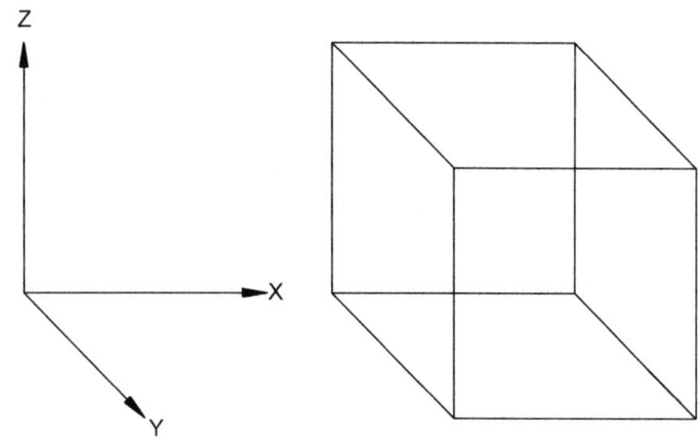

Fig. 4 Coordinate system

accordingly. This is why the laminated composites are most likely to be delaminated at the edges if those are subjected to out of plane loading.

It is important to mention that while the theory of thin plate omits the effect of the strain component in the thickness direction on bending, the vertical shear forces Q_x and Q_y are not negligible. In fact, they are necessary for equilibrium of the plate element. Vertical s_{hear} force can be expressed in the terms of the stress component.

$$Q_x = \int_{-h/2xz}^{h/2} dz,$$

$$Q_y = \int_{-h/2yz}^{h/2} dz$$

Determination of the τ_{xz} and τ_{yz} is not possible by Hooke's law because of the assumption of classical plate theory. The differential equations of equilibrium for a plate element under a general state of stress serve well for this purpose; however, if the faces of the plate are free of any tangential external load, then τ_{xz} and τ_{is} are zero for Z = ± *h*/2. The shear stresses τ_{xz} and τ_{yz} are as follows:

$$\tau_{xz} = -\int_{-h/2}^{h/2}\left(\frac{\partial\sigma_x}{\partial x} + \frac{\partial\tau_{xy}}{\partial y}\right)dz = \frac{E\left(z^2 - \frac{h^2}{4}\right)}{2\left(1-^2\right)}\frac{\partial}{\partial x}\nabla^2 w,$$

$$\tau_{yz} = -\int_{-h/2}^{h/2}\left(\frac{\partial\sigma_y}{\partial y} + \frac{\partial\tau_{yx}}{\partial x}\right)dz = \frac{E\left(z^2 - \frac{h^2}{4}\right)}{2\left(1-^2\right)}\frac{\partial}{\partial y}\nabla^2 w$$

where ^2w () is the Laplace operator, given by

$$\nabla^2 = \frac{d^2w}{dx^2} + \frac{d^2w}{dy^2}$$

In these equations, it is observed that the stresses τ_{xz} and τ_{yz} vary according to a parabolic law. Magnitude of these stresses can be reduced if the thickness of plate is reduced. These stresses can also be minimised if plate material is having lower elastic modulus. Laminated composites can be designed with desired modulus with different combinations of reinforcements. It was the fundamental basis of this research to avoid the delamination at Y-joint of CRMC. Final ply thickness and ply sequence is derived on the basis of finite element analysis results.

4 Finite Element Analysis

FEA of the CRMC is carried out by using ANSYS. For the purpose of analysis, a number of cases have been identified and FEA has been carried out for each of these cases. The number of load cases considered for analysis is as follows (Table 1).

External thermal protection is the hoop layer winding by the insulation material to protect the CRMC from kinetic heating during the flight. With external thermal protection layer, there is increment in stresses like σ_{yy} and σ_{yz} which has been observed at NE seating zones, and other all parameters like helical stress and hoop stress are comparatively not changed. According to Professor Paul A. Lagace and John Charles Brewer (with reference to PHD thesis produced by John Charles Brewer at MIT Boston), the delamination will start at the corner of laminate if the following condition is violated:

$$\left(\frac{\sigma_{yy}}{S_{YY}}\right)^2 + \left(\frac{\sigma_{YZ}}{S_{YZ}}\right)^2 + \left(\frac{\sigma_{XY}}{S_{XY}}\right)^2 \leq 1.$$

σ_{yy}—Average interlaminar tensile or compressive stress.
σ_{YZ}, σ_{XY}—Average interlaminar shear stress.
S_{YY}—Interlaminar tensile or compressive strength.
S_{YZ}, S_{XY}—Interlaminar shear strengh.

The left-hand side portion of above equation is calculated by putting interlaminar stress and strength. This calculated value is termed as delamination factor. This is an indicative parameter for initiation of delamination of composite laminates subjected to loads. The delamination will initiate if delamination factor is greater than or equal to one under acceptable loading condition.

With external thermal protection layer, CRMC is violating the above conditions marginally by considering interlaminar shear strength equal to 50 Mpa and interlaminar compressive strength is equal to 100 Mpa (resin compressive strength = 100 Mpa) (Fig. 5).

The results shown in Table 2 is giving the indication of delamination at NE seating during the PPT of CRMC if external thermal protection layers are wound over the CRMC. The similar outcomes are observed to CRMC after PPT.

Table 1 Categorization of finite element analysis

Condition	Cases (considered for analysis)			
	Without external thermal protection	With external thermal protection layer	Remedial action-1 with external layers	Remedial action-2 with external layers
Without crack	Yes	Yes	Yes	Yes
With crack at NE seating	No	Yes	No	No

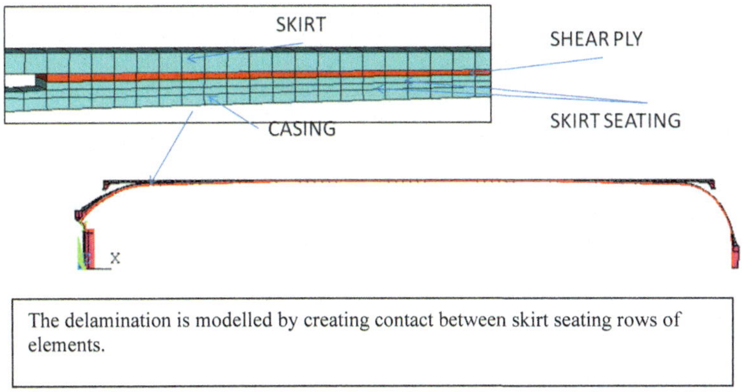

Fig. 5 FE model

Table 2 Results' comparison for different states of CRMC

Parameter	With external thermal protection layers		Without external thermal protection layers
	Without crack	With crack at NE seating	Without crack
σ_{yy} (Mpa) radial stress at NE seating	− 18	− 8	− 37
σ_{yz} (Mpa) at NE seating	− 48	− 25	− 42
σ_{xy} (Mpa) at NE seating	− 16	− 8	− 18
Delamination factor	**1.0564**	0.4944	0.97

4.1 Proposed Remedial Action-1

In order to avoid the delamination problem, a modified configuration has been finalised based on the iterative design through analysis method. Number of iterations has been carried out with respect to decreased delamination factor by decreasing the seating thickness and other performance parameter are altered marginally with in acceptance limit of factor of safety. These analyses were carried out on ANSYS finite element analysis software. Three-dimensional eight-node solid element with structural and layered solid capabilities is used. Based on performance parameter optimization and manufacturing processability, the configuration is finalised. In this configuration, thickness of NE and IE seating zone has been reduced to 3.5 mm from 8 mm as shown in Fig. 6.

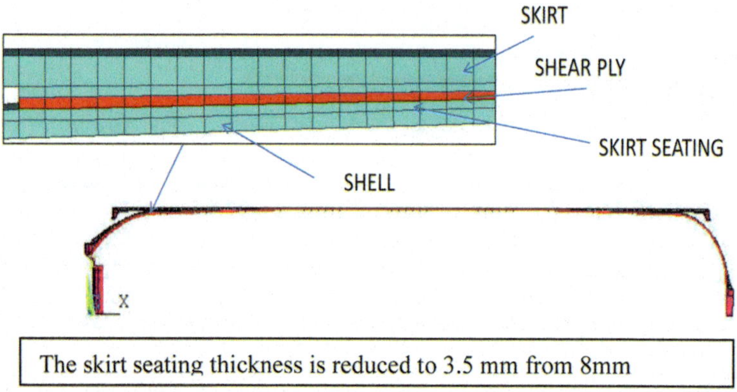

Fig. 6 FE model of remedial action-1

4.2 Proposed Remedial Action-2

In order to avoid the delamination problem one modified configuration has been modelled on FEM and analysed. In this configuration, the material between the shear ply and skirt is replaced by E-glass. Keeping seating thickness to 3.5 mm, as shown in Fig. 7.

The results shown in Table 3 are indicating that the delamination factor has been reduced drastically by reducing the thickness of seatings. This factor is decreased further by replacing the material between shear ply and skirt from carbon epoxy to E-glass epoxy.

The finite element results are having the similar kind of trend as the theory of plates describes. The interlaminar stresses are decreasing by reducing the seating thickness as well as by providing low stiffness material like E-glass in place of carbon as reinforcement. However, these results may affect by the geometry of product, loading conditions also.

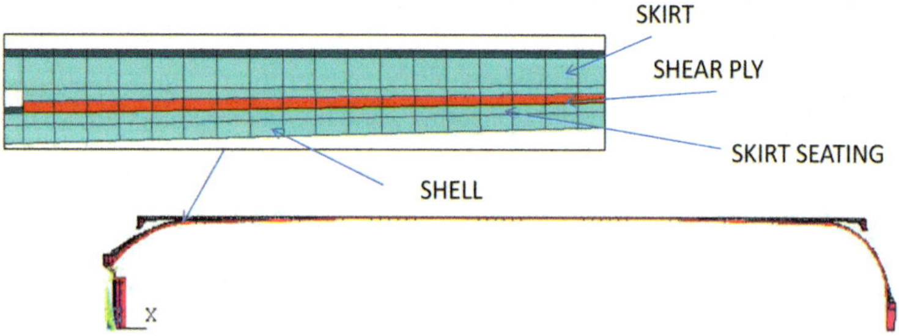

Fig. 7 FE model of remedial action-2

Table 3 Results comparison between remedial actions

Parameter	Remedial action-1	Remedial action-2
σ_{yy} (Mpa) at seating	− 25	− 22
σ_{yz} (Mpa) at seating	− 34	− 7
σ_{xy} (Mpa) at seating	− 6	− 6
Delamination factor	0.5393	**0.09**

5 Conclusion

As a remedy, the remedial action-2 configurations are accepted and CRMC is fabricated accordingly. This problem of delamination occurring during the PPT gets rectified. This configuration is combination of E-glass and carbon fibre as reinforcement in order to get desired stiffness of composites to avoid the delamination.

However, the ratio of E-glass and carbon epoxy and their ply sequence to be used can be determined by proper analysis only as the interlaminar stress is dependent on geometry and loading conditions of product also.

Acknowledgements The authors would like to thank Director Advanced systems Laboratory (DRDO), colleagues, and members of different technology centres, to carry out the research work of delamination study of CRMC.

References

1. Jayatilake IN, Karunasena W, Lokuge W (2016) finite element based dynamic analysis of multilayer fibre composite sand wich plates with interlayer delamination
2. Kutlu Z, Chang FK (1992) Modelling compression failure of laminated composites containing multiple through the width delamination
3. Li DH, Liu Y, Zhang X (2015) An extended layer wise method for composite laminated beams with multiple delaminations and matrix cracks
4. Maimi P, Camanho PP, Mayugo JA, Turon A (2011) Matrix cracking and delamination in laminated composites
5. Short GJ, Guild FJ, Pavier MJ (2001) The effect of delamination geometry on the compressive failures of composite laminates
6. Suemasu H, Sasaki WT, Aoki Y (2008) A numerical study on compressive behaviour of composite plates with multiple circular delaminations considering delamination progression
7. Wang XW, Lezica IP, Harris JM, Guild FJ, Pavier MJ (2005) Compressive failure of composite laminates containing multiple delaminations

Application, Product Realisation and Testing

Design, Development, Optimization and Standardization of Indian Mini Satellite (IMS-2) Bus-Structure Using Composites

Archana Shukla, G. Srinivasa Rao, M. N. M. Patnaik, M. V. Peereswara Rao, C. S. Varghese, T. S. Govindaraja, and D. Poomani

1 Introduction

The modular construction facilitates standardization [1] of the main bus systems with standard mechanical and electrical interfaces for the payload. This allows the flexibility of design and qualification of the payload module independently. In IMS-2 bus, the main module is designed to accommodate the spacecraft main subsystem elements like propellant tanks, power elements, AOCE packages (Attitude and Orbit Control Electronics), data receiving and transmission packages, on-board data collection elements. This imposes a challenge to provide a structural bus which is volume efficient and provides the required strength and stiffness at same time. Though to take care of this challenge, the previous standard structural buses were designed with a central load-carrying composite cylinder with all the other structural members built around it.

IMS-2 bus is to be used for Lower Earth Orbit (LEO) missions payloads and to be launched by ISRO-PSLV launcher. Hence, the structure design should consider the above-mentioned requirements in addition to incorporating the assembly modularity of the main and payload modules (Fig. 1).

The ISRO's 400–450 kg IMS-2 bus has two separate modules as main and payload platforms. In this 400–450 kg satellite bus, about 200–250 kg mass is assigned for the main module and about 200 kg mass for the payload module. The payload platform interfaces with the main platform at four locations using titanium brackets. These brackets are specially designed to transfer the loads from payload bus to main bus effectively. Titanium brackets meet both the structural load and thermal requirements

A. Shukla (✉) · G. Srinivasa Rao · M. N. M. Patnaik · M. V. Peereswara Rao · C. S. Varghese · T. S. Govindaraja · D. Poomani
U R Rao Satellite Centre, ISRO, Bangalore 560017, India
e-mail: sarchana@ursc.gov.in

M. R. Manohara Babu et al. (eds.), *Recent Advances in Composites Engineering and Research*, Springer Proceedings in Materials 39,
https://doi.org/10.1007/978-981-99-8807-5_6

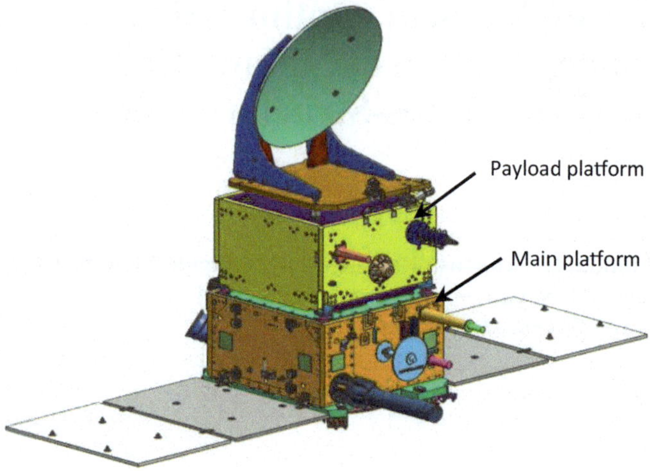

Fig. 1 IMS-2 modular bus structure. *Image source* ISRO website

of the payload. The global flatness of the payload module interfaces is maintained as 0.1 mm.

In order to develop and qualify this bus, the detailed design, analysis and testing are performed.

2 Design Requirements and Constraints

The structure design [2] is primarily based on the requirements and constraints imposed by PSLV launch vehicle. The design load requirements of PSLV launch vehicle are given below.

2.1 *Static and Dynamic Loads*

The quasi-static and sine loads are specified by the launcher which are given in Tables 1 and 2.

[Center of Gravity (CG) of the spacecraft: 647 mm from the separation plane].

The qualification loads are 1.25 times the flight loads.

Table 1 Quasi-static loads

S. No.	Longitudinal 'g'	Lateral 'g'
1	− 7 g	± 1.5 g

Table 2 Dynamic loads

Particulars	Longitudinal		Particulars	Lateral	
	Amplitude			Amplitude	
Freq. (Hz)	Qual. level	Accpt. level	Freq. (Hz)	Qual. level	Accpt. level
5–11.5	3.4 mm	2.3 mm	5–7	3.4 mm	2.3 mm
11.5–38	1.8 g	1.2 g	7–30	0.67 g	0.45 g
38–60	1.8–0.75 g	1.2 g–0.5 g	30–100	0.45 g	0.3 g
60–100	0.75 g	0.5 g			
Sweep rate	2 Oct/Min	4 Oct/Min	Sweep rate	2 Oct/Min	4 Oct/Min

2.2 *Stiffness, Strength and Buckling Requirements*

Spacecraft structural design is generally driven by stiffness. The stiffness requirements specified by launcher and the required design strength margins are given in Table 3.

3 Material Selection

Spacecraft structure design is primarily stiffness based. The design [4] is based on the improved structural efficiency for a given mass. Traditionally aluminum alloys are used in design and construction of spacecraft structures. The basic reason to use aluminum alloys is its lightweight and having better mechanical performance parameters such as specific stiffness (E/ρ) and specific strength (σ/ρ) compared to other metals. Honeycomb sandwich construction decks with aluminum face sheet and aluminum honeycomb core are used to minimize the mass and increase the reliability of the process.

Titanium brackets are used for connecting payload module to mainframe module. Table 4 provides the mechanical properties of metals used. Table 5 provides the mechanical properties of honeycomb core.

Where

Table 3 Stiffness, strength and buckling requirements

1	Stiffness	First global lateral frequency	> 18 Hz
		Global longitudinal frequency	> 40 Hz
2	Strength	Strength margin to material failure	> 0.1
3	Buckling	Local buckling	> 0.25
		Global buckling	> 2.0
4	Fasteners/Inserts	Margin of safety	> 2.0

Table 4 Material properties for metals

Material	E (GPa)	μ	G (GPa)	σ (MPa)	ρ (kg/m^3)
AA 2024	72	0.33	27	270	2800
Titanium	111	0.3	43	910	4500

Table 5 Material properties of honeycomb core

Material	μ_{lw}	G_{lt} (N/m^2)	G_{wt} (N/m^2)	ρ (kg /m^3)
Low-density core	0.30	1.85E + 08	0.89E + 08	32
High-density core	0.30	9.85E + 08	3.51E + 08	130

'E' denotes material Young's modulus.

'G' denotes material shear modulus.

'σ' denotes material strength.

'ρ' denotes material density.

And 'μ' denotes the Poisson's ratio for the material.

4 Structural Configuration and Design

4.1 Main Platform

Generally, spacecraft structural configuration consists of a central cylinder and structural panels built around it. The central cylinder provides the major stiffness and is the main load-carrying member. However, considering volume constraints and subsystem assembly requirements having central thrust cylinder is not feasible here. The overall configuration is a cuboid (box) structure with six honeycomb panels interconnected by structural angles. The required stiffness is arrived by proper design of panels and connecting angles.

Panel thickness can be arrived at as per stiffness requirements. To transfer launch loads to the interface ring, bottom deck and other structural elements are strengthened adequately. Design safety margins for all structural joints are established under all static and dynamic load environments (Fig. 2).

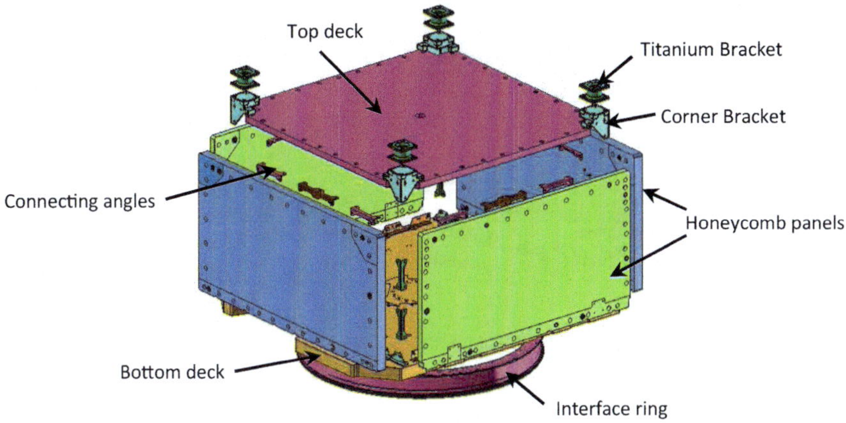

Fig. 2 Exploded view of a typical IMS-2 main platform

4.2 Payload Platform

The payload platform is the module which contains the assembly of various payload elements. This module can be assembled separately and brought to the main platform for the overall spacecraft assembly.

4.3 Interface to Payload Platform

Payload module is assembled to the main platform at four corners using titanium brackets. Payload module to main platform assembly is shown in Fig. 3. Flatness measurements at the assembly interfaces are carried out, meeting design specifications.

Fig. 3 Payload module to main platform assembly

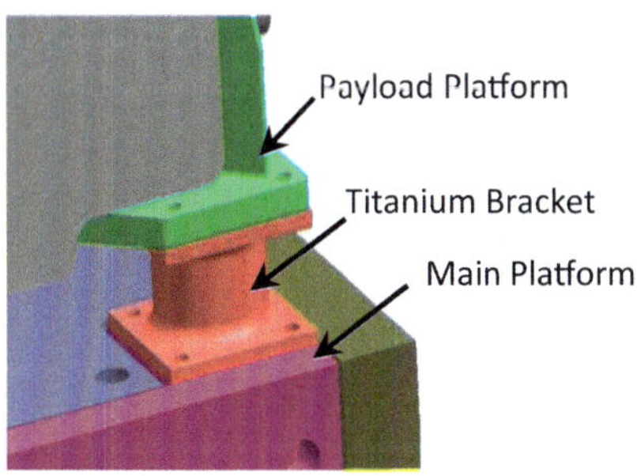

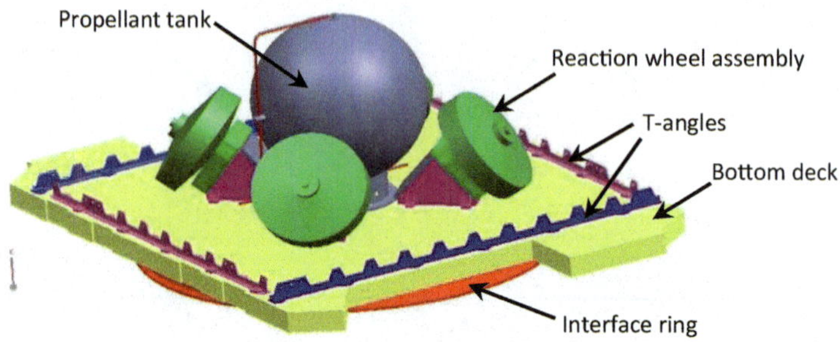

Fig. 4 Bottom deck with propellant tank and reaction wheels

5 Major Spacecraft Assemblies

5.1 Bottom Deck Assembly

Bottom deck is an aluminum sandwich construction panel which accommodates the propulsion elements and reaction wheels as shown in Fig. 4. One mono-propellant tank is housed at the center of bottom deck having an embedded ring. The tank is connected at 18 number of interfaces. The bottom deck is sufficiently stiffened at interface ring and equipment decks' interface locations.

5.2 Bottom Deck to Equipment Deck Assembly

Bottom deck is connected to the vertical decks using specially designed T-shaped angles as shown in Figs. 5a, b.

5.3 Equipment Decks' Assembly

One equipment deck is connected to another equipment deck using specially designed corner brackets as shown in Fig. 6. Upper portion of the bracket is connected to titanium bracket of the payload module. Along the length of the equipment decks, L-angles are used for mutual connections.

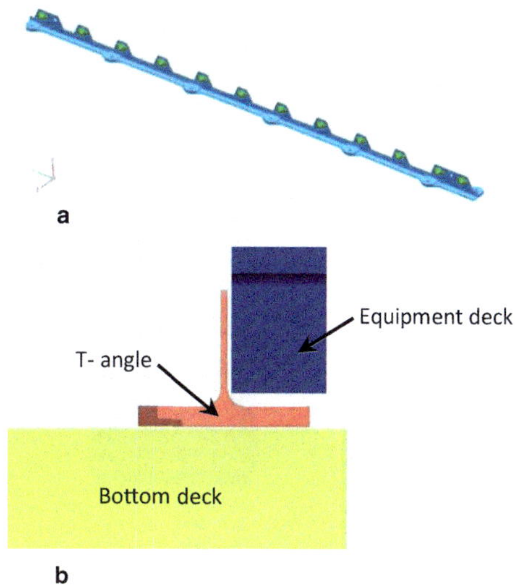

Fig. 5 **a** T-shaped angle. **b** Bottom deck to equipment deck assembly

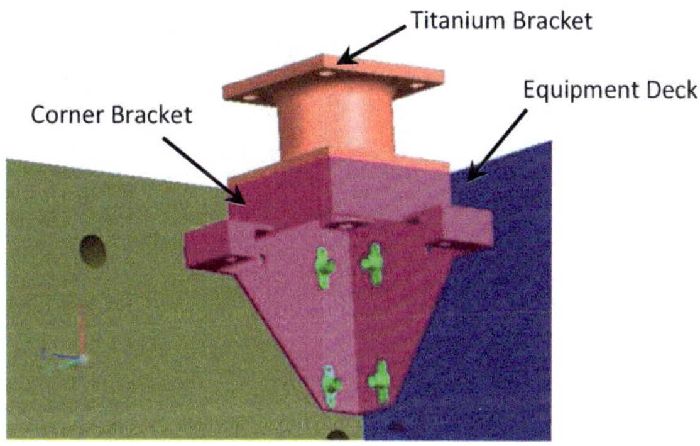

Fig. 6 Equipment deck to equipment deck assembly

6 Finite Element Analysis

The finite element (FE) model (Refer Fig. 7) of the spacecraft is generated in PATRAN software. Finite element model of the spacecraft has over nine-lakh degree of freedom. Detailed finite element model is developed using shell elements (CQUAD4), beam elements (CBAR) and other similar elements. The lower end of the interface ring (also a connecting interface to the launch vehicle) of the spacecraft is taken as the separation plan.

Free vibration analysis of the spacecraft is carried out by constraining all the six degrees of freedom of all the nodes corresponding to the base of the interface ring of the spacecraft (Refer Fig. 7). This forms the boundary condition for FE analysis. Important subsystems like battery, solar array panels, inertial reference unit (IRU) and propellant tanks are modeled and attached. Subsystem FE models, like reaction

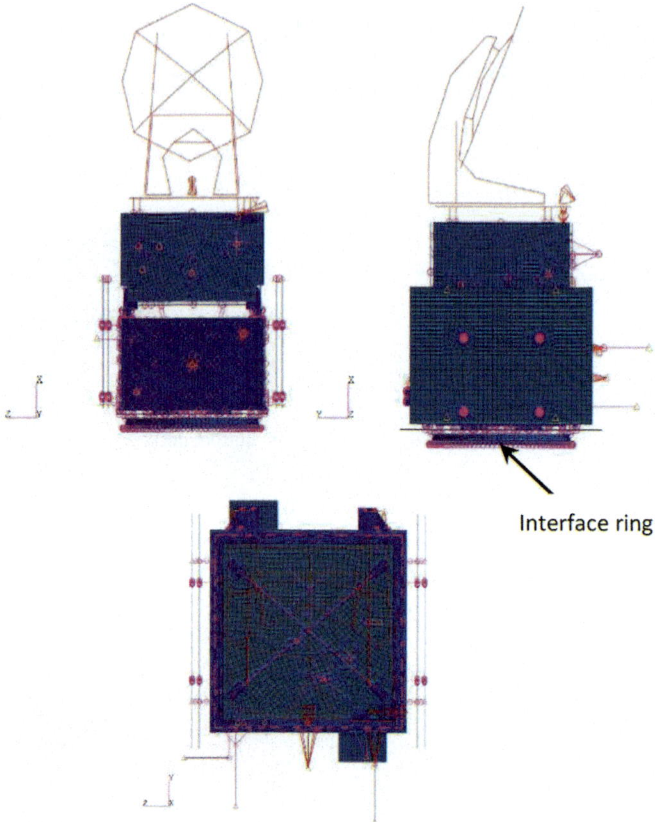

Fig. 7 FE model of IMS-2 spacecraft

wheels, sensors, telemetry, tracking and command antenna (TTC), are also assembled to the respective locations. Subsystems having mass more than 2 kg are simulated as lumped masses at their respective CG locations and mounted at the corresponding decks using beam elements, simulating their frequencies > 120 Hz. Other subsystems with mass less than 2 kg are smeared over their respective panels. FE model of payload module is assembled at its interface location to the main platform. An Eigen value problem is solved on the simulated mathematical model, of overall 450 kg spacecraft mass and its associated stiffness. Table 6 shows the important compliance parameters obtained with respect to the design specifications provided.

Table 6 Important compliance parameters

S. No.	Design requirement	Specs (Hz)	Obtained (Hz)
1	First global lateral frequency	> 18	22.7
2	Global longitudinal frequency	> 40	51.8

Fig. 8 Fabricated structure of main module

Response analysis is performed for the complete FE model, by giving 1-g acceleration in all the three directions, at the spacecraft base. The acceleration responses at all the subsystem interfaces are estimated and are meeting the design requirements. Static and dynamic forces are calculated at all the critical locations, such as the joints between interface ring to bottom deck, bottom angle to bottom deck, and at the payloads' interfaces. Safety margins are assured before clearing the design.

For the preprocessing and postprocessing, i.e., for the FE modeling and output result processing, PATRAN software is utilized. NASTRAN software is utilized for solving the numerical equations. The NASTRAN analysis outputs are read into PATRAN software, and the results are presented.

7 Structure Assembly

Equipment decks are attached to each other by using thin L-angles. Refer Fig. 8 for the fabricated main structure module and Fig. 9 for the integrated modular spacecraft assembly.

8 Static and Dynamic Tests

8.1 Static Test

As a part of static qualification, the static load tests [3] were done on the payload joint as this is a novel joint. The objective of the static test is to qualify this joint suitable for IMS-2 Bus. The two load cases considered for the tests are:

Fig. 9 Fabricated modular spacecraft assembly

Load case1 (LC-1)

Longitudinal: – 12.5 g

Load case2 (LC-2)

Longitudinal: – 4 g

Lateral: 6 g

The payload mass considered is 200 kg. The CG height taken is 400 mm from the titanium bracket interfaces. Figures 10 and 11 show the longitudinal and lateral test setups for the same. Strain gauges are mounted at the important locations, which are arrived on the basis of FE stress analysis.

Fig. 10 Longitudinal loading test setup

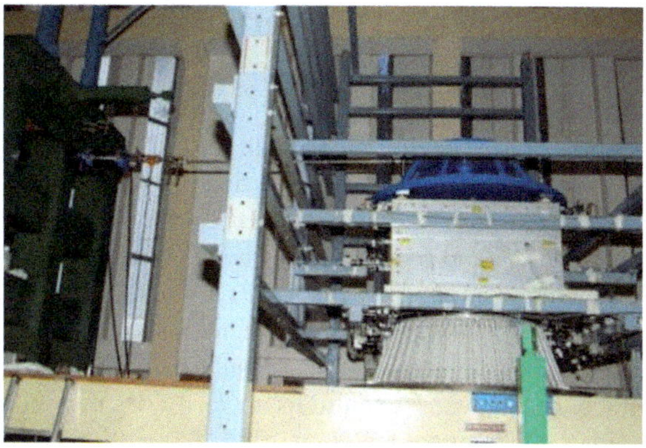

Fig. 11 Lateral loading test setup

Maximum stress and displacements are within design limits and negligible permanent set is observed. The low-level sine vibration test qualifies the whole bus structure. Hence, IMS-2 bus structure is demonstrated to be integral and safe.

8.2 Dynamic Test

After the static qualification test, the dynamic test [3] is carried out on the whole integrated satellite. The main objective of the tests is to demonstrate satisfactory performance of the spacecraft and to qualify the spacecraft for the specified sine and acoustic loads.

It also serves as a workmanship test for already qualified elements and assemblies. Proto-flight sine vibration sweeps between 5 and 100 Hz is carried out. The fundamental mode frequencies and vibration responses are captured during the vibration test. A comparison is made between FE analysis and test natural frequencies, which shows a good correlation as mentioned in Table 7.

Table 7 Comparison of FE analysis and test natural frequencies

S. No.	Mode	Analytical frequency (Hz)	Test frequency (Hz)
1	Global lateral	22.7	26.47
2	Global longitudinal	51.8	50

9 Test and Analysis Correlation

The detailed structural analysis is carried out for the static test configuration of IMS-2 spacecraft structure along with test adaptor and payload simulator. The test load cases comprise of pure compressive load (specified as − 12.5 g longitudinal for the payload assembly) and combined compressive and lateral loads (specified as -4 g longitudinal and 6 g lateral for the payload assembly). Displacement pattern, stress and strain contours and the forces at the structural interfaces are obtained for these load cases. Maximum displacement, stresses, strains and interface forces are noted for the above load cases. These results helped in understanding the structural behavior under the static test conditions, to verify whether the test simulation closely matches with the theoretical results.

The strains and deflections were correlated with FE model and the FE model was updated accordingly.

10 Conclusion

The IMS-2 bus is qualified for the quasi-static loads as per PSLV load specifications, i.e., longitudinal limit load of7 g (compression) and lateral limit load of ± 1.5 g. The designed global lateral and global longitudinal frequencies are 22.7 Hz and 51.8 Hz, respectively. The tested global lateral and global longitudinal frequencies are 26.47 Hz and 50 Hz, respectively. Hence the design is validated by dynamic tests and found good match. Maximum stress and displacements are well within design limits and negligible permanent set observed. The realized structure mass is 50 kg, which is only 11% of overall spacecraft mass, hence a good mass saving by the use of presented design.

References

1. Fukuzawa K, Abe N, Nakamura S, Suzuki R, Kimura S, Akioka M (2003) 100–300 kg class satellite standard bus system design and applications. In: 21st International communications satellite systems conference and exhibit, ICSSC 2003—Yokohama, Japan
2. Aborehab A, Kassem M, Nemnem F, Kamel M, Kamel H (2020) Configuration design and modeling of an efficient small satellite structure. Eng Solid Mech 8
3. Madry S, Martinez P, Laufer R (2018) Innovative design, manufacturing and testing of small satellites. http://www.springer.com/series/5495
4. Madey JM, Bazlmann RC, Goddmd Space Flight Center Greenbelt Md (1969) Design techniques for small scientific satellite structures. NASA Technical Note. NASA TN D-4988

Composite Capsule Tube for Underwater System

Rajesh Addanki, Ayush Mahajan, Ramakrishna Dinavahi, and Manoj Kumar Buragohain

1 Introduction

The capsule tube is a protective casing for the launch vehicle that will be launched from underwater environment. The capsule has to function for various load conditions in the underwater environment till the launch vehicle ejects out of the capsule. This composite capsule tube is designed with carbon fibre as the primary load bearing material with epoxy as the matrix ingredient. The primary technology adopted is filament wet winding where in 90% of the composite tubes are wound using filament winding. Three-dimensional models are configured and analysed by FE.

2 Capsule Configuration

The capsule is configured primarily with carbon/epoxy composite material along with aluminium alloy end parts. The composite tube is joined with other metallic sections with nose cap at the fore end and base dome at the aft end. Overall length is about 5 m with cylindrical diameter of about 0.5 m which is considered for study, and the configuration with nose cap and base dome is presented in Fig. 1.

R. Addanki (✉) · A. Mahajan · R. Dinavahi · M. K. Buragohain
Composite Product Development Centre, Advanced Systems Laboratory, DRDO, Hyderabad, India
e-mail: rajeshaddanki@rediffmail.com

M. R. Manohara Babu et al. (eds.), *Recent Advances in Composites Engineering and Research*, Springer Proceedings in Materials 39,
https://doi.org/10.1007/978-981-99-8807-5_7

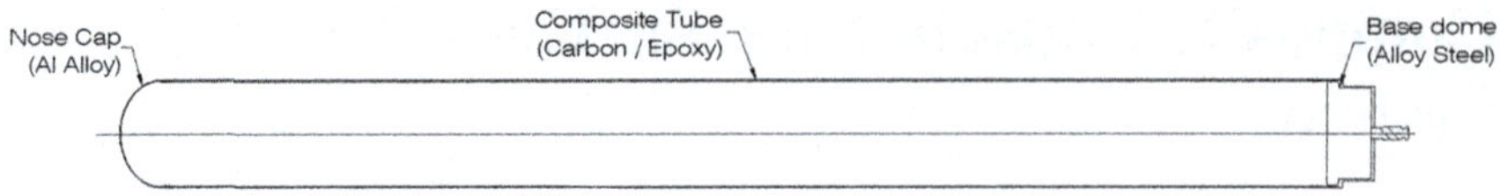

Fig. 1 Configuration of capsule

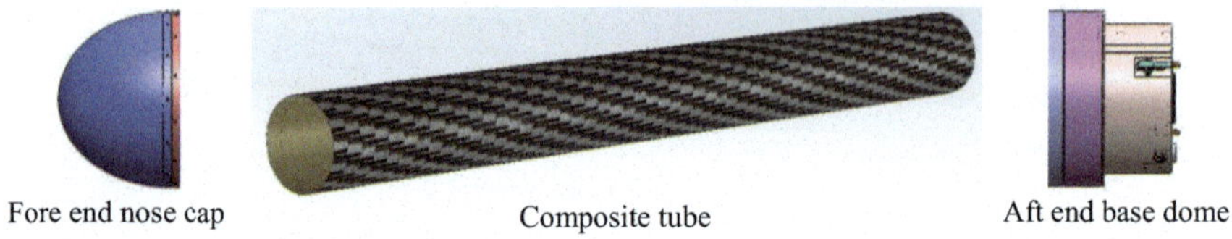

Fig. 2 Parts of composite capsule

3 Metallic Parts

The nose cap is designed as hemispherical dome to house electrical packages which are not a part of this study. The base dome is an axisymmetric cylinder with flat end at the rear for housing the propulsion system and is attached to the composite tube. The base dome is in turn connected to the launch vehicle and helps in pushing the vehicle out of capsule body. Aluminium alloy is considered for nose cap and alloy steel base dome configurations. The detail parts are presented in Fig. 2.

4 Joint Configuration

The metal-to-composite interfaces are joined by fastening. The joints are configured using blind rivets of steel material. The fasteners are configured in two rows maintaining pitch distance of at least 2d (where d is the fastener diameter) and also staggered at a distance of at least 4d. The joint configuration is represented in Fig. 3.

The design allowables considered are:

- Design factor of safety on UTS: 1.25.

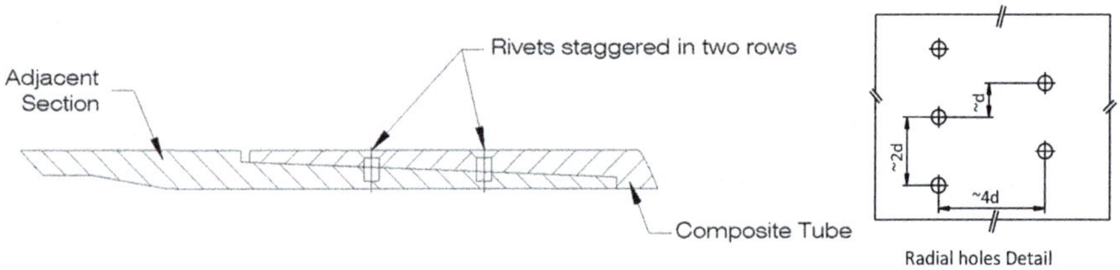

Fig. 3 Composite to metal joint

- Design factor of safety on YS: 1.125.
- Allowable stress: Min [UTS/1.25, YS/1.125].

5 Design Calculations for the Joint

The design considerations include:

- Size of the fastener, their quantity, and pitch distance according to overall load/force acting on the component.
- No of rows and distance between rows.
- Circumferential distance available for fastener location.

During the design phase, the possible failures considered are shear, tear, and bond failures [1, 2].

5.1 *Shear Failure of Fasteners*

The shear failure is calculated by equating the compressive axial force to the shear force acting in the fasteners at the interface of two joining materials and is given in Eq. 1.

$$F = n \times t \times \pi/4 \times d^2, \tag{1}$$

where F is maximum axial force, n is total no. of fasteners used, ι is shear stress in a fastener, and d is diameter of a fastener. The shear stress acting on fastener is calculated and compared with the material value for attaining the margin availability.

5.2 *Tearing Failure of Adjacent Section*

With the composite plies weakening in bearing strength, this failure condition is computed. Considering the available distance between fasteners in a single row as a worst possible case, the tearing stress induced in the section is as calculated in Eq. 2.

$$T_s = \frac{\text{Maximum compressive force}}{\text{Tearing Area}}. \tag{2}$$

The computed value is compared to bearing strength of the material for margin calculations.

Table 1 Process details of the composite tube

Material	Carbon/epoxy, E- Glass/epoxy composite, and nitrile rubber
Process	Filament winding of hoop plies, lay-up of unidirectional meridional doilies, rubber
Possible failure modes	Excessive stress in composite plies

5.3 Bonding Failure

Overall bond area is also important while joining the composite section to the metallic section with adhesive. Hence, bonding as a lap joint, the stress induced is as calculated in Eq. 3.

$$\mathrm{Bs} = \frac{\text{Maximum axial load}}{\text{Bonding Area}} \tag{3}$$

This in turn is compared to lap shear strength of the materials for finding out factor of safety.

6 Composite Tube

Composite tube is the only composite part in the capsule assembly and is designed with carbon/epoxy composite material. As the capsule is operated in sea environment, nitrile-based insulation rubber is considered for layer of leak proofness. Filament winding is chosen as technology for composite tube manufacturing, and based on the fabrication technique, the process details of the composite tube are given in Table 1.

7 Ply Design

The basic design method is to orient the fibres in the direction of the principal stresses (i.e. circumferential and longitudinal), as typically in the case of composite tube. The major assumptions in composite ply design are [2]:

- Fibre and matrix strains are equal.
- Tensile and compressive deformations are equal.
- Shear stresses developed in the fibre matrix interfaces are low.
- Structure obeys Hook's law.
- The fibres are straight and continuous.

The primary loads considered for capsule design are structural loads which are taken care by the netting system of the fibres and the leak proofness function is taken care of by the rubber layer.

The design allowables for composites are considered as:

- Design pressure: 1.25 × pressure.
- Design axial force: 1.25 × axial force.

However, fibre damage during winding, voids, batch-to-batch property variations, etc. to be taken into account during ply design as a translation efficiency which in turn is the reduction of fibre stress at failure in a laminate as compared to that in a coupon level.

7.1 Netting Model

In a filament wound composite structures, the fibres are essentially in tension, and for all practical purposes, the interlaminar shear stresses can be ignored during initial design. The netting model can be very conveniently used during the preliminary design of the composite shells. With local meridional slope as 0°, local component radius r, and local angle of winding α, the equations are derived from the netting model [3] representing in Fig 4.

Now, considering static equilibrium in the axial direction, the thicknesses are calculated from the Eqs. 4–6 and are as given below.

$$\pi r^2 p = 2\pi r t_\alpha \sigma_f \cos^2 \alpha \tag{4}$$

or

$$t_\alpha = \frac{pr}{2\sigma_f \cos^2 \alpha}, \tag{5}$$

where

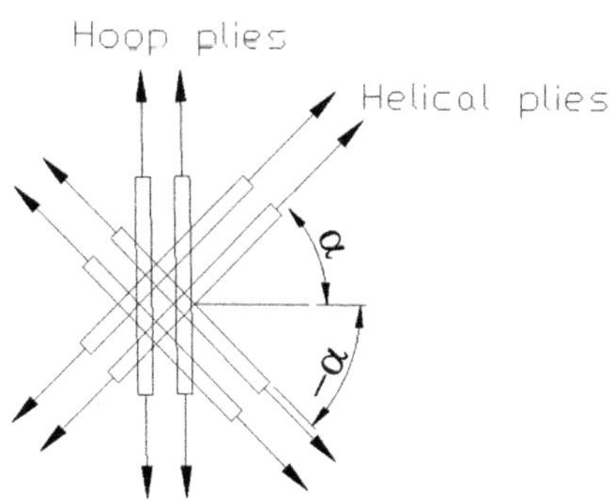

Fig. 4 Netting model

t_α is required helical thickness.
p is internal pressure.
σ_f is tensile strength of fibre.

To counteract the radial stresses, hoop windings are provided. Then, considering static equilibrium in the radial direction, the required hoop thickness is computed.

$$t_{90} = \frac{pr(2 - \tan^2\alpha)}{2\sigma_f}, \tag{6}$$

where t_{90} is required hoop ply thickness.

7.2 Thickness Computation

A technological constraint to be considered here is bandwidth of the fibre, volume fraction with resin pick up in wet winding, quantity of spools, etc., for the distribution of plies so as to ensure only the minimum required thickness per ply.

Helical or hoop thickness per ply for a given angle of winding and cross-sectional radius can be theoretically estimated by Eq. 7.

$$t_{\alpha/90} = \frac{n_\alpha a}{W_\alpha V_{f\alpha}}, \tag{7}$$

where

$t_{\alpha/90}$ is helical/hoop ply thickness.
$V_{f\alpha}$ is fibre volume fraction (0.50–0.55 for carbon–epoxy).
a is 0.225 mm^2, for T-700, 12K, carbon fibres.
n_α is number of spools.
W_α is bandwidth.

7.3 Ply Sequence

The composite part is designed with helical and hoop plies with rubber layer interspersing in between them, and the ply sequence is given as [89°/R/65°/-65°/8 9°$_6$]$_s$. A constant 45° helical angle ply is designed to cater different load cases and for ease of manufacture. About 90° ply refers to hoop ply with larger helical angles and R refers to rubber ply.

8 FE Analysis

A numerical solution technique-finite element analysis is carried out for the composite capsule assembly. Material idealisations are done using different material models and by discretizing the composite shell into certain number of zones. Each zone is assumed to have a constant angle of winding such that to each zone, unique three-dimensional stiffness properties (viz. E_1, E_2, E_3, G_{12}, G_{23}, and G_{31}) and the Poisson's ratios (viz. ν_{12}, ν_{23}, ν_{31}) are allocated for the helical/hoop thicknesses.

The constituent relations [2, 4] are given in Eq. 8.

$$\{\sigma_i\} = [C_{ij}]\{\varepsilon_j\} \tag{8}$$

where
$\{\varepsilon\}^{\mathrm{T}} = [\varepsilon_1, \varepsilon_2, \varepsilon_3, \varepsilon_{23}, \varepsilon_{31}, \varepsilon_{12}]$.
$\{\sigma\}^{\mathrm{T}} = [\sigma_1, \sigma_2, \sigma_3, \sigma_{23}, \sigma_{31}, \sigma_{12}]$.

Using orthogonal plane of symmetry for composites, the reduction of stiffness matrix is given in Eq. 9.

$$[C_{ij}] = \begin{pmatrix} C_{11} & C_{12} & C_{13} & 0 & 0 & 0 \\ C_{12} & C_{22} & C_{23} & 0 & 0 & 0 \\ C_{13} & C_{23} & C_{33} & 0 & 0 & 0 \\ 0 & 0 & 0 & C_{44} & 0 & 0 \\ 0 & 0 & 0 & 0 & C_{55} & 0 \\ 0 & 0 & 0 & 0 & 0 & C_{66} \end{pmatrix} \tag{9}$$

For the basic lamina, the coefficients C_{ij}'s are functions of the elastic constants E's, ν's are given in Eqs. 10 and 11.

$$\left.\begin{aligned} C_{11} &= E_{11}(1 - \nu_{23}\nu_{32})/DET \\ C_{22} &= E_{22}(1 - \nu_{13}\nu_{31})/DET \\ C_{33} &= E_{33}(1 - \nu_{12}\nu_{21})/DET \\ C_{12} &= E_{22}(\nu_{12} + \nu_{13}\nu_{32})/DET \\ C_{13} &= E_{33}(\nu_{13} + \nu_{12}\nu_{23})/DET \\ C_{23} &= E_{22}(\nu_{32} + \nu_{31}\nu_{12})/DET \\ C_{44} &= G_{23} \\ C_{55} &= G_{13} \\ C_{66} &= G_{12} \end{aligned}\right\} \tag{10}$$

where

$$DET = (1 - \nu_{12}\nu_{21} - \nu_{13}\nu_{31} - \nu_{23}\nu_{32} - \nu_{21}\nu_{13}\nu_{32} - \nu_{31}\nu_{12}\nu_{23}$$
$$n_{ij}E_{ji} = n_{ji}E_{ij}$$

$$\Rightarrow \nu_{ij} = \frac{E_{ij}}{E_{ji}}_{ji} \tag{11}$$

The stiffness constants for a lamina with fibres at an angle α to the axis are estimated as in Eqs. 12–15.

$$\left.\begin{array}{l}
C'_{11} = C_{11}m^2 + 2(C_{12} + 2C_{66})m^2n^2 + C_{22}n^4 \\
C'_{12} = (C_{11} + C_{22} - 4C_{66})m^2n^2 + C_{22}(m^4 + n^4) \\
C'_{22} = C_{11}n^4 + 2(C_{12} + 2C_{66})m^2n^2 + C_{22}m^4 \\
C'_{13} = C_{13} \\
C'_{23} = C_{23} \\
C'_{33} = C_{33} \\
C'_{16} = (C_{11} - C_{12} - 2C_{66})m^3n + (C_{12} - C_{22} + 2C_{66})mn^3 \\
C'_{26} = (C_{11} - C_{12} - 2C_{66})mn^3 + (C_{12} - C_{22} + 2C_{66})m^3n \\
C'_{66} = (C_{11} + C_{22} - 2C_{12} - 2C_{66})m^2n^2 + C_{66}(m^4 + n^4)
\end{array}\right\} \tag{12}$$

where m is $\cos\alpha$ and n is $\sin\alpha$

Then, the stiffness values for the complete 2D element are obtained by Eq. 13.

$$[C'_{ij}] = \frac{1}{N\sum_{i=1}^{N}[C_{ij}]} \tag{13}$$

The engineering constants for the orthotropic solid are then obtained after computing $[S]$ from Eq. 14.

$$[S] = [C']^{-1} \tag{14}$$

And, using the relation of $[S]$ with constants can be computed from Eq. 15.

$$[S] = \begin{pmatrix}
1/E_1 & -\nu_{21}/E_2 & -\nu_{31}/E_3 & 0 & 0 & 0 \\
-\nu_{12}/E_1 & 1/E_2 & -\nu_{32}/E_3 & 0 & 0 & 0 \\
-\nu_{13}/E_1 & -\nu_{23}/E_2 & 1/E_3 & 0 & 0 & 0 \\
0 & 0 & 0 & 1/G_{23} & 0 & 0 \\
0 & 0 & 0 & 0 & 1/G_{31} & 0 \\
0 & 0 & 0 & 0 & 0 & 1/G_{12}
\end{pmatrix} \tag{15}$$

With the material parameters, the strains are computed and further the stresses are calculated [4].

Table 2 Load cases

Sr. No.	Load scenario	Load case
1	Transportation load	Lateral compressive load with acceleration
2	Combined loads including emergency dive	Axial compressive load combined with bending and internal cum external pressure

8.1 Load Scenario

The capsule is designed for various load cases considering its operation and handling aspects. With the underwater operational loads and transportation of capsule are considered as the critical load scenarios and are listed in Table 2 which are studied in detail, analysed.

A 3D model with 360° sector is generated using ANSYS [5]. The composite is modelled using solid shell 190 element and metal, rubber are modelled using 8-noded structural solid elements.

8.2 Load Case-1: Transportation Load

The weight of the capsule assembly is simulated along with acceleration loads in this load scenario. Resting of capsule on 120° sector supports was done so that the CG lies in between the support locations and total load with weights of nose cap, composite tube, and base dome was distributed accordingly.

The following are the boundary conditions applied on to the model and the same are represented in Fig. 5.

8.2.1 Scenario 1 Plots

FE analysis [5] is carried out with 360° sector meshed model and the displacements and stresses are plotted from Figs. 6, 7, 8, 9, 10 and 11.

8.3 Load Case-2: Combined Load

In this load case, the external pressure of about 10bars is considered with an equivalent axial compressive load (combining bending load) of 100kN. The test environment also is considered in this analysis wherein the capsule is rested on supports and pressurised inside a chamber along with application of other loads. The test configuration with load details and boundary conditions are presented in Fig. 12.

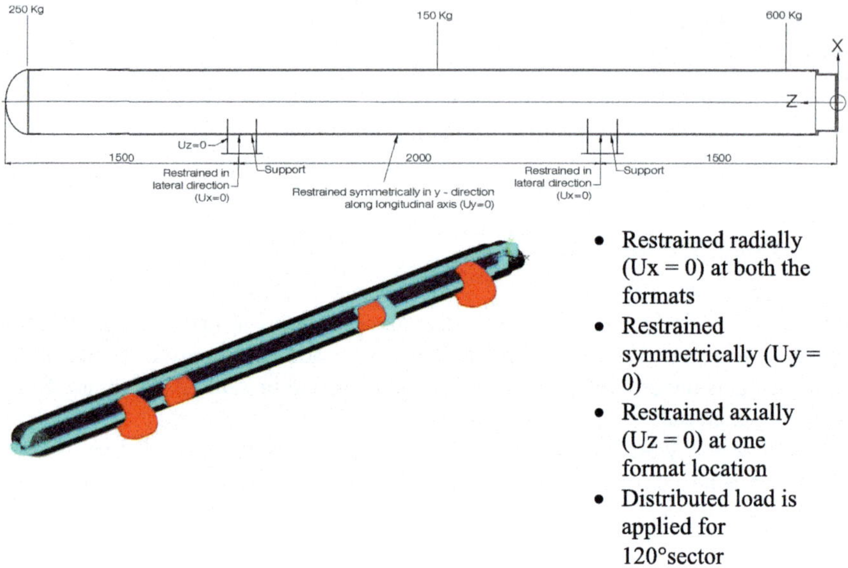

Fig. 5 FE model with constraints

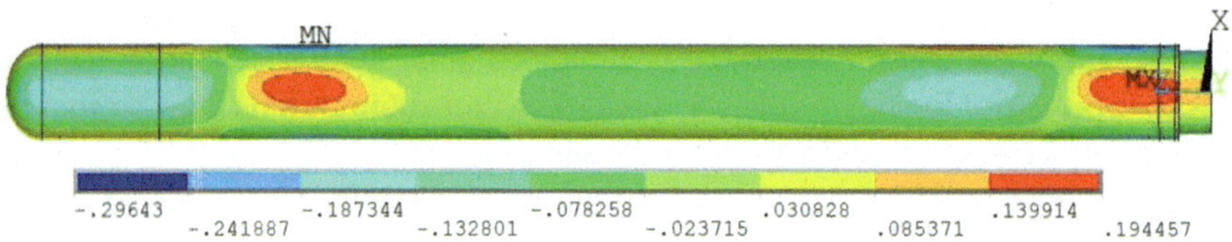

Fig. 6 Longitudinal displacement

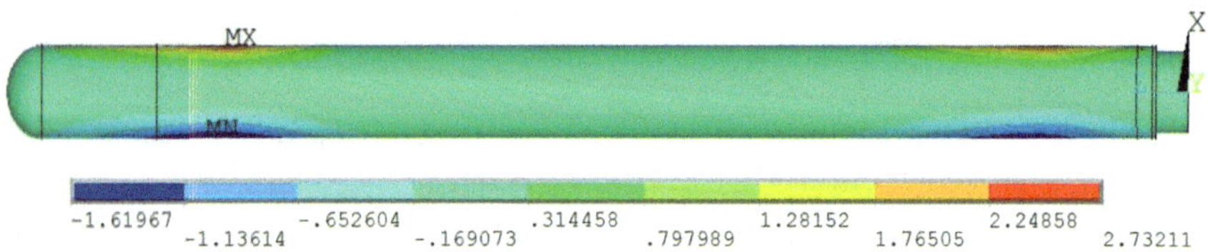

Fig. 7 Radial displacement

8.3.1 Scenario 2 Plots

FE analysis is carried out with 360° sector meshed model, and the displacements and stresses are plotted from Figs. 13, 14, 15, 16, 17, 18, 19 and 20. This load case is being the combined load case, and the stress patterns of composite and metal are plotted.

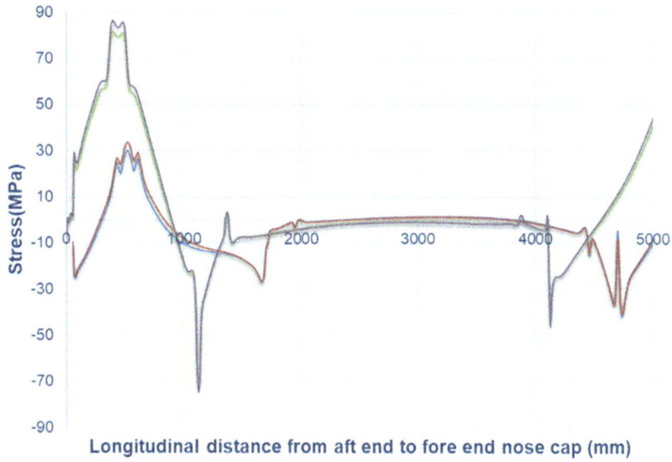

Fig. 8 Helical stress plot

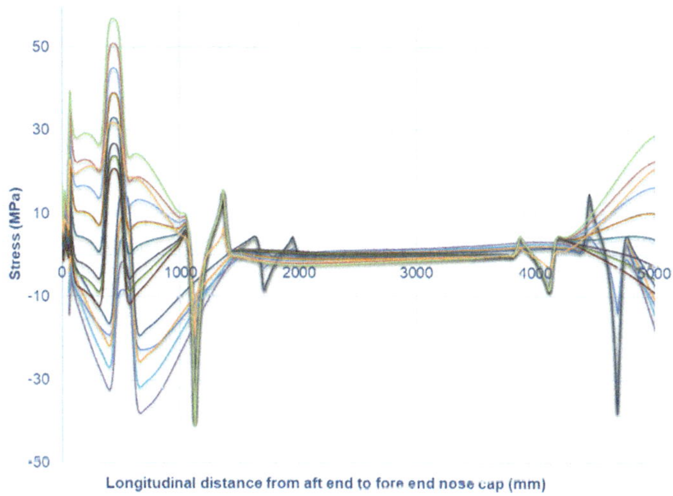

Fig. 9 Hoop stress plot

One of the most important analyses to be carried out for longer systems is buckling. The buckling of the capsule under axial compression is analysed by extending short column theory of isotropic material and treating the capsule as a short column with hinged support as given in Eqs. 16–18.

$$P_c = \frac{\sigma_c A}{\left(1 + a\left(\frac{1}{k}\right)^2\right)}, \tag{16}$$

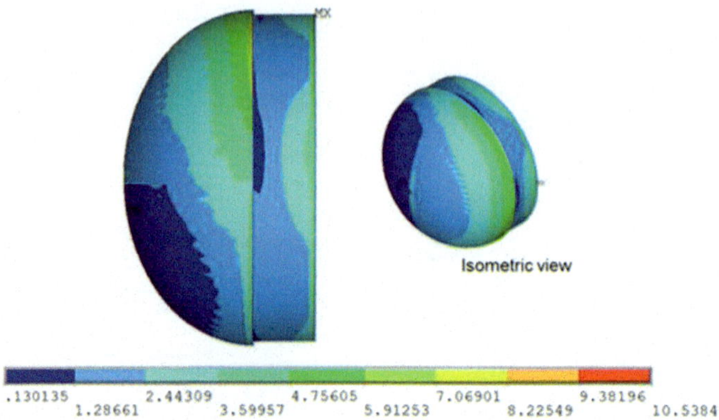

Fig. 10 Von-Mises stress in nose cap

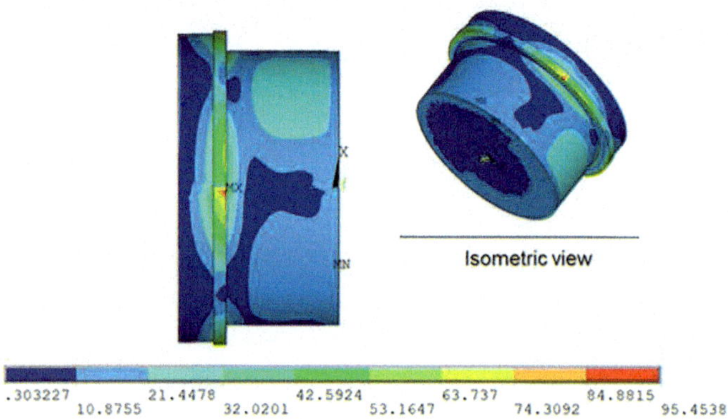

Fig. 11 Von-Mises stress in base dome

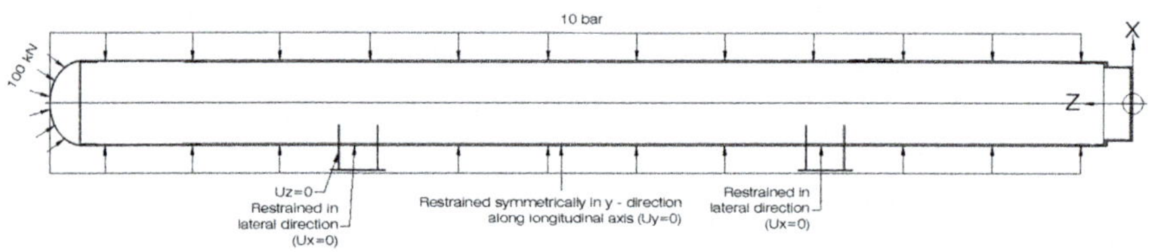

Fig. 12 Capsule assembly with loads and constraints

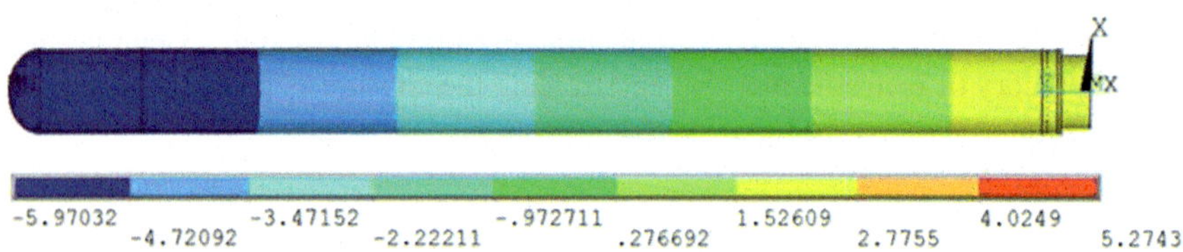

Fig. 13 Longitudinal displacement

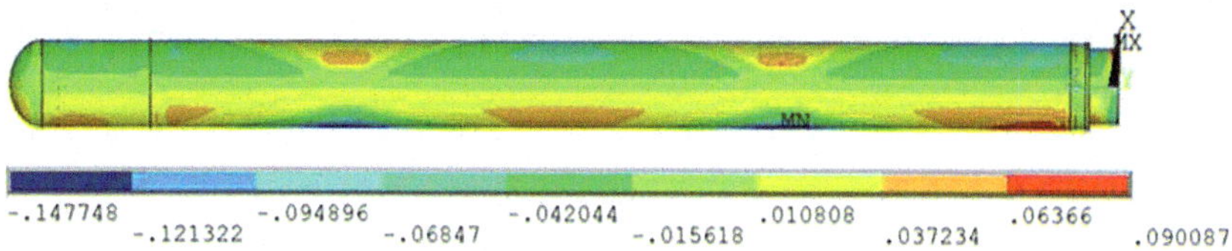

Fig. 14 Radial displacement

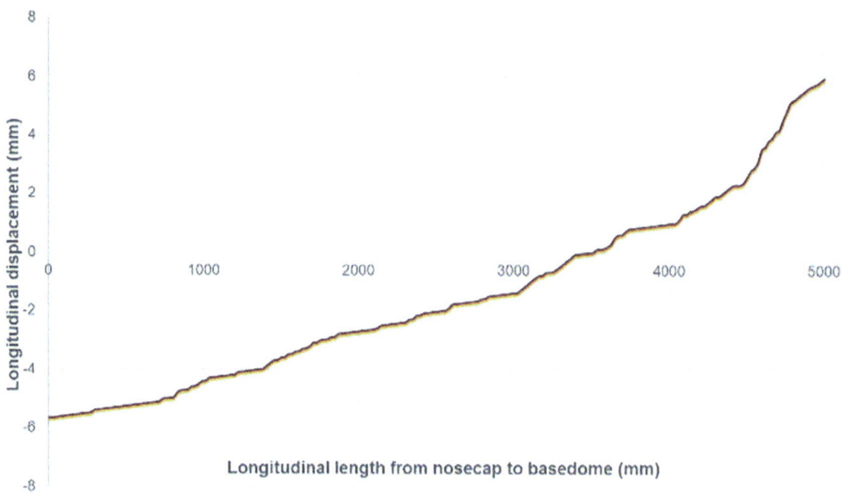

Fig. 15 Longitudinal displacement

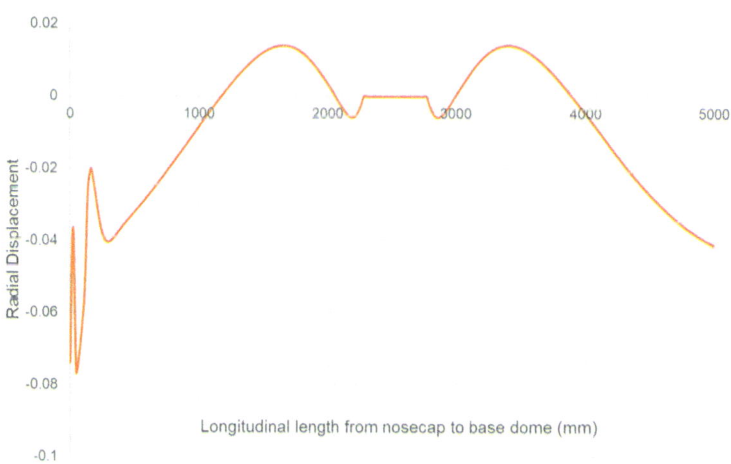

Fig. 16 Radial displacement

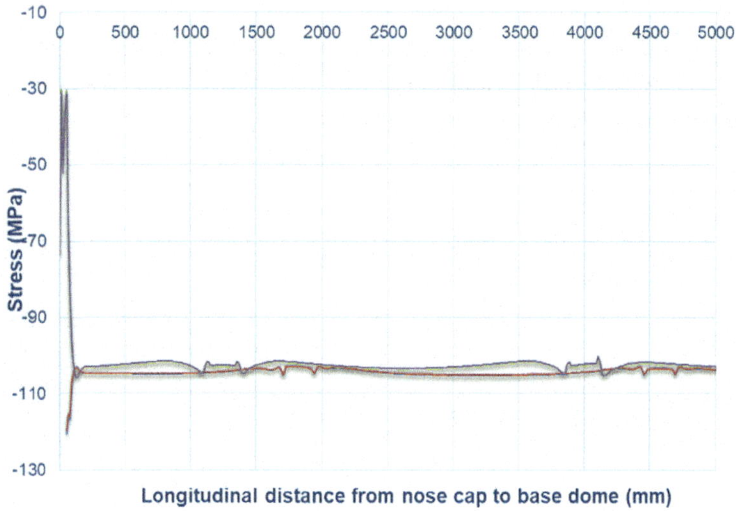

Fig. 17 Helical stress plot

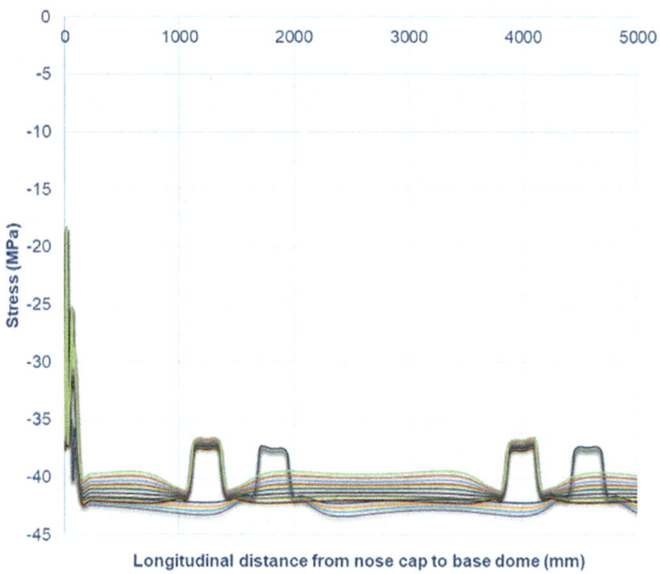

Fig. 18 Hoop stress plot

$$\text{Buckling factor} = P_c/P, \tag{17}$$

where

P_c Critical buckling load (Rankine).

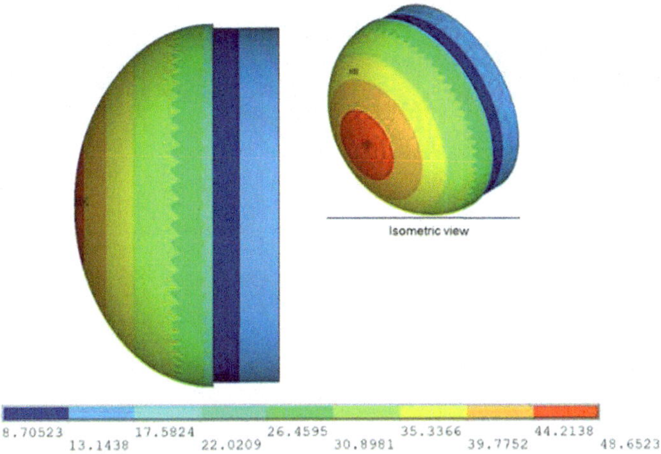

Fig. 19 Von-Mises stress in nose cap

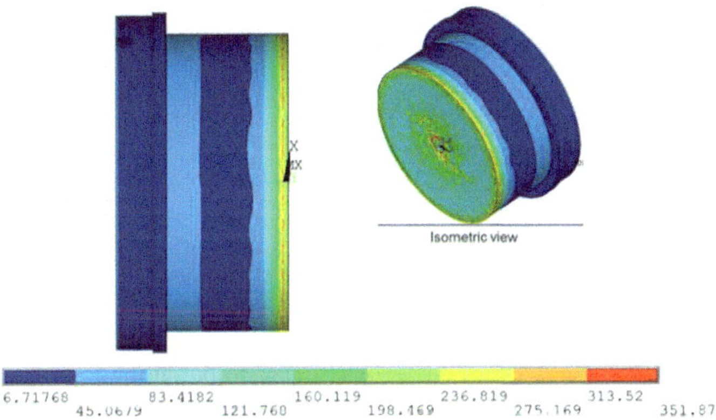

Fig. 20 Von-Mises stress in base dome

σ_c Buckling stress.
P Axial load.
A Area of cross section of the capsule.
a Material constant, given by

$$a = \frac{\sigma_c}{\pi^2 E}.$$

l Length of the capsule.
k Least radius of gyration, given by

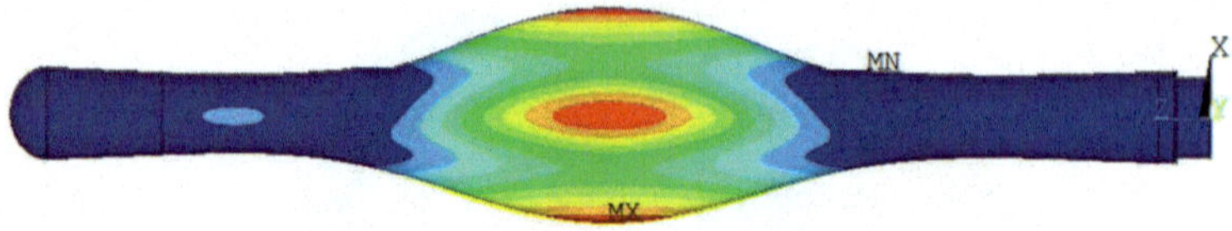

Fig. 21 Buckling analysis plot

$$k = \sqrt{\frac{I}{A}}.$$

E Transverse modulus.
I Minimum section modulus.

Buckling of the capsule is also checked by using the orthotropic from of the classical buckling equation:

$$\sigma_{cr} = \left\{C[E_l E_t]^{1/2}\right\}/(R/t), \tag{18}$$

where

C Coefficient of buckling, considered is 0.48.
E_l Longitudinal modulus.
E_t Transverse modulus.
R Outer radius of the capsule.
t Thickness of the composite capsule.

However, Eigen buckling analysis is carried out by FE and the results are presented in Fig. 21.

On observation, the buckling factor was found to be on the lower side. Hence, composite stiffeners of 2 mm thick were modelled and analysed.

The stiffeners of 50 mm width with 2 mm thick of same carbon–epoxy material are considered for analysis. The number of stiffeners considered is 4 and 6, respectively, distributed equidistantly along the length of composite part. Post-analysis results are presented in Figs. 22, 23, 24, 25, 26 and 27.

9 Results and Concluding Remarks

9.1 Load Case-1

- The maximum displacement in longitudinal direction observed is 0.3 mm along the length of the capsule.
- The maximum radial displacement observed is 2.73 mm (perpendicular to the length of the capsule)

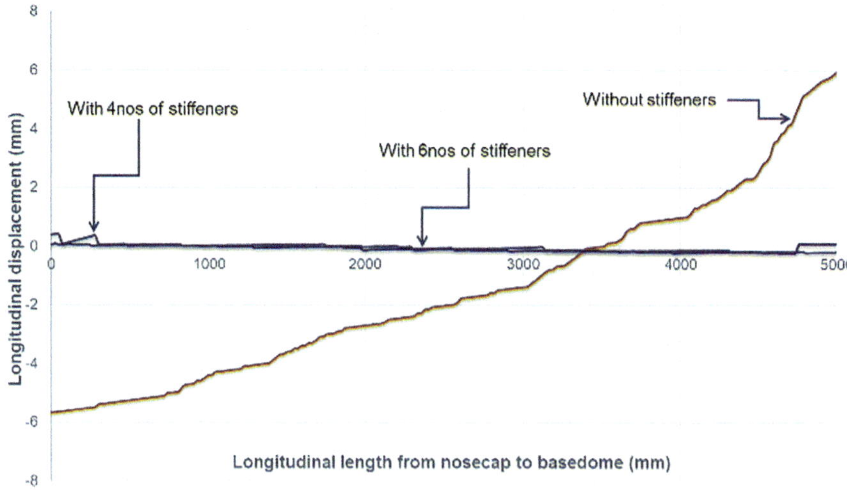

Fig. 22 Longitudinal displacement

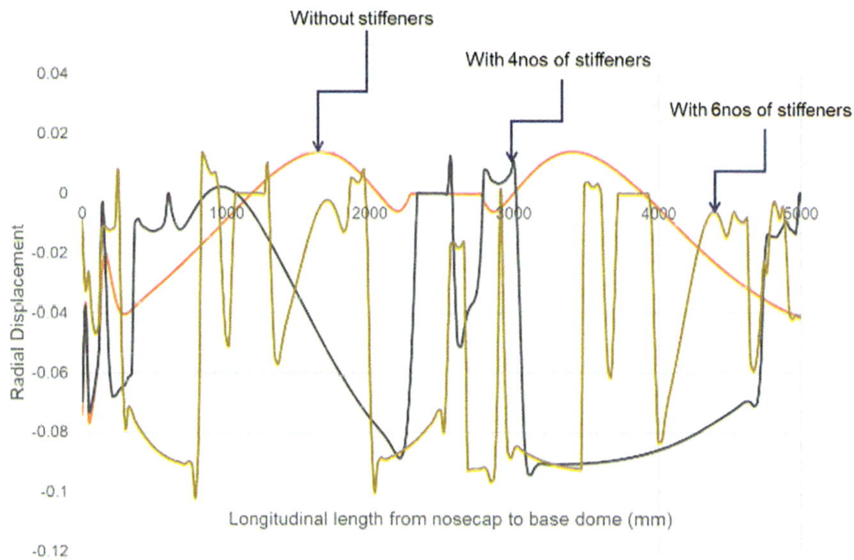

Fig. 23 Radial displacement

- The displacement variation can be further reduced with provision of locking rings at the aft end during transportation and can be further minimised when the capsule is tied with ratchet belts.
- The helical layers are apart from the neutral axis and slight bending is observed from variation in the stress pattern and the maximum stress is 89 MPa.

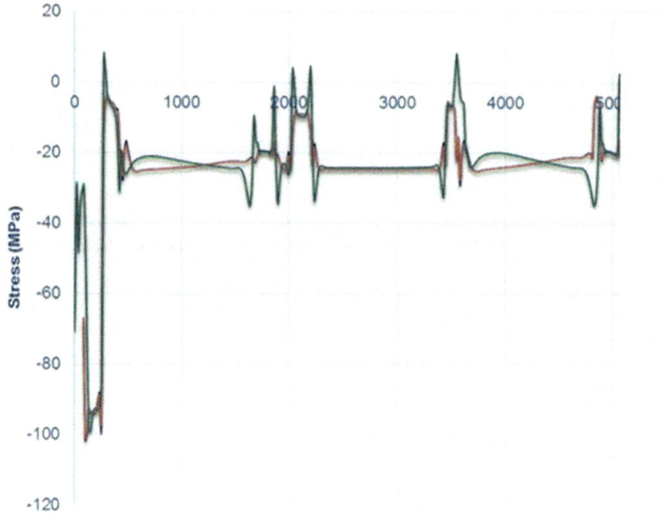

Fig. 24 Helical stress plot

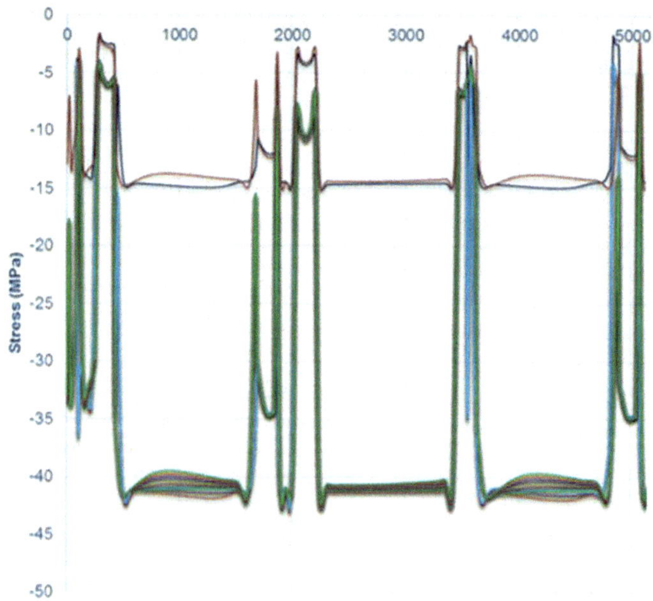

Fig. 25 Hoop stress plot

Fig. 26 Buckling analysis plot (with 4 stiffeners)

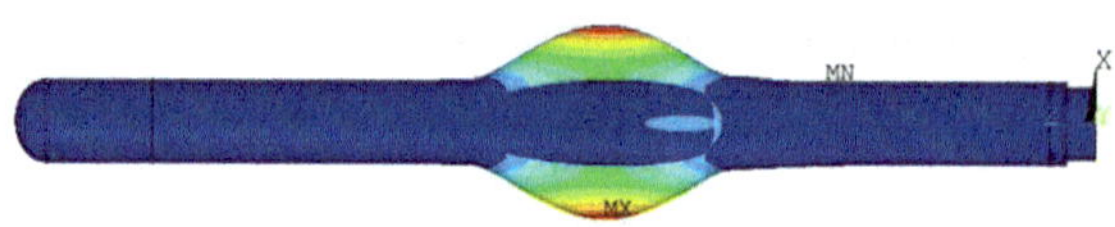

Fig. 27 Buckling analysis plot (with 6 stiffeners)

- The hoop stress observed is about 70 MPa and the variation is due to the loading pattern and model constraints.
- Von-Mises stress in nose cap is 11 MPa and in the base dome is 96 MPa.
- Maximum Von-Mises stresses in the metallic parts are highly localised and further reduced with the passengers mounted in the nose cap.

9.2 Load Case-2

- The displacement of the capsule in longitudinal direction is varied from -5.97 mm to 5.27 mm. This variation is all along the length and due to the presence of different material combinations.
- The radial displacement observed is 0.1 mm and the waviness in the curve all along the length is due to constraints and the joining sections.
- Negative stresses of about 45–125 MPa are observed in hoop and helical stress plots which are due to the combined load phenomenon of external pressure, axial loads.
- Von-Mises stress in nose cap is 49 MPa and 351 MPa is observed in base dome which is within the material stress limits.
- The longitudinal and radial displacements have reduced to 0.8 mm and 0.1 mm, respectively, due to the presence of stiffeners.
- Helical and hoop stress plots show minor stress variations at the stiffener locations.
- The buckling factor observed is 0.63 without stiffeners and the buckling factor improved from 2.81 with four numbers of stiffeners and to 4.12 with six numbers of stiffeners. Also, the mode has shifted from centre to base dome.

9.3 Conclusions

- Stress pattern in metallic parts is lower than the material strength values of 450 MPa for aluminium alloy and 1200 MPa for alloy steels.
- The test data of carbon/epoxy composite infer longitudinal tensile strength of 2000 MPa and compressive strength of 800 MPa along fibre direction. The results obtained in both the load cases are benign to the material capability.
- Providing stiffeners at intermittent zones along the axis of the composite capsule has improved the buckling load ability.

Acknowledgements The authors extend their sincere gratitude to Project Director and Director ASL for supporting in completion of this project.

Declaration of Competing Interest The authors declare that they have no known competing financial interests or personal relationships that could have appeared to influence the work reported in this paper.

References

1. Lubin G (1982) Hand book of composites. Van Nostrand Reinhold Publishers, New York
2. Jones RM (1999) Mechanics of composite materials, 2nd edn. Taylor & Francis group
3. Peters ST, Humphrey WD, Foral RF (1991) Filament winding composite structure fabrication, 2nd edn. Society for the Advancement of Material and Process Engineering (SAMPE)
4. Roark RJ (1997) Formulas for stress strain. Mc-Graw Hill International Student Edition.
5. ANSYS, Inc., Mechanical, Release 17.0

Analysis and Repair of Delaminated Continuous Helical Ply in a Composite Pressure Vessel

Rajesh Addanki, Ayush Mahajan, Surjeet Kumar Sinha, and Manoj Kumar Buragohain

1 Introduction

Composite pressure vessel in general consists of a cylindrical shell with dish end domes and in turn is fitted with two or more openings, and schematic representation is presented in Fig. 1.

The vessel is processed by filament winding incorporating helical, hoop and doily layers. The helical layers with non-geodesic winding are provided from one opening end to the other opening end through continuous winding. Hoop windings cover the cylindrical region, and on the dished domes, doilies are interspersed in between helicals. For the vessel to connect to other sections, metal joints are fitted to the cylinder and to the domes at the openings.

Specific layup is followed for the composite pressure vessel covering the domes and cylinder regions, and the typical ply sequence is presented in Fig. 2.

2 Manufacture

To carryout filament winding, for the configured pressure vessel, an axisymmetric mandrel is necessary. Metallic shaft is prepared with foam and Plaster of Paris (PoP) overlay and subsequently machined to the inner profile of the casing. The foam stacking and the machined profile are presented in Fig. 3, over which the winding and doily layup are performed.

R. Addanki (✉) · A. Mahajan · S. K. Sinha · M. K. Buragohain
Composite Product Development Centre, Advanced Systems Laboratory, DRDO, Hyderabad, India
e-mail: rajeshaddanki@rediffmail.com

M. R. Manohara Babu et al. (eds.), *Recent Advances in Composites Engineering and Research*, Springer Proceedings in Materials 39,
https://doi.org/10.1007/978-981-99-8807-5_8

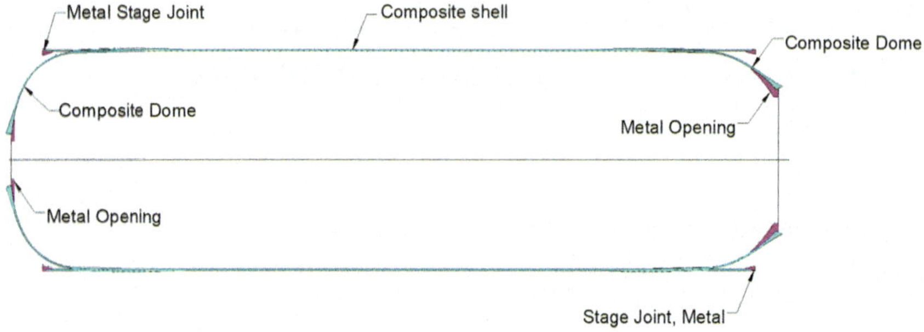

Fig. 1 Configuration of pressure vessel

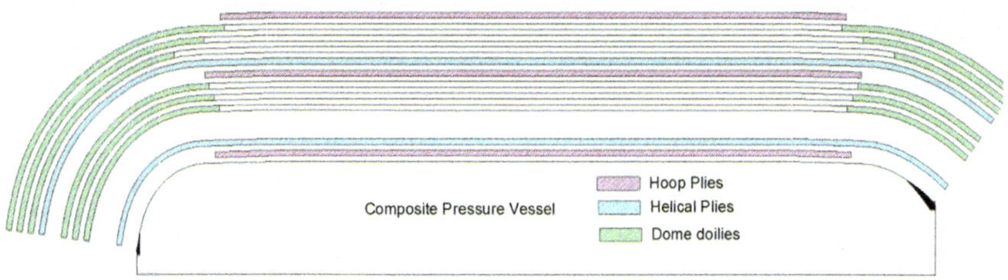

Fig. 2 Pressure vessel ply sequence

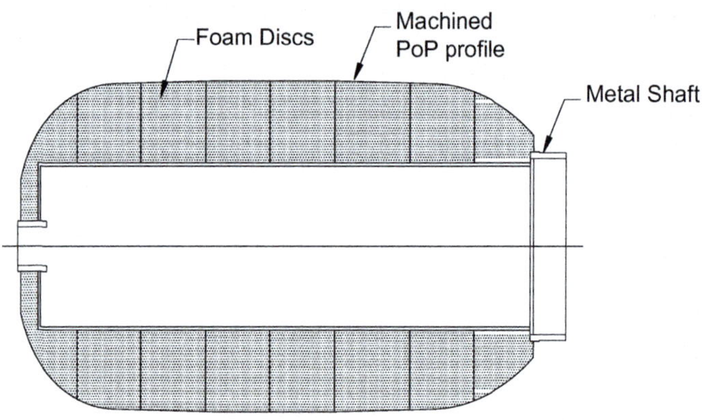

Fig. 3 Winding mandrel for pressure vessel

Composite materials are being very sensitive, care and caution to be exercised at each step/level and all activities [1–4]. Helical layers, hoop layers are done over the mandrel as per machine programming in continuous fashion, and doilies are placed manually over the dome regions as per designated zones. After the vessel manufacture, it is cured and shaft is extracted by chipping off foam, PoP. The process and extracted vessel are represented in Fig. 4.

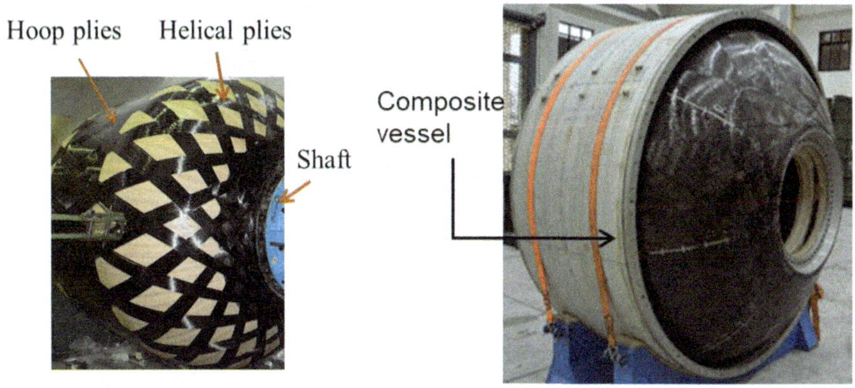

Fig. 4 Filament winding process and extracted pressure vessel

3 Failure Mechanisms

The defects in the composite structures include porosity, inclusions, delaminations, non-uniform dispersion of fibre, matrix, etc. The major contribution in usage and service life of the component amongst all the defects is delamination and is mostly introduced during manufacturing.

With the presence of delamination, primary failure to be considered is fibre buckling which is based on the stability of the fibres in the softer matrix. With the lower fibre volume fractions, the failure happens in extension mode, i.e. out of phase, and if the laminate fibre volume fraction is higher, in phase failure, i.e. shear mode failure occurs. In unidirectional composites or composites with considerably low transverse strengths, Poisson effects are predominant creating failures by transverse tension and also waviness in the structure.

Some factors that influence delamination are:

- **Fibre orientation**: Misorientation of fibres causes loss in strength and further changes the loading patterns from direct tension/compression to shear loading.
- **Stacking sequence**: Fibres close to the neutral axis should be sufficiently stacked for better load transfers avoiding unanticipated loads.
- **Fibre waviness**: The longitudinal shear modulus increases with increase in amplitude of the fibre waviness. Also micro and kinking of fibre arise.
- **Matrix cure**: One most important parameter is proper curing of the matrix to attain the required glass transition temperature. Else, unreacted sites in the network will rise to stress risers and also can lower the properties.

The interactions with normal and shear stresses in delaminated composite stack are represented in Fig. 5.

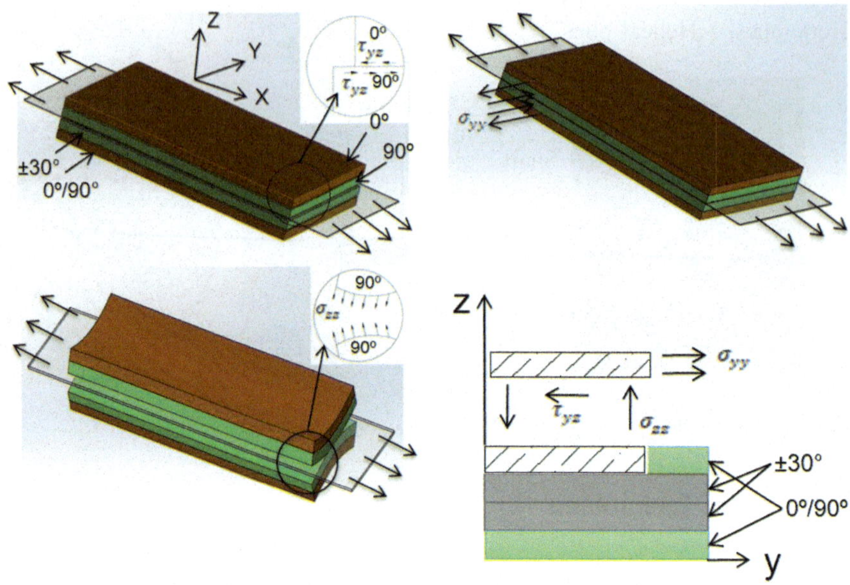

Fig. 5 Stress interaction in a delaminated composite laminate

3.1 Damage Mechanics

Delaminations result from interlaminar stresses, and considering initiation using interlaminar stresses, different criteria have been employed in the literature. Ye [5] studied from Hashin–Rotem failure criteria.

When $\overline{\sigma}_{ZZ} > 0$,

$$\left(\frac{\overline{\sigma}_{ZZ}}{T}\right)^2 + \left(\frac{\overline{\tau}_{LZ}}{S}\right)^2 + \left(\frac{\overline{\tau}_{TZ}}{S_T}\right)^2 = 1. \tag{1}$$

When $\overline{\sigma}_{ZZ} \leq 0$,

$$\left(\frac{\overline{\tau}_{LZ}}{S}\right)^2 + \left(\frac{\overline{\tau}_{TZ}}{S_T}\right)^2 = 1, \tag{2}$$

where z denotes the direction of laminate thickness, and T and S are the transverse tensile and shear strengths, respectively. S_T is the shear strength in the plane perpendicular to the fibres. In the above equation, the stress components are defined as the averaged stresses over a distance, from the free edge. Hosseini-Toudeshky et al. [6] had employed the quadratic failure criterion to predict the delamination initiation.

$$\left(\frac{\sigma_n}{\sigma_n^0}\right)^2 + \left(\frac{\tau_{sn}}{\tau_{sn}^0}\right)^2 + \left(\frac{\tau_{tn}}{\tau_{tn}^0}\right)^2 = 1, \tag{3}$$

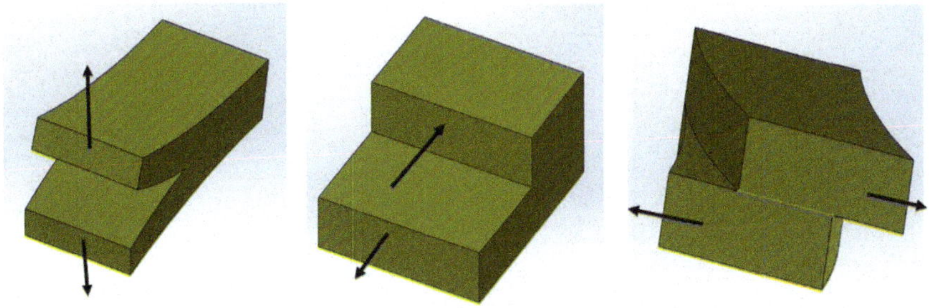

(a) (b) (c)

Fig. 6 **a** Interlaminar tension mode G_I, **b** interlaminar sliding shear G_{II} and **c** interlaminar scissoring shear G_{III}

where σ_n^0, τ_{sn}^0 and τ_{tn}^0 are the mode I, II and III strengths, respectively. The mode strengths are found out by testing the samples as shown in Fig. 6.

The delamination growth criteria can be evaluated from power law

$$\left(\frac{G_I}{G_{IC}}\right)^{\alpha} + \left(\frac{G_{II}}{G_{IIC}}\right)^{\beta} + \left(\frac{G_{III}}{G_{IIIC}}\right)^{\gamma} \geq 1, \tag{4}$$

where

$$G_I = \frac{1}{2\Delta A} \cdot F_Z(w_2 - w_1),$$
$$G_{II} = \frac{1}{2\Delta A} \cdot F_X(u_2 - u_1),$$
$$G_{III} = \frac{1}{2\Delta A} \cdot F_Y(v_2 - v_1),$$

where F_x, F_y, F_z are the forces exerting at the crack tip and ΔA is the crack area due to delamination. *u*, *v* and *w* are crack tip displacements in *X*, *Y* and *Z* directions, respectively.

4 Problem Identification

Visual inspection was carried out after the vessel extraction and bulges at about three different locations on the inner surface were noticed. After thorough inspection via ultrasonic and radiography, these bulges were notified as delaminations. The delamination is the macro-level mechanism that occurs in a laminate where in the interface, resin layer damages and causes the adjacent fibre/fabric layers to separate. Delamination reduces the strength and stiffness of the plies and thus limits the life of the pressure vessel and further causes stress concentration in primary composite

Fig. 7 Delaminated zones in the pressure vessel

load bearing plies leading to further growth of delamination resulting failure of the structure.

Three major delamination zones were identified having maximum size of 500 × 300 mm with air gap of 12 mm. The delaminated zones are presented in Fig. 7.

4.1 Mapping of Defect

All the three delaminated zones are at the location of dome to cylinder junction which would have occurred during the processing of the vessel enriching with composite ply stacking. From the radiographs, it was noticed that the inner most plies were pulled towards the inner surface creating an air gap between the major ply stack and delaminated stack. The exact thickness is mapped in the maximum delaminated area considering the total shell thickness. The air gap variation is mapped from the radiograph and is represented in Fig. 8.

The maximum-to-minimum air gap is mapped in the composite vessel, and henceforth, the number of plies to be removed are determined. The outermost plies are intact as well as the inner most plies are intact as a different set. It was observed that the delaminated stack (innermost plies) consists of helical hoop and unidirectional plies with [30°/90°/0°] angles.

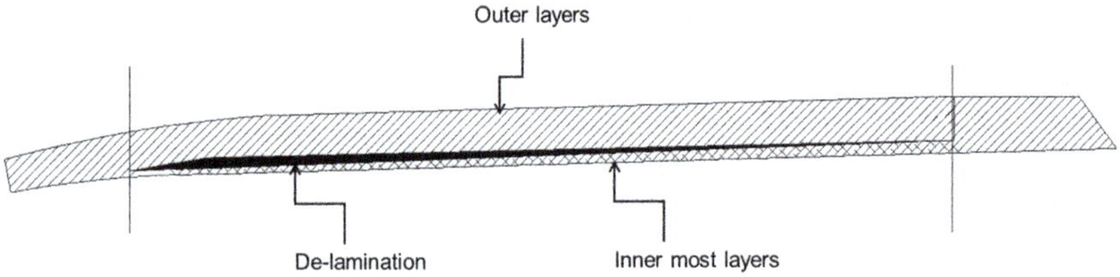

Fig. 8 Thickness mapping of defect

5 FE Analysis

The major load cases that the composite vessel will be subjected to are internal pressure and handling. FE analysis for the study of delamination observed is carried out by using ANSYS [7] for the pressure load case where it is the critical one amongst all. Complete model and also local model analysis were carried out to capture the ply closing behaviour.

The vessel is modelled in 3D using layered solid 186 elements incorporating the layer properties of the carbon/epoxy composite. The dome regions, delaminated zones [6] are also modelled with inputting the material data as tabulated in Table 1.

Hexagonal fine mesh size is chosen for all the elements considering their aspect ratios and the zone of delamination into account. The detailed mesh of the model along with loads and constraints is presented in Fig. 9.

A clear-cut gap as mapped is modelled for depicting the delamination and the ply data is inputted accordingly. A bulge formation is modelled featuring 12mm delamination gap. Total 360° sector model is considered for analysis and node-to-node contact element is incorporated. Section data with ply inputs are provided for analysis and the model is constrained at the openings simulating the operational conditions.

Table 1 Engineering constants of carbon/epoxy composite

E_{11} (GPa)	E_{22} (GPa)	E_{33} (GPa)	G_{12} (GPa)	G_{23} (GPa)	G_{13} (GPa)	ν_{12}	ν_{23}	ν_{13}
65	65	3.5	4.5	4.5	2.5	0.25	0.25	0.09

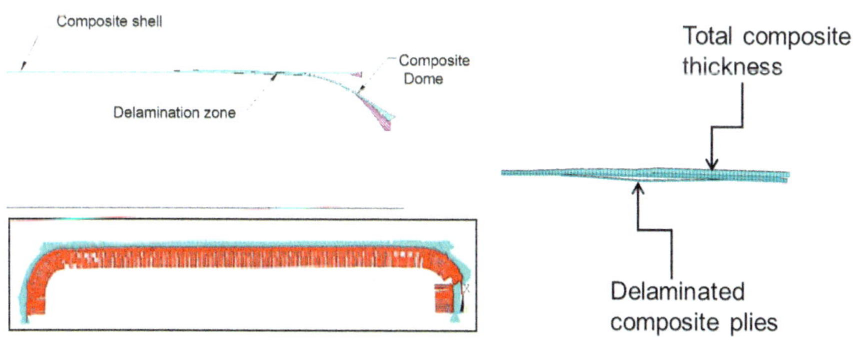

Fig. 9 FE model with load and constraints

5.1 FE Results and Discussions

The FE analysis was carried out and the major stress results around the delaminated zone are discussed. Incremental pressure load is provided to capture the gap closure phenomenon.

- From FE analysis, it was found that the delaminated gap closes at very lower pressure of 1.5 MPa and is represented in Fig. 10a.
- However, the stress in the composite plies has increased with the loading pressure increase.
- Further, it shows that the stress variation is observed around the delaminated zone increasing up to 1600 MPa as represented in Fig. 10b.
- The stress variation around the delaminated zone is observed from the coloured patches of Fig. 10b.
- The overall helical stress increase in the vessel is noticed at the delaminated zone only as shown in Fig. 10c
- However, overall stress pictures from Fig. 10a–c infers that the composite plies have failed and created the local stress variations.

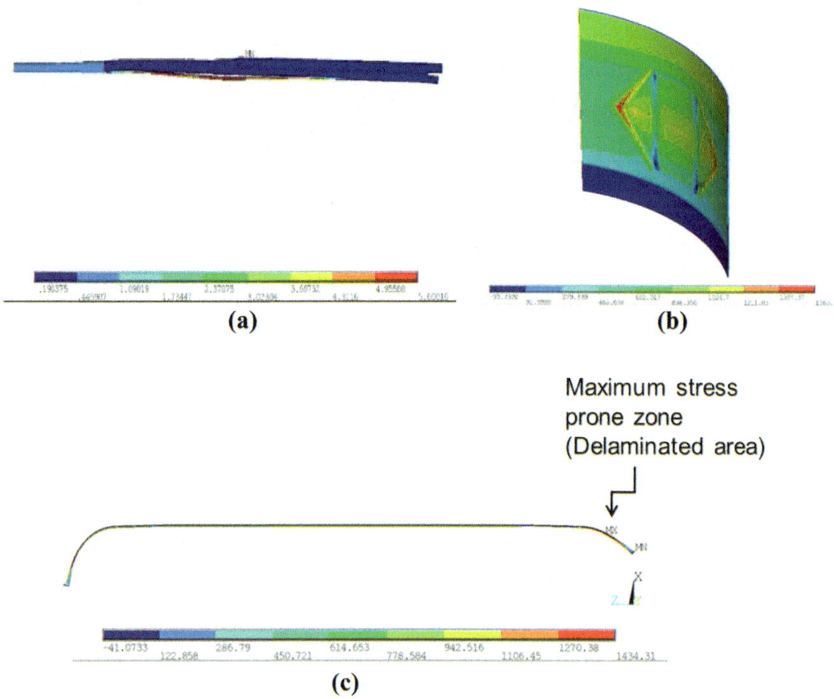

Fig. 10 **a** Dilation pattern in the delaminated plies, **b** stress variation around the delaminated zone, **c** helical ply stress variation in the overall composite vessel

6 Repair Work

Considering the intended use of the article, the delaminations can cause adverse effects during the article service life. The delamination leads to a redistribution of structural load paths which, in turn, precipitates structural failure. Hence, delamination indirectly affects the final failure of the structure thus affecting its life. Therefore, delamination is known as the most prevalent life-limiting damage growth mode. A repair is attempted at the delaminated zones of the pressure vessel wherein the helicals, hoops and doilies are laid and cured.

To remove the air pocket drilling/cutting to be performed from inside, henceforth the inner plies would be cut. Removal of a substantial amount of plies in the damage portion is a high skilled affair and many trials were carried out not to disturb the healthy plies in the structure. During fibre removal, it was ensured that no fibre beyond the air gap is being cut. Hence, a laminate is prepared depicting the similar thickness of the composite vessel and by introducing a defect. A machine with variable rpm and with different cutter sizes of 5mm–50mm is identified for cutting. Figure 11 shows the flow process.

The main objective of this model preparation is to provide an overall insight of different works, which is to be carried out on the delamination before starting with the actual vessel. This has helped to choose the suitable machining tools, the size of segments and also provided the necessary precautions that has to be followed while removing the segments without hampering the inside surface of the casing. Small pieces were cut using special cutting machine with rpm adjustment, and finally, the cavity is filled. The process of cutout making and refilling is represented Fig. 12.

Effort is put for replacement of the damaged fibre reinforcements in the structure to restore the original mechanical properties. In our study, the fibre orientation in the scooped out portion was chosen close to the design/manufactured part orientations and the composite layup sequence was formulated accordingly with carbon fibre, fabric and epoxy resin system for cutout filling. Carbon/epoxy composite plies of $[0°/\pm30°/90°]_s$ were laid in the scooped out region in more than 3–4 sectors. Filling at the corners was done by fibre and resin mix pieces ensuring of filling of layers to the maximum extent of the cavity. Since a bonded-on repair constitutes a discontinuity of the original plies, and therefore a stress raiser, structural repair schemes normally require extra plies to be provided in the repair area. Finally, the levelling of the cutout region is done after overall filling of $[90°/\pm45°/90°]$ plies. The detail

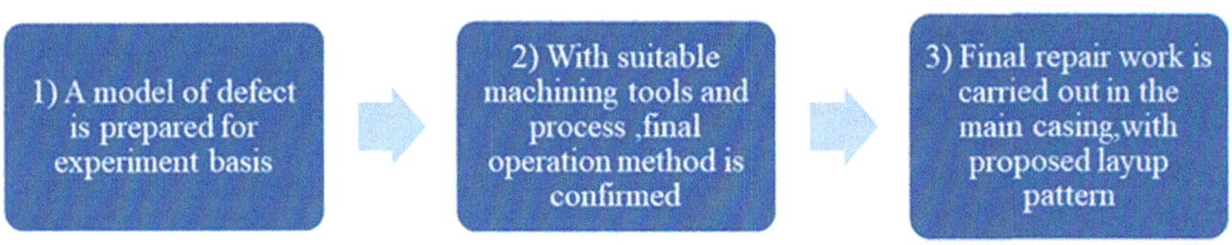

Fig. 11 Flowchart depicting repair work process

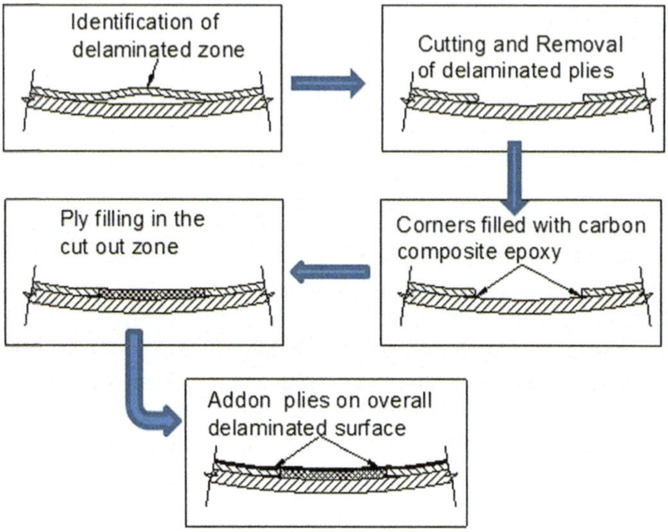

Fig. 12 Repair scheme of delaminated zone

pictorial representation of filling the cutout with angle plies (a), with overall add-on circumferential plies (b) and subsequent curing (c) is presented in Fig. 13.

The vessel after repair work, bagging is done and cured as per the desired curing cycle of resin system. Later, the inner surface was thoroughly cleaned and no fibre cut marks were found depicting proper bonding of layers. The vessel is made vertical and closure plates are assembled on either ends. The vessel is then filled with water for hydropressure test as a part of qualification trials. An intermittent leak proof layer of rubber is provided to avoid water contamination into the composite layers.

(a) (b) (c)

Fig. 13 Filling of composite layup and curing

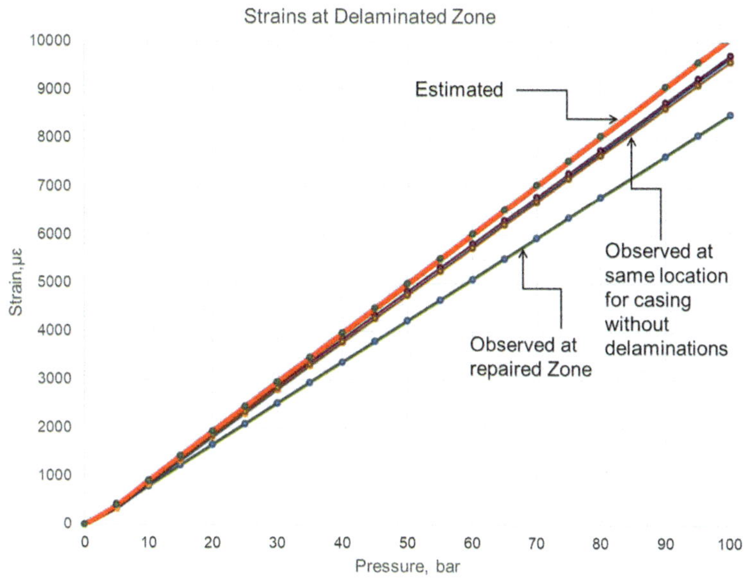

Fig. 14 Pressure test strain plot

7 Testing and Analysis

The vessel is analysed along with repair scheme and compared the results with and without delaminations. The pressure vessel after repair work was hydropressure tested successfully. The strains and dilations are measured and compared with other tests and found sufficient margins which are available. Also, the influence of all the three delaminations on the maximum pressure-carrying capability was studied and found satisfactory. The strain results of pressure test are plotted in Fig. 14.

The strains are plotted against pressure and the observed values are found to be in close correlation with the estimates. For every incremental of 10bar, the strains are plotted at the delaminated zone and at the same location for casings without delamination. Maximum strain value of 9729με was observed for pressure vessel without delamination and strain value of 8756με at delaminated zone was recorded. A variation of about 2.76% is observed at the dome regions compared to the finite element analysis estimates. At the delaminated zone, about 12.49% variation is observed which is due to the angle variation in the hand layup and also due to the overall add-on plies. This shows that the load distribution has happened though the composite filling and the add-on plies. However, not much change in the dilation pattern is noticed as compared to the previous tests.

Acknowledgements The authors extend their sincere gratitude to Project Director and Director ASL for supporting in completion of this work.

Declaration of Competing Interest The authors declare that they have no known competing financial interests or personal relationships that could have appeared to influence the work reported in this paper.

References

1. Smith RA, Composite defects and their detection, Mater Sci Eng III
2. Luke Everett Turoski (2000) Thesis on "Effects of manufacturing defects on the strength of toughened carbon/epoxy prepreg composites." Montana State University, Bozeman
3. Thirkettle A et al (2005) Modelling delamination caused by defects in large diameter composite cylinders. In: European conference on spacecraft structures, materials & mechanical testing
4. Kim CU et al (2002) Structural analysis and strain monitoring of the filament wound motor case. J Compos Mater (36)20
5. Ye L (1988) Role of matrix resin in delamination onset and growth in composite laminates. Compos Sci Technol 33:257–277
6. Hosseini-Toudeshky H, Hosseini S, Mohammadi B (2010) Buckling and delamination growth analysis of composite laminates containing embedded delaminations. Appl Compos Mater 17:95–109
7. ANSYS, Inc., Mechanical, Release 17.0

Best Practices in Additive Manufacturing of Continuous Fiber Composite Parts

Raghunath Rao Komati and Rohit Shivaji Rasale

1 Introduction

Manufacturing can be classified as conservative, subtractive and additive. In conservative manufacturing, the initial stock is reshaped to the final shape with conservation of mass, like in forging and compression molding. In subtractive manufacturing, the material is removed from the raw stock, like turning to get the final shape. Whereas in the case of additive manufacturing, we build the layers one over the other from bottom to top in a sequential manner, to get the final product, like in filament winding. Mostly we get the finished product in additive manufacturing, but in some cases, final machining is needed. 3D printing, a disruptive technology, is the only process which does not need tooling, while all other methods need tooling during manufacturing. Depending on the material of the printed part, 3D printing is classified as metal, polymer, ceramic and composite. This paper is about the 3D printing of polymers and composites only. In polymers, it is specifically related to the nylon-based onyx polymer. In composites, it focuses on the combination of the onyx matrix with one of the three fibers (i.e., carbon, glass, kevlar and their variants) at a time. The basic operations in 3D printing are the generation of CAD model, .STL file, print file and printing of component. This paper specifically deals with the best practices that we followed during the generation of mechanical properties of materials, model file optimization, analysis and optimization of fiber orientation, error compensation due to orientation of the part, optimization of parameters related to software and hardware, printing, inspection and testing of components. A Markforged X7 3D printer was used for printing composite components.

Komati Raghunath Rao (✉)
Technology Director, Advanced Technology Group (ATG), ASL, Hyderabad, India
e-mail: raghunathraok.asl@gov.in

Rasale Rohit Shivaji
JRF, ATG, Advanced Systems Laboratory (ASL), Hyderabad, India

M. R. Manohara Babu et al. (eds.), *Recent Advances in Composites Engineering and Research*, Springer Proceedings in Materials 39,
https://doi.org/10.1007/978-981-99-8807-5_9

2 Literature Survey

Systems engineering approach and methodology in additive manufacturing were explained in references [1–3]. Proper planning, design, methods and applications of 3D printing were explained in these references. ASME Y14.46 [4] explains the unique definition of product geometry, process and data. Markforged additive manufacturing literature [5] was extensively used for the actual development of 3D-printed continuous fiber composite parts. Some articles in references [6–9] explained current trends in the 3D printing industry.

3 Best Practices

3.1 Data Generation for Mechanical Properties

Mechanical properties are required for the analysis of composite parts. First, drawings were generated as per sizes mentioned in ASTM standards for tensile, flexural and impact tests. Subsequently, we printed test specimens on two different occasions with the same batch of material, to assess mechanical properties and variation of properties, with time by using Markforged X7 3D printer. Then we inspected the printed samples for dimensional accuracy. Finally, we carried out the testing of samples as per ASTM standards. A total of 54 samples were tested in fiber direction only. No post or oven curing was done as the complete polymerization took place in printing itself. In the following bar charts, the first two specimens in each group were tested in the year 2020 and the remaining were tested in year 2022. Some degradation is observed in properties with time, in this limited number of samples. Onyx, high strength and high temperature(HSHT) glass and carbon samples were printed and tested, and results are shown in Charts 1, 2 and 3. We used minimum values of these properties in the analysis of actual components.

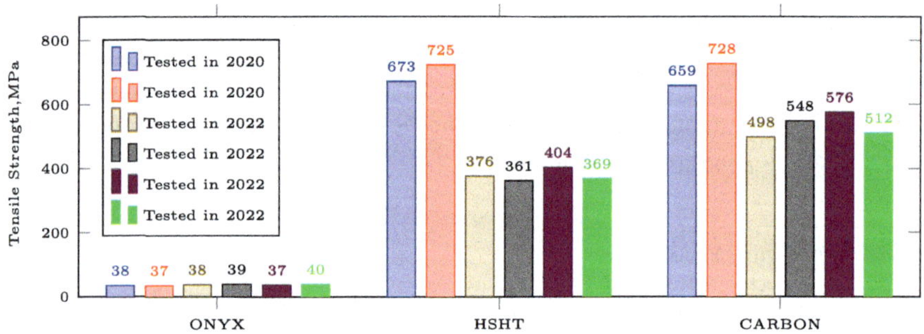

Chart 1 Tensile strength of 3D printed specimens

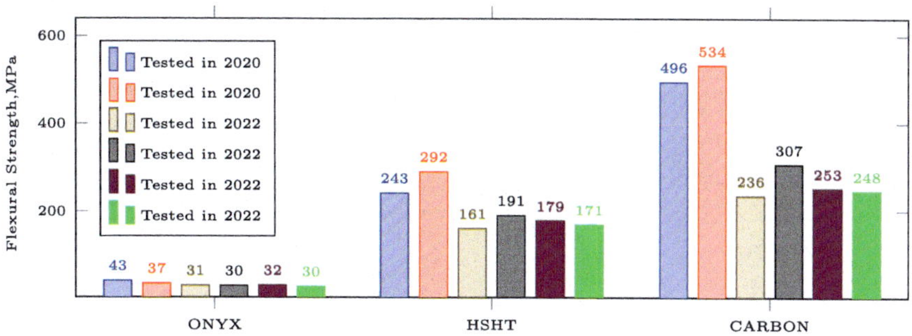

Chart 2 Flexural strength of 3D printed specimens

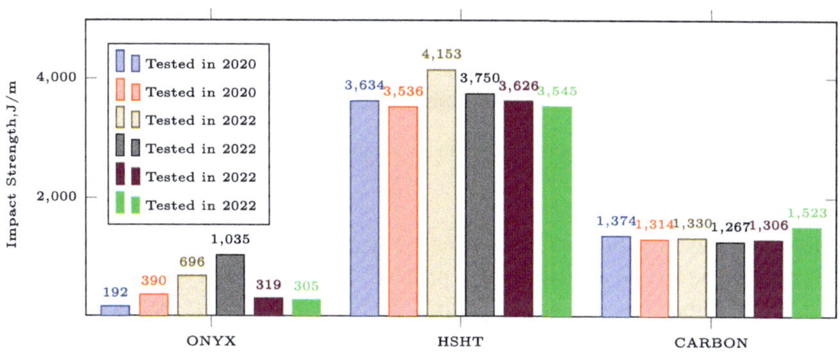

Chart 3 Impact strength of 3D printed specimens

3.2 Model File Generation

First, the CAD model was generated in SpaceClaim, an add-in module of ANSYS 2021R2. Sometimes 2D drawings, .STEP files and other native file formats like .sldprt were also used, to generate CAD models. Then various tools in SpaceClaim were used for optimization. Finally, .STL file was generated. Here the model size, complexity and level of detail dictate the size of the .STL file. If the file size is huge, then the printing time increases enormously. Without compromising the functional requirement, print time was optimized by adjusting the size of the element in triangulation. Other file formats, .AMF, .VRML are also available for supporting multi-color and multi-material printing.

3.3 Analysis and Optimization of Fiber Orientation

In the case of metal 3D printing which is an isotropic material, the lattice structure inside the object, only can be engineered. Whereas in the case of composite 3D

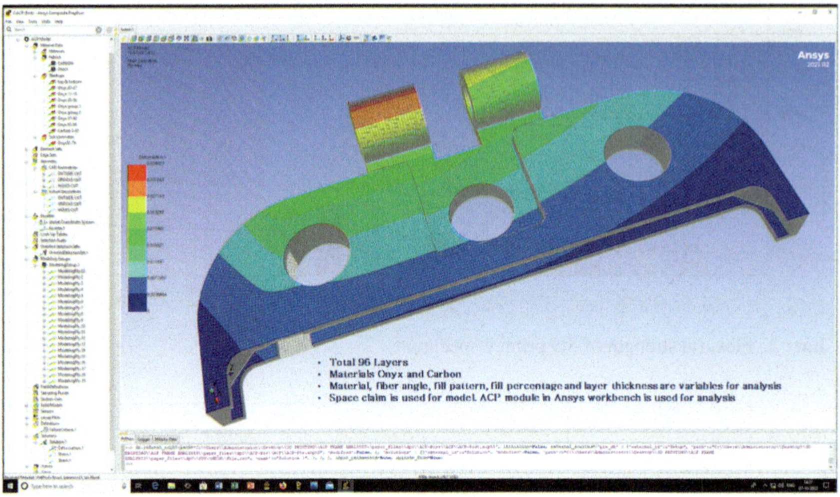

Fig. 1 FEM analysis of representative functional component

printing, particularly in continuous fiber 3D printing, there is a lot of scope for optimization of the weight and stiffness of printed components. The first parameter is an alignment of fiber direction with load direction in the component. The second parameter is the fill pattern inside the component either rectangular, triangular or hexagonal. The third parameter is fill density. The fourth parameter is layer thickness. The fifth parameter is the type of material. Above-mentioned five parameters were considered, in optimizing the weight and time of the printed product. Analysis was carried out for several different print layup patterns by changing the above-mentioned five parameters in Ansys workbench 2021R2, after making primitive models in SpaceClaim, and finally, an optimized printable layup pattern was used for final printing. Figure 1 shows the deformation of a typical component in the Ansys ACP post. Here carbon fiber was chosen, as the high specific stiffness was required in the application. The layer thickness was 125 μm for carbon and was fixed. The component comprised holes, slots, channels and other complex features with constraints on thickness; hence 100 percent fill pattern was chosen. The layup pattern was optimized using FEM. A rectangular pattern was chosen to get maximum fill density.

3.4 Parameters Related to Slicing Software

From the analysis, material, fiber direction, fill pattern, fill density and thickness of the layer were chosen. On-premises Eiger software was used to generate the print file/path file for the movement of the print head in the printer. If continuous fiber is the recommended choice while designing, then the layer thickness shall be as per

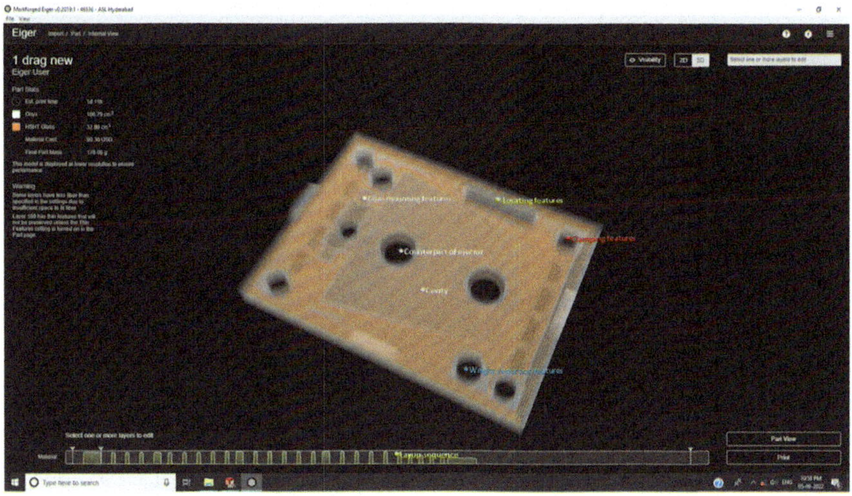

Fig. 2 Details of mold halve in slicing software

pre-setting thickness in the 3D printer. It is 125 μm for parts with continuous carbon fiber and 100 μm for parts with other fibers like kevlar, glass and their variants. If onyx is recommended choice while designing, then we have the freedom to choose the thickness of the layer from 50 μm, 100 μm, 125 μm and 200 μm. The surface finish and print time are usually inversely proportional to the layer thickness. Print time is proportional to fill density. Figure 2 shows the mold halve with HSHT and onyx materials. It shows print time, quantity of material, layup pattern and the number of layers. It also shows the locating features, ejection features, clamping features, weight reduction and component cavity on the mold halve. It was printed and used for the production of functional composite parts.

3.5 *Printer Parameters in 3D Printing Machine*

Quality in terms of conforming to shape, size and tolerances, conforming to internal layup pattern and surface finish depends on many machine-dependent parameters. These include belt tension, axis (X, Y and Z) positional accuracy while executing traverse commands, nozzle wear, nozzle offset, bed clearance and bed leveling. In addition to the above, the path of plastic and fiber from the spool to the print head should be clean and clear to facilitate uniform movement of material. Moisture content in the plastic also plays a critical role in the quality of the part. To achieve the best quality, these parameters need to be set as per original equipment manufacturer(OEM) instructions. Acceptable components were printed by following the guidelines of OEM.

Fig. 3 Typical complex features in printed component

3.6 *Printable Complex Features*

3D printing machines can print thin features of less than a millimeter to a maximum size equal to build volume. The geometry includes all definable features in the CAD software. The shape includes all Booleans of primitives in the CAD modeling software. Multiple features like waveguides, support structures for PCBs, complex contoured ducts, features for metal inserts and honeycomb structures with variable void volume and grid patterns were incorporated in a single model and were printed successfully. Some of these features were impossible to produce in traditional manufacturing processes. Figure 3 shows the printed component without any support material while printing, which demonstrates the capability of 3D printing.

3.7 *Printing of Critical Local Features*

Because of the limitation in build volume (330 mm × 270 mm × 200 mm), splitting is necessary for the parts, larger than the build volume. In some cases, though the build volume accommodates the part, because of fiber orientation requirements, the part needs splitting. In some cases, the part may be a replacement of a metal part in an assembly. To achieve smooth assembly in the above-mentioned circumstances, parts must be printed within the designed tolerance band, particularly at interface features. In actual printing due to the orientation of the part, a mean shift was observed, which caused the printed part in out of the designed tolerance, though the model was within the designed tolerance and also the printer was capable. To overcome this problem, the mating features were printed separately with the same orientation, and the error was compensated in the model to achieve the designed fit between mating parts.

3.8 Metrology and NDT

A dimensional inspection of printed test samples, prototypes and functional parts was carried out to assess the capability of the 3D printer in terms of accuracy and precision. Accuracy and precision of linear dimensions were excellent as long as the nozzle offset, wear and bed level were within the range, recommended by OEM. But we found some deviation in the accuracy of oriented complex features, like inclined holes, slots and cavities due to inherent issues of the printer. This deviation was compensated in the CAD model file by printing localized features alone with the same orientation and then measuring the deviation. Thermography was carried out on printed samples to ascertain the continuity of the material. The dielectric constant, moisture content, fiber volume fraction and Tg of materials were carried out, for some functional applications of printed components.

4 Results and Discussions

4.1 Results

- A facility for 3D printing of continuous fiber composite was established.
- Parts were printed for functional applications in addition to tools and prototypes and used successfully.
- Carried out the material characterization of materials and used them in analysis.
- Analyzed the functional parts using tested mechanical properties for different inter and intra-layup patterns to optimize weight and printing time using Ansys Mechanical Enterprise 2021R2.
- Some complex features were incorporated into a single model and printed to demonstrate the capability of 3D printing.

4.2 *Discussions*

- Quality dependence on software and hardware parameters was discussed.
- Prior printing of primitives to compensate for errors in inclined features was discussed.
- Few NDT methods were carried out to ascertain the quality of functional parts.
- This paper focuses only on the totality of product development using continuous fibers. As per the authors' best knowledge and with testing of the limited number of samples, in the fiber direction, the mechanical properties of 3D-printed continuous fiber composite parts were parred with aluminum alloy or even superior in certain cases.
- Standards are yet to be established for processing parameters and acceptance criteria.

5 Conclusion

Though higher mechanical properties in fiber direction, sometimes it is not possible to orient the reinforcement to all features, in the desired direction. Nevertheless, within the build volume, with proper orientation of the job and proper concentric reinforcement around the open contours, parts were printed and used successfully for functional applications. If any 3D printer is capable of printing open source continuous fibers with thermosetting matrix, along with multidirectional traverse, then it really will be useful in many applications of the aerospace industry

Acknowledgements The authors would like to express their sincere thanks to the people who cooperated while the procurement of the printer and while inspection, testing and qualification of materials, functional parts and assemblies. Author1 would like to thank the hierarchy of the organization for the allocation of author2, who was associated with this work.

References

1. Parandoush P, Lin D (2017) A review on additive manufacturing of polymer-fiber composites. In: Department of industrial and manufacturing systems engineering, Kansas state university, Manhattan, KS66506, vol 182. USA Composite structures, pp 36–53
2. Sonmez G, Yong CK (2016) Additive manufacturing of composites: current developments and use in the near future. In: Editor FS (ed) UTWENTE 3D printing research papers 2015/2016, Conference 2016, LNCS, vol 9999. Springer, Heidelberg, pp 1–13. 10.10007/1234567890
3. Badiru AB, Valencia VV, Liu D (2017) Additive manufacturing handbook, product development for the defense industry. CRC Press Taylor and Francis Group
4. Product Definition for Additive Manufacturing, Engineering Product Definition and Related Documentation Practices, An American National Standard, ASME Y14.46-2022
5. Industrial Printer User Guide Version 1.01/June2019 of Markforged X7 3D printer
6. AIRTECH Largescale 3D printing, Resins and tooling https://www.airtech3d.com

7. Gibson I, Rosen D, Stucker B, Khorasani M (2021) Additve manufacturing Technologies. Springer.
8. JEC composites magazine-issues. https://www.jeccomposites.com/?s=additive+manufacturing+of+composites
9. Composites world (CW), Additive manufacturing of composites. https://www.compositesworld.com/search?q=additive manufacturing of composites

VARTM Process Automation Using Fiber Optic Sensors

Irfan Khan, Mayur Godbole, and Makarand Joshi

1 Introduction

1.1 VARTM Process

The Vacuum Assisted Resin Transfer Molding (VARTM) is one of the prominent processes for composite fabrication, especially in defense application where reliability, strength, and product quality are at higher priority. It is based on Liquid Composite Molding (LCM) process technique. Additionally, we can say, it is one of the variant of RTM process in which resin impregnation is carried out by applying vacuum instead of using high injection pressure. In VARTM process, a vacuum bag replaces the upper mold, and the resin is infused under atmospheric pressure. However, there are some variants of VARTM such as vacuum injection molding which infuses resin under pressure. VARTM process is widely used for manufacturing both large and small sized complex components. One of the chief advantages of VARTM over RTM is the ability to produce high fiber content product. This is due to the fact that, as fiber content is increased in RTM, there is a tendency of fiber wash due to resin impregnation under pressure. Hence, VARTM is generally suitable for realization of large and thick composites structures. Although, the process apparently looks simple to implement, it requires detailed process design (involving placement of resin and vacuum lines) and a process monitoring setup to confirm resin arrival at each and every location in the mold.

I. Khan (✉) · M. Godbole · M. Joshi
Composite Research Centre, Research and Development Establishment (Engrs), DRDO, Pune 411015, India
e-mail: irfankhan.rde@gov.in

M. R. Manohara Babu et al. (eds.), *Recent Advances in Composites Engineering and Research*, Springer Proceedings in Materials 39,
https://doi.org/10.1007/978-981-99-8807-5_10

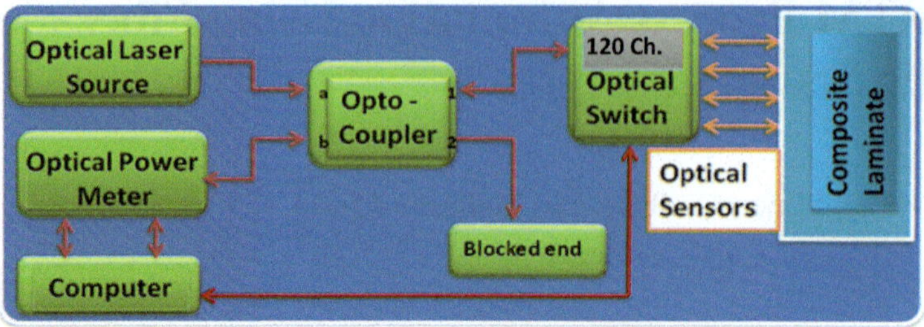

Fig. 1 Block diagram for optical sensors-based process monitoring setup

1.2 Process Monitoring Setup for VARTM

Improper process control may result in incomplete fill, wastage of resin, excessive cycle times, under cure, etc. In complex structure having different size and thickness, it's difficult to ensure proper resin reach at all critical locations during infusion process. Even after making at most technical efforts, practically there remain certain regions in the fabricated structures with resin rich areas or resin dry areas. Process monitoring setup is one of the essential elements for fabrication of composite product using VARTM technique. Performing process monitoring improves quality of product and also reduces manufacturing cost by reducing excess resin flow. It consists of optical-based instruments to monitor resin flow front by using fiber optic sensors. Being very immune to electromagnetic interference, ease of handling, and refrain to change structural property of product, these fiber optic sensors are embedded in large number inside the reinforcement so as to provide feedback for complete resin fill within the fabric layers. These feedback sensors ensure that the product is free from sub-surface voids with full saturation [1]. Optical bare fibers are initially cleaved at 90° using cleaver so as to act as an optical sensor. These sensors work on the principle of change in refractive index (u). Change in refractive index at the end of embedded sensor causes drastic drop in optical power value, ensuring resin reach at that particular location. For executing the process monitoring system, available optical instrumentation setup such as DFB laser source of 1550 nm wavelength, optical circulator, or bidirectional coupler, JGR make optical switch (1×120) and JDSU make optical power meter have been used [2]. Schematic diagram of the optical setup is shown in Fig. 1.

1.3 Requirement to Automate Process Monitoring Setup

To carryout effective process monitoring on large structure composite, it was required to embed huge numbers of optical sensors to ensure complete saturation at all critical locations. Thus, manual scanning as well as monitoring of such large number of embedded sensors within stipulated time was quite difficult. However, navigating the channel numbers manually through push button may lead to wear and tear of the hardware as well as it leads to time lag during switching from first channel to last channel. During this manual scanning, many sensors reading were missed out due to system time lagging, and also it includes human error while tabulating the data manually.

These problems were overcome by automating the existing process monitoring optical setup. In view of this, a LabView-based customized software GUI is developed to carry out all process monitoring task automatically. This GUI effectively performs scanning of embedded sensors (auto/manual mode), updating of sensor status, and indication for responded sensors. By setting of optical power threshold value in software, output from batch of sensors leads to create ON/OFF type control action and actuate solenoid valves of next stage resin line. It also automatically logs the sensor data with respect to elapsed time and moreover minimizing overall execution time delay as well as human error. Further, this software configuration is utilized to implement the logic-based control actions for VARTM automation.

1.4 Motivation for VARTM Automation

As discussed, VARTM is considered as the most prominent process for world class composite manufacturing technique. This process is frequently used to develop huge structures like windmill blades and marine application because of being low weight, stiffened, durable, easy to repair, stronger, and reliable [4, 5]. While implementing this process various uncertainties like variation in resin flow front due to race-tracking or human error arises that leads to incomplete resin saturation. Large number of labor force required during huge structure fabrication process that increases the possibility of accidental misconduct or misinterpretations. In traditional manual mode, resin line pipes were used to pinch by binding three fold turn that causes temporary deformation and act as obstacle to drop resin flow pressure. Even after deploying trained labors to open/close the designated resin lines, most of the time misunderstanding occurs that causes major disturbances in sequential resin infusion process. Manually monitoring and maintaining of adequate resin level within resin tray often causes overflow or lack of resin that allow air bubbles to get inside fabric layer. Performing process monitoring independently with other VARTM parameters can miss out certain important monitoring candidate that leads to unconditional failure of product.

Thus, in order to resolve such issues related to VARTM process, it is required to automate overall system. For this, along with process monitoring setup, various monitoring and controlling parameters of VARTM process are consolidated on LabView-based developed GUI software. The automation process automatically monitors and makes necessary adjustments for uniform resin flow in the vacuum sealed job to avoid any variation in resin flow front [6]. This was done by incorporating the control action based on sensor input from process monitoring setup for auto-actuation/opening and closing of resin lines through solenoid valves, monitoring of reinforcement compaction, resin level, hardener mixed resin temperature, etc. Addition to this the most important was real-time data acquisition for embedded sensors with respect to elapsed time, various trial parameters, date, time, etc.

The overall automation setup is equipped with computer-based compact DAQ controller, resin dispensing lines with pinch type solenoid valves, sensors for vacuum, level, and temperature. Using this, a customized GUI program is developed for real-time sensors data acquisition as well as control action implementation based on optical sensors input. In this GUI, we can easily scroll and scan the complete channels/sensors with desired scanning rate in auto or manual mode [3].

2 Components for VARTM Automation Process

VARTM setup comprises various component configuration as shown in schematic Fig. 2. Before placing fabric plies, a button size load cell is fixed in mold cavity, and its data cable is drawn out through cable slot on mold. Both load cell and its cable are encapsulated with release agent as well as sealed vacuum bag. Later, rest of the components as shown in figure is configured to execute the process.

2.1 *Pinch Type Solenoid Valves*

Solenoid valves play an important role while implementing automation process with sequential stage wise operation. It is allowed to sequentially actuate through computer controlled GUI, based on response from embedded optical sensors. For this application, customized pinch type solenoid valve is deployed to control resin flow sequence. These pinch type solenoid valves are in lined with flexible silicon tubes, as shown in Fig. 3.

Resin replenishment line, i.e., main line from resin reservoir to resin dispensing tank is connected using blue PU pipes (having ID:10 mm and OD:12 mm). In order

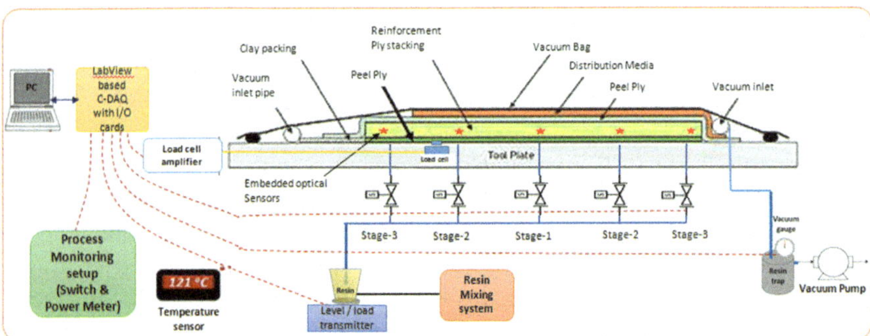

Fig. 2 Schematic configuration diagram for VARTM automation system

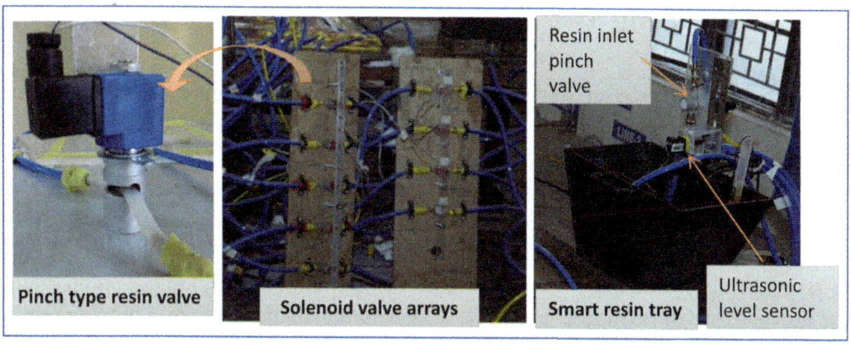

Fig. 3 Customized pinch type solenoid valve and resin tank configuration

to stop the resin flow at high pressure and such huge diameter PU pipe, a customized high torque solenoid valve had been used with proper fixture assembly. For mold resin lines feeding, comparatively small size PU pipes (having ID: 6 mm & OD: 8 mm) without external pressure are used. Being various advantages like small, compact design, safe operating voltage: 24 VDC, Valve Type: 2-Way, available in both modes: Normally-open and Normally-closed, Medium: air and resin fluids, easy to clean (replace tubes), low power consumption: 2.5 Watts, Response Time: 5–25 ms., high cycle life, unobstructed flow path, laboratory grade silicone tubing (easily replaceable), these valves had been deployed for all resin line opening stages.

2.2 *Smart Resin Tank*

Smart resin tank is responsible for maintaining the desired resin level range as well as its temperature during infusion process. It is instrumented with ultrasonic level sensor (instead pressure transmitter or load cell can also be deployed) and temperature sensor (2 wire thermocouple). Addition to this customized pinch high torque pinch type solenoid valve used as inlet and outlet (drain) valves. Resin level as well as temperature data is acquired on the computer controlled GUI in the form of numerical and graphical representation. If the pressure of the resin tanks falls below threshold value, it actuates the resin replenishment solenoid valve so as to fill the resin till upper level limit and later de-energies the valve. Similarly, if the temperature of the hardener mixed resin rises beyond 35° Celsius, the software triggers the output to actuate the resin drain valve so as to drain the complete batch at certain level.

2.3 Resin Mixing System

This is a pneumatic-based system for mixing resin and hardener in a specific ratio as specified by the data sheet of the resin. It consists of two feeder pumps with programmable mixing system unit, which delivers the resin hardener mixed ratio with pressurized flow. However, for this small automation trial, a pressured resin reservoir is used to simulate the functioning of mixing system. The equivalence of functioning between the pressurized resin reservoirs and mixing system was achieved in terms of the parameter flow rate. Previously explained resin tank is connected to the downstream of resin reservoir. The flow between resin reservoir acting as mixing system and resin tank is controlled by a solenoid valve whose trigger point is the low level indication of resin in the resin tank.

2.4 Button Size Load Cell

In order to monitor the fabric compaction, load cell of compact static low profile having range of '10–50 kg.' with conditioner is deployed. This button size load cell is located in between mold and fabric layup. This parameter is continuously monitored throughout the infusion process so as to maintain reasonable vacuum pressure with respect to resin flow. It has been observed that the vacuum pressure drops at some extend after the resin injecting process started.

2.5 Resin Trap with Vacuum Gauge

Resin trap is a metallic tightly sealed container placed at the vacuum line for the purpose of filtering the vacuum lines through vacuum pump. Vacuum pressure transmitter, Model: S-10, Output: 4–20 mAmps, range of '0–1 bar absolute' is incorporated with resin trap unit so as to continuously monitor and display the active pressure of incoming vacuum line on computer controlled GUI. This resin trap basically accumulates the excess resin that comes in vacuum line due to complete fabric fill or due to race-tracking.

Fig. 4 Trial laminate under vacuum with optical sensors

3 Trial Laminate for VARTM Automation

3.1 Preparation of Trial Laminate for VARTM Automation

The process of fabricating laminate by VARTM technique is done by systematic approach. For demonstrating VARTM automation process, trial laminate had been prepared by considering all essential parameters. This trial had been done on flat plate mold with resin lines slot with replica of deck section of composite bridge having various thickness level. The procedure for preparation of trial laminate is cleaning of mold surface plate using acetone and applying vacuum sealing tape to mark layup area, waxing and PVA coating application within inside region of vacuum sealing tape, peel ply placement on the layup area of the mold, positioning of distribution media 'mesh' on the layup area of the mold, embedment of optical sensors in various thicknesses of fabric preforms, clay and sealant tape dams are formed so as to stop and avoid resin flow to form race-tracking during the infusion, on the top of the complete layup, peel ply and then distribution media, i.e., mesh is placed, flexible plastic cauls plate with attached PU pipe for resin line is placed on mesh, vacuum bag with sealing tape is applied to get leak proof vacuums on the whole layup setup area. Breathers used on sharp edges and corners of mold to avoid vacuum bag damage under pressure. All resin lines as well as vacuum lines were scanned and tagged with certain nomenclature so as to release it sequentially during stage wise infusion process. Vacuum pressure sensors deployed to monitor vacuum level and leak drop inside the sealed fabric preforms. Complete setup is prepared for resin infusion as shown in Fig. 4.

3.2 Optical Sensors Embedment During Fabric Layup of Trial Laminate

For demonstration of VARTM automation, Glass Fiber Reinforced Plastic (GFRP) fabric is laid on mold surface within layup area in quasi-isotropic sequence. Implementing the fabrication process, dust free enclosed as well as air conditioned fabrication bay was setup. Optical bare fiber prepared as a sensor is embedded at different

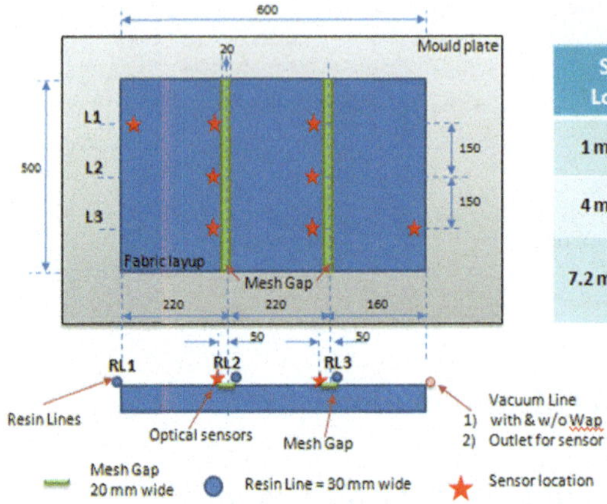

Sensor Location	SL-1	SL-2
1 mm / 2 ply	1L1-2, 2L3-2	3L2-2, 4L3-2
4 mm / 9 ply	5L1-9, 6L2-9, 7L3-9, 8DRSL3-9	9L1-9, 10L2-9, 11L3-9, 12DRSL1-9
7.2 mm / 16 ply	13L1-16, 14L2-16	15L2-16, 16L3-16

Total Sensors in Deck : 16 Nos.

Fig. 5 Schematic of fabric layup, mesh gap, and sensor location

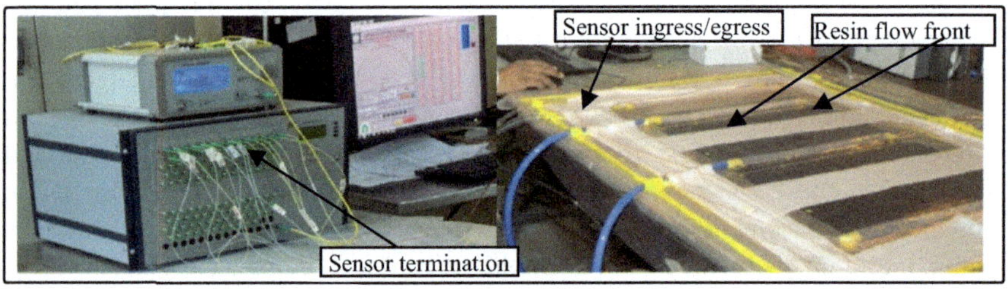

Fig. 6 Process monitoring experimental setup with trial laminate

thickness during fiber layup sequence. This section will elaborate the detailed sensor layout at different thickness location. Optical sensors have been embedded at various thickness layers so as to monitor flow of resin through fabric preforms thickness. Figure 5 shows a schematic of sensor positioning layout.

Bare optical fiber end, inside layup is initially cleaned using an IPA solution. About 40–50 mm length coating is removed from end-tip using a cutter tool. Retaining about 10 mm of core-cladding region, rest is cleaved at 90° by using a cleaver tool. This cleaved end-tip of 10 mm is treated as a sensor part which is embedded in laminate and other end is connected to optical switch through pigtails as shown in Fig. 6.

3.3 Targeted and Achieved Properties of Trial Laminate

Composite laminate fabricated using VARTM automation technique was targeted for specific properties such as Thickness: 20 mm, Density: 1.9 gm/cc, Resin content: 40%, Fiber content: 60%. Later, achieved properties were more or less closer to targeted. Achieved Thickness: 20.76 mm, Density: approximately 2 gm/cc, Resin content: 38%, Fiber content: 58–60%, and void content: 1.8–2% approx.

Secondly, for trial laminate fabrication, Fiber used for layup: Glass fabric (Chomarat make) and Resin-Araldite-LY1564, Hardener-Araldite-3486. Also, this technique can be adopted for those fiber and resin systems where fabric and resin compatibility is ensured, infusion under room-temperature and resin is compatible with hardware used in infusion system.

4 Demonstration of VARTM Automation Scheme

4.1 Development of LabView-Based GUI for VARTM Automation

A customized Graphical User Interface (GUI) has been developed to incorporate all essential components of VARTM automation. This GUI is developed on NI's LabVIEW software (version 2010) platform. Computer controlled GUI receives the main source of input from optical instrument setup so as to monitor and control the overall process performance. Additionally, it also accepts input data from vacuum pressure gauge, resin tank level transmitter, resin tank temperature sensor, load cell, resin mixing system, solenoid valve ON/OFF status, etc., to provide single platform system monitoring. Based on the optical sensor information, it is capable of executing the control actions like resin lines opening/closing and raising alarm on change in magnitude of threshold value. Developed GUI is user friendly and provides the information about: current location/channel number of optical switch, optical power meter reading of existing sensor, current date/time information, total elapsed time of overall process, active status of solenoid valves, number of sensors sensed/responded so far, vacuum pressure level, resin tank level with temperature for monitoring purpose, etc.

Initially, at the time of infusion, the VARTM automation software is configured for number of sensor classified as monitoring and decision making. Monitoring sensors are just indication for progress of infusion while decision making sensor respond to some control action based on process requirement. The configured GUI, shown in Fig. 7, starts scanning all connected sensors at predefined scanning rate. Also it indicates the level of resin and vacuum in terms of actual value as well as percentage range. LED indicators of sensors remain unlit at initial stage, interpreting that no sensors responded yet. Before starting the infusion process, all connected sensors were scanned, and its corresponding optical power values (μW) were noted. Embedded sensors which initially being in air (before infusion), reflects optical power value (about 50–300 μW) based on the length of its fiber. Thus, refractive index of the medium (i.e., air) is measured for each sensor and plotted. Once the infusion process commences, all sensors connected at the output channels of optical switch were scanned through the GUI in auto or manual mode. During infusion, resin flows through mold's resin lines and then fabric preforms under the driving force of vacuum.

Resin arrival at the tip of the embedded sensor interprets to change in medium (from air to liquid). Thus, optical light almost get refracted and power meter reads the power value near to zero (about 1–2 μW), which means drop of total optical power. This ensures that, fiber preforms are impregnated with resin at particular layer/thickness or location. As power meter reading drops below the threshold value configured in GUI, it drives an event and the channel indicating LED glows green to convey that the sensor has sensed the resin arrival at particular location, as shown in Fig. 7.

On soft panel GUI, 'Data saving' button is provided so as to save sensor information manually with respect to total elapsed time. Also 'Channel rest' button is provided for resetting any particular or complete channel status while using it for multiple batches. Process data acquired is saved in separate text file (.txt format) at desired location or file path. Addition to this, saved data sheet provides the information about trial location, test name, trial date/time, etc. Thus, we can have ready data sheet for sensor responding time after complete infusion trial.

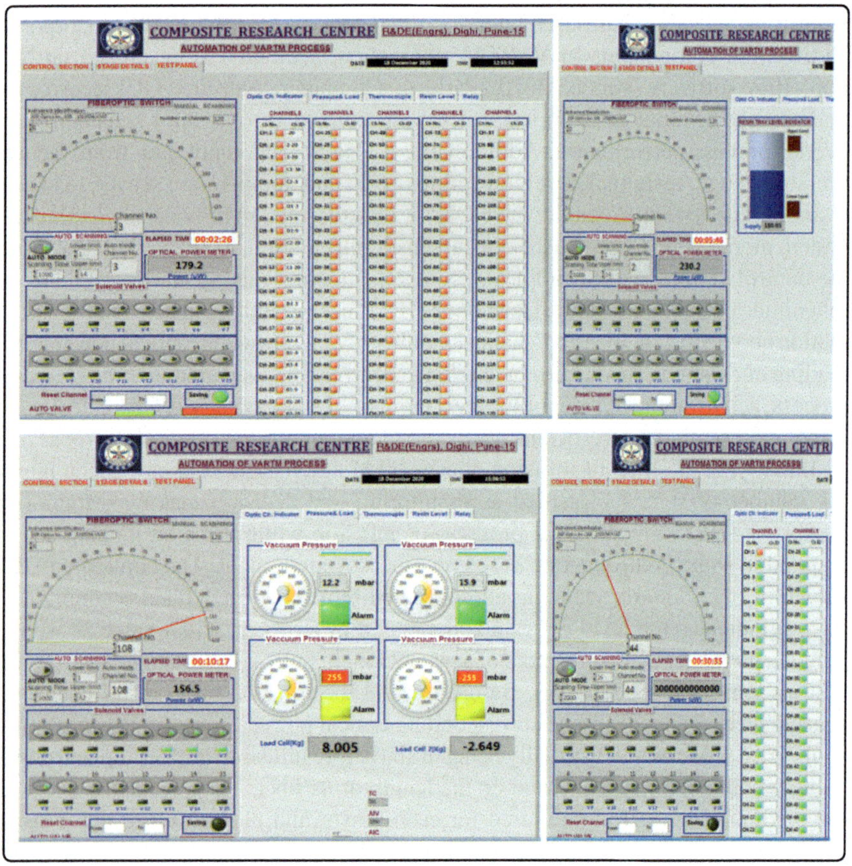

Fig. 7 LabView-based GUI showing various control and monitoring parameters

4.2 Data Acquisition for VARTM Automation Trial

Optical sensors embedded with sequential identification tagging play an important role for monitoring and controlling the resin flow during infusion. Thus, it is important to scrutinize and list up the number of sensors as per the desired stages. The detailed sensor mapping for various stages with respect to resin lines on mold and job is given in Table 1. In this, first column is nomenclature with 'Stage No.' having different characterized stages and resin lines opening sequences. Second and third column specifies sensor details, which is further classified with monitoring and decision making based on their locations. Fourth column indicates the power value of each sensors with and without vacuum application. Last one is acquired time data during consequent stages for each sensor with respect to elapsed time. Addition to monitoring and decision making sensors few more sensors embedded at vacuum line side trim area, to ensure complete fill of laminate.

Table 1 Sensor mapping and data acquisition for VARTM automation trial laminate

Stage No.	Channel No.	Sensor No.	Power		Time (MM:SS)	
			Without vacuum	With vacuum	Simul^{n} time (Min)	Experiment time
Stage-1 Resin line R1	1	13L1-14	125	104	4	00:03:14
	2	14L2-14	149	126		00:02:42
	3	**5L1-9**	102	51	8	00:07:25
	4	**8DRSL3-9**	116	106		00:07:36
	5	**6L2-9**	193	172		00:08:05
	6	**7L3-9**	122	100		00:07:52
	7	1L1-2	201	197	4	00:03:47
	8	2L3-2	178	168		00:03:52
Stage-2 Resin line R2	9	15L2-14	155	140	10	00:09:50
	10	16L3-14	138	135		00:10:15
	11	**9L1-9**	104	101	12	00:11:09
	12	**12DRSL1-9**	72	67		00:10:44
	13	**10L2-9**	103	97		00:11:23
	14	**11L3-9**	176	171		00:11:55
	15	3L2-2	124	114	10	00:10:00
	16	4L3-2	181	177		00:10:06
Stage-3 Resin R3	17	Trim-1	146	131	17	00:18:08
	18	Trim-2	193	185		

The sensor numbers are indicated in bold because they are placed in mid thickness of the laminate and these are decision making sensors to carryout further control action like opening or closing of solenoid valves

Table 2 Data compilation for VARTM automation trial laminate

Stages	Sensor location	Simulation Time (Min)	Infusion Time (Min)
1	upper region	4	3
	Bottom region	4	4
	Mid region	8	8

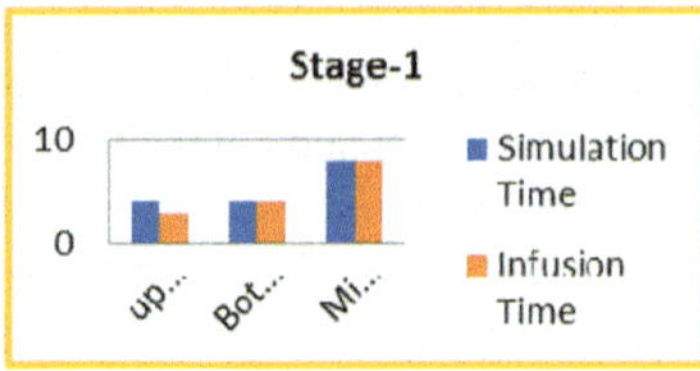

Stages	Sensor location	Simulation Time (Min)	Infusion Time (Min)
2	upper region	10	10
	Bottom region	10	10
	Mid region	12	12

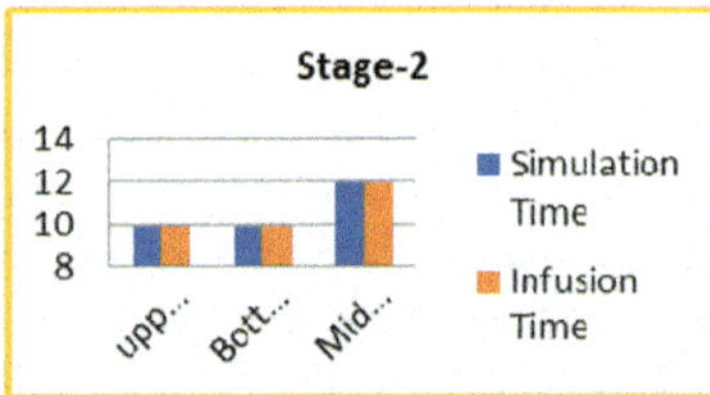

Stages	Sensor location	Simulation Time (Min)	Infusion Time (Min)
Complete Fill	upper region	14	16
	Bottom region	14	16
	Mid region	17	18

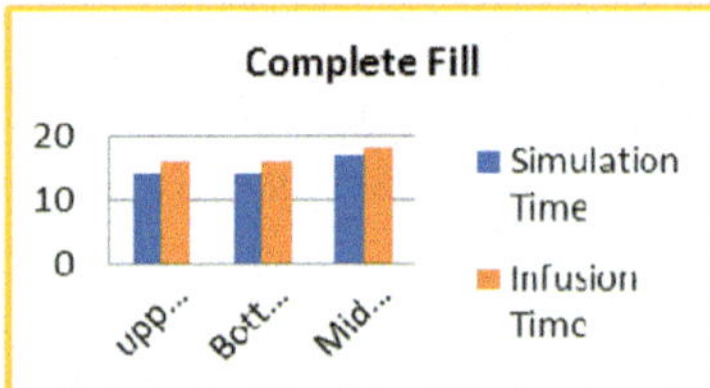

During VARTM automation trial demonstration, data pertaining to each sensor had been acquired successfully using LabView GUI data acquisition. Later, the acquired data had been compiled so as to compare with simulation values as tabulated and plotted in Table 2.

Acknowledgements This paper has been materialize after performing various laminate level trials using VARTM technique. Major portion of the paper is focused on implementation of process automation for VARTM technique adopted for delivering good quality composite product. To prepare this exclusive report, authors convey deep thanks for valuable help rendered by Mr. Jayanta Gope, Mr. Pramod Vadawkar, and Felix Barla. Last but not the least, authors also thank to all colleagues, who had directly or indirectly contributed to complete this task successfully.

References

1. Keulen CJ (2012) Process and structural health monitoring of composite structures with embedded fiber optic sensors and piezoelectric transducers. University of Victoria
2. Gope JK, Godbole M, Process monitoring process of vacuum assisted resin transfer moulding (VARTM) using single mode optical fiber (SMF) and fiber bragg grating (FBG) sensors. Research and Development Establishment (Engineers)
3. Khan I, Godbole MG (2019) Technical report on VARTM automation for composite development (Using LabView software). DRDO-RDEE-CRC-VACD-01-2019, R&DE(E), Pune-15

4. Mouritz AP, Gellert E, Burchill P, Challis K (2001) Compos Struct 53:21
5. Kim D, Hennigan D, Beavers K (2001) Int J Naval Archit Ocean Eng 2:21 (2001)
6. Swan S, Yuksel T, Kim D, Gurocak H (2015) Automation of the vacuum assisted resin transfer molding process for recreational composite yachts. In: Article in polymer composites October 2015. School of Engineering and Computer Science, Washington State University, Washington

Microwave Assisted Curing of CFRP Composite Laminates for Aerospace Applications

Sandhya Rao, Sachin V. Revgade, N. M. Manjunath, and M. Rajendra Prakash

1 Introduction

Carbon fiber reinforced plastic (CFRP) is the one of the main composite material used in the field of aerospace and defence. Conventional curing of CFRP using hot-air ovens or autoclaves is tedious, time-consuming, and energy-intensive. One of the alternative and environmental friendly methods that may be adopted is microwave curing as it is reported to be rapid and energy-efficient [1, 2]. Glass being transparent to microwaves, curing of glass–epoxy is considerably straight-forward and simple, whereas curing of carbon-epoxy composites could lead to safety issues. This is because of the conductive nature of carbon fibers and electrical discontinuities cause sparks, arcing and damage the microwave sources [2]. Further, they also affect the laminate quality resulting in poor consolidation after curing [3]. These in turn affect the cure status (lower glass transition temperature) as well as the composite mechanical properties. In situ temperature monitoring is also a significant issue associated with microwave processing be it GFRP or CFRP. Researchers are trying novel techniques like shielded thermocouples, FBG sensors, pyrometers, thermal imagers, etc. The heating methodology is unique in case of microwave curing since it results in volumetric heating whereas in case of conventional heating heat flows in the form of heat flux, mainly by conduction and convection [4, 5]. Heating CFRP composites using microwaves is not yet a very common process, although microwave heating of GFRP composites is popular [6–12]. When manufacturing CFRP components with microwaves, the potential advantages may be realized only with certain adaptations in the technology [7].

S. Rao (✉) · S. V. Revgade · N. M. Manjunath · M. R. Prakash
Advanced Composites Division, CSIR-National Aerospace Laboratories, Bengaluru 560017, India
e-mail: sandhya.rao@nal.res.in

M. R. Manohara Babu et al. (eds.), *Recent Advances in Composites Engineering and Research*, Springer Proceedings in Materials 39,
https://doi.org/10.1007/978-981-99-8807-5_11

In the present work, a custom designed and developed dual frequency microwave system (DFMPC) has been successfully used to cure carbon-epoxy composites by employing microwave amenable silicon carbide-based molds. The results showed that the cure status of the DFMPC microwave cured unidirectional (UD) and bidirectional (BD) CFRP was equivalent or superior to its thermally cured counterpart. The weight fractions were comparable, and the measured mechanical properties were also comparable or superior. The process cure time was reduced from hours to minutes with the benefit of significant energy savings. Most importantly, the curing was very uniform as revealed by the process thermal images.

2 Experimental Details

2.1 *Materials and Methodology*

The UD and BD carbon fabrics were cut in the dimensions of 200 × 200 mm. Microwave amenable molds based on SiC composite formulation were used as alternative molds in place of metal molds. LY 5052/5052 CH epoxy resin system supplied by Ms. Huntsman, USA, was used as the matrix material.

The 4mil 160 GSM T-300 grade UD fabric was supplied by M/s. Bhor Chemicals, Mumbai. The 150 GSM 2 × 2 twill weave T-300, 1 k grade BD fabric was supplied by M/s. Marktech Composites, Bengaluru.

For the purpose of laminate curing using microwave radiation, a custom-built Dual frequency Microwave System for curing Polymer Composites (DFMPC) which has 2 operating frequencies was used. The DFMPC is shown in Fig. 1.

Fig. 1 DFMPC microwave system used for curing process along with inside chamber view

2.2 Processing Procedure

The CFRP composite fabrication technique used for microwave curing is hand layup followed by room temperature vacuum bag molding (RTVBM) process using the custom-made MW susceptible molds of size (260 × 260 × 10) mm. The RTVBM process is detailed elsewhere [7]. After an initial 2 min vacuum application for uniform dispersion of the matrix phase, the layup was soaked at room temperature (RT) for 90 min to attain the matrix gel time. This was followed by Microwave assisted curing with the application of vacuum throughout the entire MW cure schedule. The microwave cure schedules were optimized for both UD-CFRP and BD-CFRP.

For mutual comparison purposes, equivalent thermally cured laminates were processed in a hot-air oven. The laminate was initially cured at room temperature and later post cured in a free standing condition as per the following post cure schedule:

50 °C for 30 min → 70 °C for 1 h → 85 °C for 2 h (ramp rate = 2 °C/min), followed by natural cooling in the oven.

2.3 Characterization of Thermally Cured and Microwave Cured CFRP

(i) Thermal Imaging: Required to monitor the temperature uniformity of the molds periodically during microwave curing CFRP and to assess the heating uniformity of the laminate after microwave cure completion. The microwave assisted cure processes were monitored intermittently using a thermal imaging camera, FLIR A325sc series.

(ii) Ultrasonic C-scan: A Meccasonic Jet Probe System-I was used for acquiring the C-scan of the vacuum assisted MW cured 2 mm CFRP laminate. Water was used as a couplant, and the ultrasonic frequency used was 2.5 MHz. The output was a color plot showing attenuation levels in different colors.

(iii) Weight Fraction: The weight fraction of carbon fibers in the carbon epoxy composite was determined as per ASTM D3171-15, Procedure A. Here, nitric acid was used for resin extraction from the composite.

(iv) Glass Transition Temperature (Tg): The Tg was determined using a Differential Mechanical Analyzer (DMA) in dual cantilever mode at a frequency of 1 Hz. The sample was heated from 30 °C to 250 °C at a heating rate of 5 °C / min in accordance with ASTM D 7028 Standard procedure. The peak value of the tan delta (δ) curve was noted as the Tg.

(v) Compression, Flexural, and Interlaminar Shear Strength (ILSS): The mechanical behavior of the CFRP was characterized by testing the Compressive, ILSS and flexural strength of the UD and BD-CFRP laminates using an INSTRON 6025 Universal Testing Machine with a 100kN load cell. The compression test specimens conformed to ASTM D6641 test standard. The specimen was loaded at a crosshead speed of 2 mm/min. ILSS of the MW cured GFRP was determined in accordance with ASTM D 2344 standard procedure. Flexural strength was determined in accordance with ASTM D 790 standard procedure. A minimum of five specimens was tested for each type of test to ensure repeatability of results.

(vi) Process Time and Energy Consumption: The curing involved laminate preparation, gelling phase, and vacuum assisted processing in the DFMPC System. The time taken for the optimized cure schedule as well as power input for specific dwell times was used to compare with conventional oven cure process.

3 Results and Discussion

3.1 Cure Schedules Followed

The cure schedules resulting in optimal properties were identified and finalized. The DFMPC MW cure schedule for UD-CFRP and BD-CFRP are shown in Figs. 2 and 3, respectively. The obtained thermal images during the process and at the end of the cure cycle are also embedded.

From Figs. 2 and 3, it is evident that the process time is less than 1 h in both the cases. The energy consumption calculated as the product of power input and dwell time was found to be 1.132 kWh for UD-CFRP and 1.04 kWh for BD-CFRP. The thermal images show the uniform heating and absence of hot-spots during microwave irradiation.

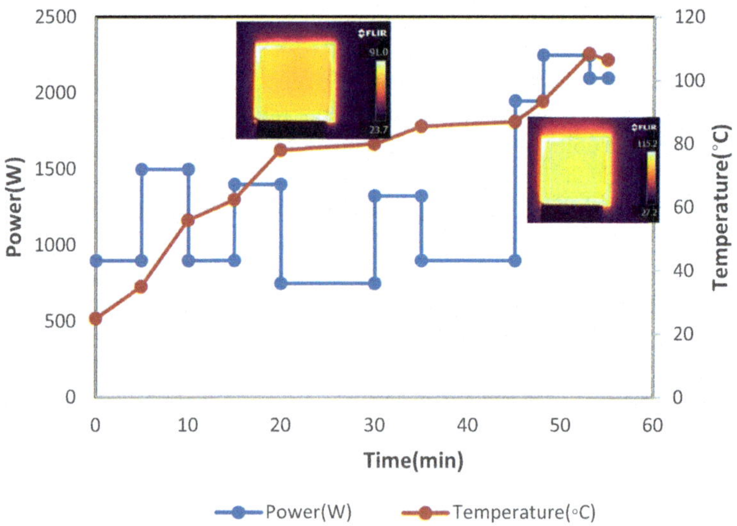

Fig. 2 DFMPC microwave cure schedule of UD-CFRP

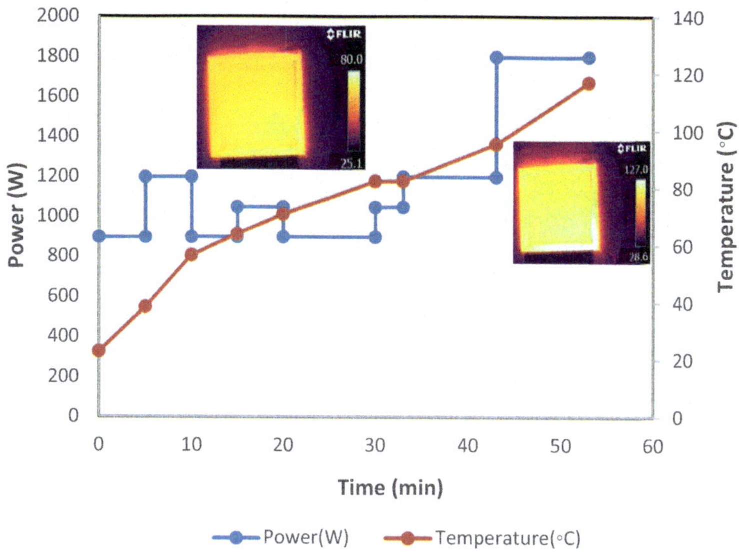

Fig. 3 DFMPC microwave cure schedule of BD-CFRP

In case of hot-air oven-based thermal curing, the composite is allowed to cure at RT for 24 h as suggested by supplier [9]. Post cure is done in the hot-air circulating oven for a cumulative period of 4 h. Even if we consider just the post cure time, then the DFMPC cure time is just 50–55 min, suggesting that the DFMPC curing is 4 times faster. The oven's energy consumption is ~3 kWh. Thus, DFMPC curing is nearly 3 times more energy-efficient.

3.2 *Non-destructive Evaluation*

The typical C-scans of DFMPC processed CFRP laminates and an equivalent thermally cured laminate are shown in Fig. 4. It was observed in general, that the laminates were defect free as revealed by the excellent attenuation distribution across the laminates.

3.3 *Fiber Weight Fraction*

The fiber weight fraction was determined for all the test laminates, both thermally cured as well as DFMPC MW cured. The values were found to be in the range of 63 ± 3%, thus revealing the process uniformity as well as proper control of DFMPC microwave cure parameters.

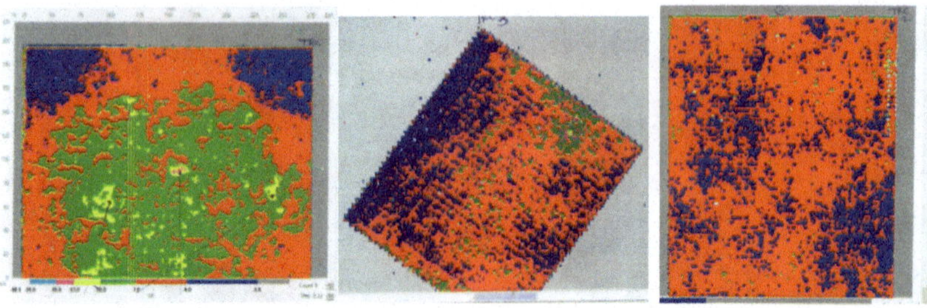

(a) (b) (c)

Fig. 4 The C scan of **a** MW cured BD-CFRP, **b** MW cured UD-CFRP, and **c** thermally cured BD-CFRP

3.4 Glass Transition Temperature (Tg)

The Tg values are shown in Fig. 5. DFMPC cured UD-CFRP showed slightly higher Tg than its thermal counterpart suggesting higher degree of curing achieved in DFMPC. In case of BD–CFRP, the enhancement in Tg was even more significant. These results suggest that volumetric heating effect of microwaves results in higher degree of curing of the matrix resin.

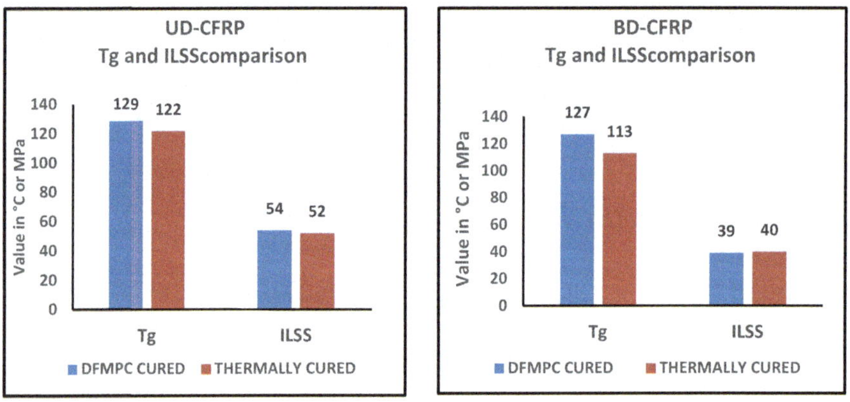

(a) (b)

Fig. 5 Comparison of Tg and ILSS of DFMPC and thermal cured CFRP **a** UD-CFRP and **b** BD-CFRP

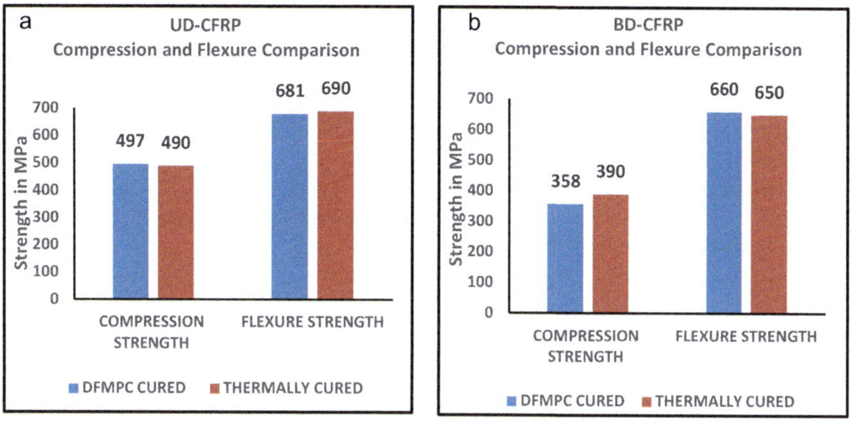

Fig. 6 Comparison of compression strength and flexural strength of DFMPC and thermal cured CFRP **a** UD-CFRP and **b** BD-CFRP

3.5 Mechanical Properties

Mechanical properties data related to compression and flexure are shown in Fig. 6, while ILSS results are shown in Fig. 5. The test results are averaged from at least 5 individual test specimens.

The mechanical properties graphs reveal that all the properties are comparable with a slight deterioration in compression strength [13]. This is attributed to the detrimental effect of microwaves on the conductive carbon fibers while being beneficial to the matrix interface as shown by the increase in ILSS of UD-CFRP. More in-depth studies are progress to ascertain this observation.

4 Conclusions

In the present work, UD and BD Carbon-Epoxy CFRP laminates were fabricated using RTVBM process and cured in a custom-developed DFMPC system. The cure schedule was optimized for rapid and energy-efficient curing. Customized microwave amenable molds were used for this purpose. Based on the properties obtained and comparison with thermally cured counterparts, the conclusions are as follows:

- Rapid and uniform curing of CFRP was achieved without hot-spots/arcing
- Fiber weight fractions were similar suggesting effective control of microwave process parameters
- Tg of DFMPC cured CFRP was higher representing higher degree of cure caused by volumetric heating effect of microwaves

- Mechanical properties were comparable or superior except compression strength which was slightly lesser which is attributed to detrimental effect of microwaves on the conductive carbon fibers.

Thus, DFMPC curing of CFRP is successfully demonstrated at laminate level.

Acknowledgements The authors thank Mr. Samiullah T H, Deputy Head, and Dr. Ramesh Kumar, Head, ACD. They are grateful to Mohan, Kavitha, and other colleagues and project assistants for the direct and indirect assistance during the progress of the present work.

References

1. Rao S (2007) Studies on radiation assisted curing of polymer matrices & composites. Ph.D. thesis, Bangalore University
2. Kwak M, Robinson P, Bismarck A, Wise R (2015) Microwave curing of carbon–epoxy composites: penetration depth and material characterization. Compos Appl Sci Manuf 75:18–27
3. Li N, Li Y, Hang X, Gao J (2014) Analysis and optimization of temperature distribution in carbon fibre reinforced composite materials during microwave curing process. J Mater Process Technol 214:544–550
4. Mooteri PS, Sridhara BK, Rao S, Rajendra Prakash M, Rao RMVGK (2006) Studies on mechanical behavior of microwave and thermally cured glass fiber reinforced polymer composites. J Reinf Plast Comp 25(5)
5. Rao RMVGK, Rao S, Sridhara BK (2006) Studies on tensile and interlaminar shear strength properties of thermally cured and microwave cured glass–epoxy composites. J Reinf Plast Comp 25(7)
6. Rao S, Rao RMVGK (2008) Cure studies on bifunctional epoxy matrices using a domestic microwave oven. Polym Test 27(5)
7. Rao S, Chiranjeevi MC (2015) Microwave curing of glass-epoxy composites using silica based moulds. J Polym Compos
8. Rao S, Chiranjeevi MC, Rajendra Prakash M (2016) Vacuum-assisted microwave processing of glass-epoxy composite laminates using novel microwave absorbing molds. Polym Compos
9. Rao S (2016) Microwave assisted processing of polymer matrix composites: a rapid & energy-efficient alternative. Indian J Adv Chem Sci S1
10. Li D, Li Y, Zhou J, Zhao Z (2020) A novel method to improve temperature uniformity in polymer composites microwave curing process through deep learning with historical data. Appl Compos Mater 27:1–17
11. Sato K, Hotta Y (2015) Carbon fiber/epoxy composite materials cured thermally and with microwave irradiation. Compos Interface 22(1):67–74
12. Nightingale C, Day RJ (2002) Flexural and interlaminar shear strength properties of carbon fibre/epoxy composites cured thermally and with microwave radiation. Compos Appl Sci Manuf 33(7):1021–1030
13. Hang X, Li Y, Hao X, Li N, Wen Y (2015) Effects of temperature profiles of microwave curing processes on mechanical properties of carbon fibre–reinforced composites. Proc Inst Mech Eng Part B: J Eng Manuf 1–9

Analytical Approach for Densification of Carbon–Carbon Composites Through Isothermal and Isobaric Chemical Vapor Infiltration (ICVI) Process

Sai Madhav and Jhon Paul

Abbreviations

Nomenclature

C	Concentration
L	Length of pore
T	Temperature
P	Pressure
ε	Porosity
$\mathcal{D}$	Diffusivity
k	First-order rate of reaction
θ	Thiele modulus
d	Molecular diameter
m	Mass of molecule
r	Radius of pore
M	Molecular weight
E_a	Activation energy
A_k	Pre-exponential factor
u	X-velocity
v	Y-velocity
μ	Dynamic viscosity
ρ	Density
S	Surface area
N	Number of molecules
k''	Surface reaction rate

S. Madhav (✉) · J. Paul
Azista Composites Pvt. Ltd, Hyderabad, Telangana 501401, India
e-mail: sai.mail.kar@gmail.com

M. R. Manohara Babu et al. (eds.), *Recent Advances in Composites Engineering and Research*, Springer Proceedings in Materials 39,
https://doi.org/10.1007/978-981-99-8807-5_12

Subscripts

R	Reactant
0	Initial or at $x = 0$
e	Effective
b	Binary
k	Knudsen

Constants

k_b	Boltzmann Constant
R	Universal gas constant
π	Pi

1 System Configuration of Industry Scale Reactor

The basic requirements of a full-scale CVD/CVI reactor are to deliver and control the flow of precursor and carrier gas into the reaction zone, provide an energy source to initiate and drive the chemical reactions, to remove the by-product gases from reaction zone. The process demands precise control of parameters like temperature, pressure, gas flow rate automatically for efficient deposition and maintaining reproducibility from one cycle to another.

CVD/CVI process model has been elaborately studied over the last few decades by many researchers. However, most of the research studies need further investigation to realize products considering the specific physical phenomena related to product and features of furnace. In the present study, to highlight the importance of this physical phenomenon, a benchmark furnace is considered to realize C–C composite products.

The current CVI system is a high-temperature hot wall reactor designed for process temperatures up to 1500 °C and process pressures ranging from 10–100 mbar. It is a top-loading reactor with a working zone of 690 mm diameter and 1100 mm length. The gas injector is placed centrally at the bottom which introduces methane (precursor) and nitrogen (carrier gas) into the reaction zone. Figure 1 shows the basic schematic diagram of a CVI reactor and important components.

The reactor is initially filled with nitrogen gas, and once the desired pressure is established, methane is introduced which will be transformed into the matrix of the C/C composite product. The reactant gas entered from injector passes through the multiple passages of preheater arrangement and distributed inside the reactor. Subsequent chemical reactions produce various intermediate products and hydrogen gas. These products and any residual methane gas will exit through the two vacuum ports at the bottom.

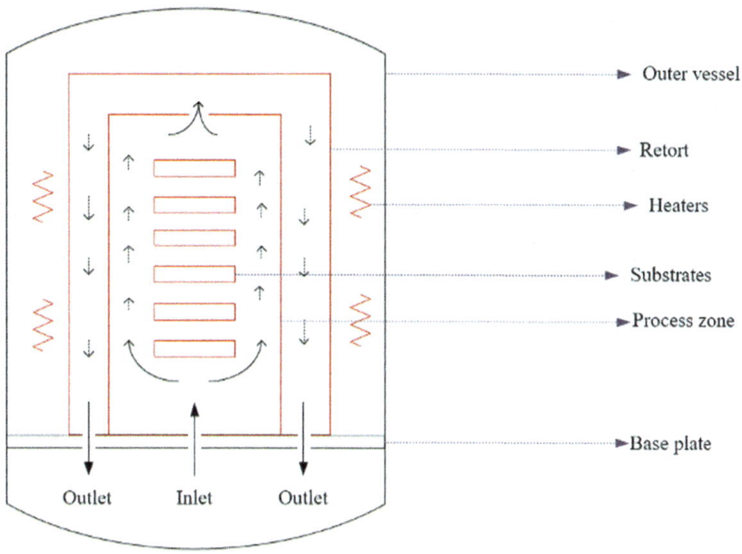

Fig. 1 Schematic of CVI reactor

2 CVI Process Model

2.1 Overview

The CVI technique is fundamentally similar to the CVD process, in which the deposition occurs due to the reaction of a carbonaceous vapor on the surface of a carbon-based substrate at high temperature and low pressure. CVI is based on the same principle but typically utilized over a large surface area to density a fibrous preform. The transport of mass, momentum, and heat plays a major role in determining the surface reaction rate and diffusion phenomenon inside each individual pore. The basic schematic of a CVI process consists of the transportation of carrier and reactant gases to the surface of the substrate through the boundary layer followed by diffusion of reactant gases inside the pore (governed by Knudsen diffusion). Subsequently, adsorption, surface reactions, formation of solid deposit, desorption, transportation of product species through boundary layer and toward the exit as shown in Fig. 2.

Numerous processing parameters, such as gas flow, diffusion behaviors, chemical reactions, and preform architecture, influence the densification process. In the present study, the effect of flow rate was considered by applying a multi-variable optimization model to predict optimum temperature and pressure of the CVI process for particular preform and process parameters.

Influence of preheater, injector, top lid, and substrate arrangement has been studied to investigate their effects on flow field. This provides better understanding of the gas flow pattern inside the reactor.

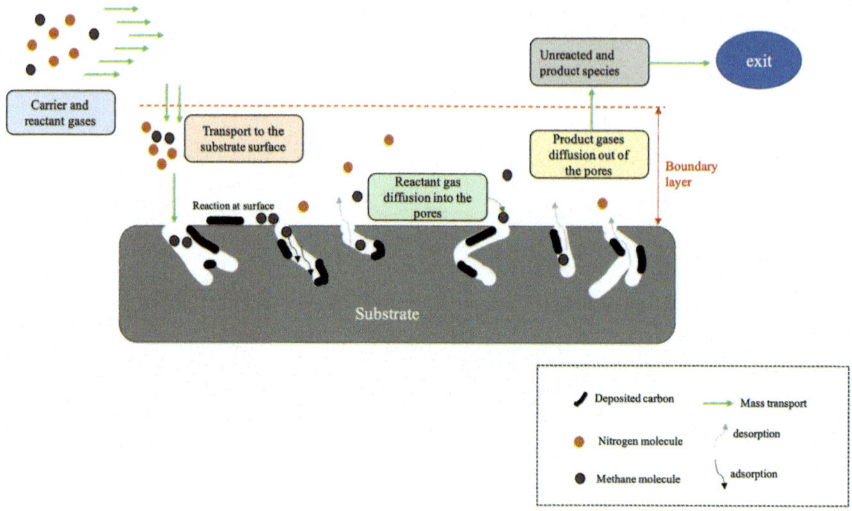

Fig. 2 CVI process schematic

2.2 *Single Pore Model*

The assumption of single pore model is that the pores are cylindrical with a single entrance for gases. In order to predict the availability of reactant gases throughout the pore, an elementary section of pore was considered to apply material balance and derive a relation between concentration of reactant and depth of pore in dimensionless form.

Consider a single cylindrical pore of length L as shown in Fig. 3. Reactant R diffuses into the pore and reacts on the surface by a first-order kinetic reaction.

$$\text{Reactant} \rightarrow^{\text{pyrolysis}} \text{product}$$

$$-r_R^{''} = -\frac{1}{S}\frac{\mathrm{d}N_\mathrm{R}}{\mathrm{d}t} = k''C_R \tag{1}$$

Let the concentration of reactant R at the entrance of the pore is C_0. The flux or movement of gases along the length of the pore decreases and reaches a minimum at the end of the pore. So the gradient of C_R reaches zero at pore end. Thus, from the above statements we can derive the boundary conditions for the single pore model to be:

$$C_\mathrm{R} = C_0 \text{ at } x = 0 \tag{2}$$

$$\& \frac{\mathrm{d}C_\mathrm{R}}{\mathrm{d}x} = 0 \text{ at } x = L \tag{3}$$

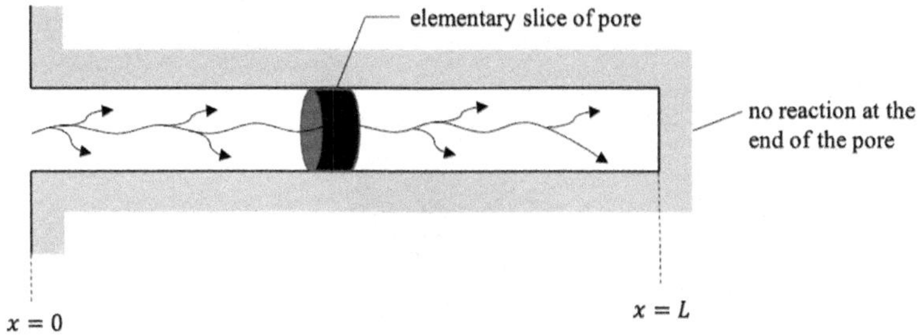

Fig. 3 Representation of a cylindrical pore

The flow of gases in and out of the section of pore is shown in Fig. 4. At steady state, applying mass balance for reactant R for this elementary section gives:

$$\text{Flow out} - \text{Flow in} + \text{Disappearance of reactant } (R) \text{ by reaction} = 0 \quad (4)$$

$$\text{Flow in} = -\pi r^2 \mathcal{D}\left(\frac{\mathrm{d}C_\mathrm{R}}{\mathrm{d}x}\right)_\mathrm{in}$$

$$\text{Flow out} = -\pi r^2 \mathcal{D}\left(\frac{\mathrm{d}C_\mathrm{R}}{\mathrm{d}x}\right)_\mathrm{out}$$

$$\text{Disappearance of } R \text{ on surface by reaction} = \left(\frac{\text{rate of disappearance}}{\text{unit surface}}\right)(S) = k'' C_R (2\pi r \Delta x)$$

Upon substituting the above three equations in Eq. (3):

$$-\pi r^2 \mathcal{D}\left(\frac{\mathrm{d}C_\mathrm{R}}{\mathrm{d}x}\right)_\mathrm{out} + \pi r^2 \mathcal{D}\left(\frac{\mathrm{d}C_\mathrm{R}}{\mathrm{d}x}\right)_\mathrm{in} + k'' C_\mathrm{R}(2\pi r \Delta x) = 0$$

$$\Rightarrow \frac{\left(\frac{\mathrm{d}C_\mathrm{R}}{\mathrm{d}x}\right)_\mathrm{out} - \left(\frac{\mathrm{d}C_\mathrm{R}}{\mathrm{d}x}\right)_\mathrm{in}}{\Delta x} - \frac{2k''}{\mathcal{D}r} C_\mathrm{R} = 0.$$

As $\Delta x \to 0$, the above equation becomes:

$$\frac{\mathrm{d}^2 C_\mathrm{R}}{\mathrm{d}x^2} - \frac{2k''}{\mathcal{D}r} C_\mathrm{R} = 0. \quad (5)$$

The interrelation between rate constants is $kV = k''S$.
So, for a cylindrical pore:

$$k = k''\left(\frac{\text{surface area}}{\text{volume}}\right) = k''\left(\frac{2\pi r l}{\pi r^2 L}\right) = \frac{2k''}{r}.$$

So, Eq. 5 becomes:

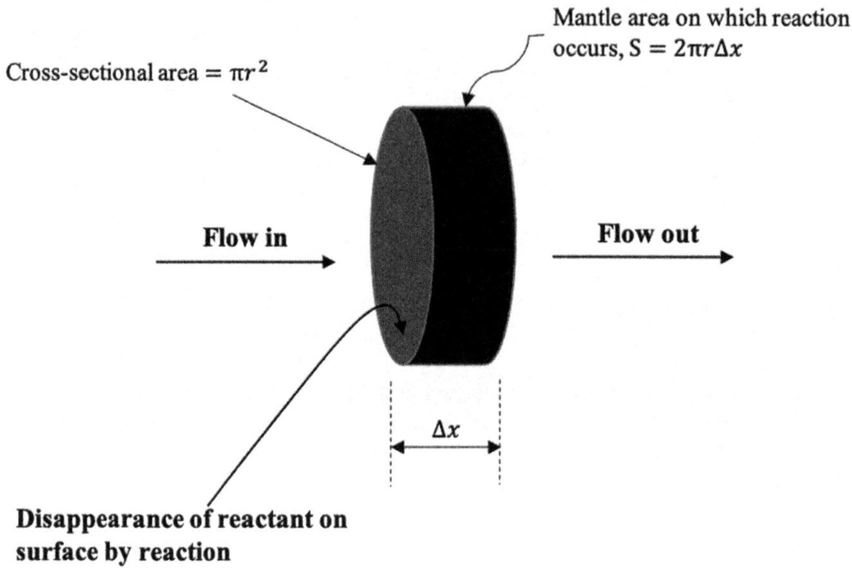

Fig. 4 Material balance for elementary slice of pore

$$\frac{d^2 C_A}{dx^2} - \frac{k}{\mathcal{D}} C_A = 0.$$

Solving the above linear differential equation and applying the boundary conditions, the solution obtained is:

$$\frac{C_R(x)}{C_0} = \frac{\cosh \theta \left(1 - \frac{x}{L}\right)}{\cosh \theta}, \quad (6)$$

where $\theta = L\sqrt{\frac{k}{\mathcal{D}}}$ = Thiele modulus.

The process and preform parameters are:

T	1423 K
P	20 mbar
L	3000 μm
ε	0.702
$\mathcal{D}_e$	34.889 cm^2/s
k	9.7335 s^{-1}

The rate of reaction is obtained from Arrhenius equation $\left(k = A_k e^{\frac{-E_a}{RT}}\right)$ for dissociation of methane and effective diffusivity is calculated from binary and Knudsen diffusivity.

Substituting the above values for θ:

$$\text{Thiele modulus } (\theta) = L\sqrt{\frac{k}{\mathcal{D}_e}} = 3000 \times 10^{-6}\sqrt{\frac{9.7335}{34.889 \times 10^{-4}}} = 0.1584$$

Then Eq. 6 becomes:

$$\frac{C_R(x)}{C_0} = \frac{\cosh 0.1584\left(1 - \frac{x}{L}\right)}{1.0126}.$$

The above curve is shown in red (Thiele = 0.1584) in the below graph (Fig. 5).

From the above plot, it can be inferred that the availability of reactant is even almost throughout the pore length indicating a high theoretical effectiveness factor required for uniform infiltration.

The main challenge on reactor level is maintaining a laminar regime inside the process zone in order to maintain uniform transport of reactant species throughout zone. This is addressed in the next section by reactor level flow modeling using a 2D Fluent analysis.

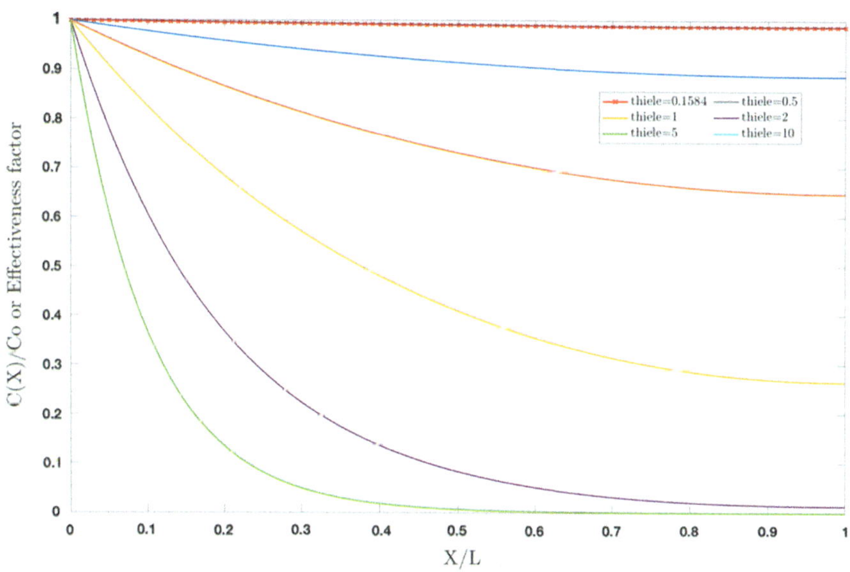

Fig. 5 Influence of Thiele modulus on pore infiltration

2.3 Multi-variable Optimization of Process Variables

From the predictions of elementary kinetic theory [1], the relation between binary diffusion, Knudsen diffusion, temperature, pressure, etc., is given by:

$$\text{Binary diffusivity} = D_{\text{b}} = \frac{2}{3d^2}\sqrt{\frac{k^3}{\pi^3 m}}\frac{T^{1.5}}{P}$$

$$\text{Knudsen diffusivity} = D_k = 97r\left(\frac{T}{M}\right)^{0.5}$$

For methane–nitrogen system substituting the appropriate values, the above equations become:

$$D_{\text{b}} = 5.698 \times 10^{-5} \times \frac{T^{1.5}}{P}\,\frac{\text{cm}^2}{\text{s}} \quad (7)$$

$$D_k = 3.637 \times \sqrt{T}\,\frac{\text{cm}^2}{\text{s}}, \quad (8)$$

where T is in Kelvin and P in kPa.

As the diffusion regime for porous substrates falls under the transition regime, the relation between effective, binary, and Knudsen diffusivities is given by:

$$\frac{1}{D_e} = \frac{1}{D_b} + \frac{1}{D_k}$$

$$\Rightarrow D_e = \frac{D_b D_k}{D_b + D_k}.$$

As $D_b = f_1(T, P) and D_k = f_2(T), therefore D_e = f(T, P)$.

The rate of reaction (k) is dependent on temperature according to Arrhenius equation $k = A_k e^{\frac{-E_a}{RT}}$

$$\Rightarrow k = g(T).$$

The above problem can be deduced into a multi-variable optimization problem in which the objective is minimizing Thiele modulus that results in uniform densification throughout the pore (Fig. 6).

The optimization is performed using optimization toolbox in MATLAB which uses **fmincon** solver to minimize the objective function. From this approach, the optimum temperature and pressure for the particular preform architecture and process parameters considered are 1001 °C and 19.9 mbar. It can be noted that these values provide an estimate at which lower Thiele modulus can be obtained for the assumed temperature and pressure ranges.

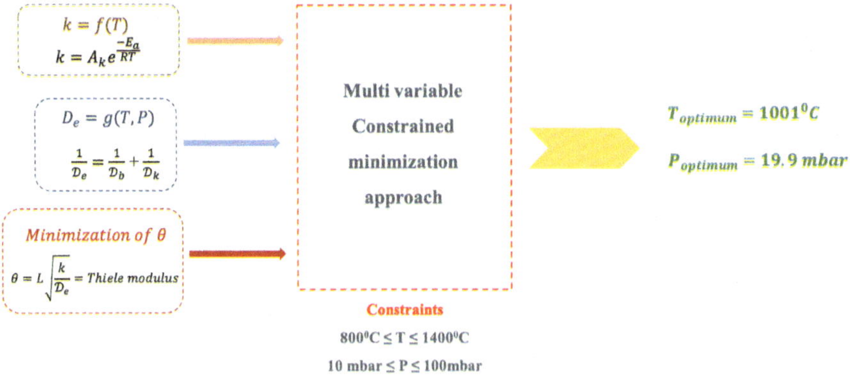

Fig. 6 Optimization problem definition

2.4 2D Steady-State Fluid Flow Analysis

2.4.1 Geometry and Grid

For the flow analysis, an axisymmetric model has been considered as the reactor is cylindrical and flow is symmetric in nature. 2D axisymmetric model reduces the computational time required for solving the governing equations in each discretized cell. The flow and symmetry axis are along x-axis. Multiple geometrical models with variations in design were made with the objective to study the flow field. For the geometry shown in Fig. 7, the substrates were arranged with a 100 mm spacing and an axial shift between alternate ones. Similarly, analysis was performed with 50 mm spacing, axial arrangement of substrates, single preheater, and top lid with multiple holes. All the design iterations were done in Design Modeler.

The discretization of computational domain was done in Ansys Workbench and is shown in Fig. 8. After a grid-independent study, the optimized mesh has a total of 57,759 elements and 182,456 nodes. In order to capture the near wall effects, precise meshing was done in the near-wall regions.

2.4.2 Solver and Results

For CFD analysis of CVD systems, it is essential to define appropriate boundary conditions for conservation equations mentioned below. In assigning the boundary conditions, it is a common practice to apply no-slip condition to the walls and define whether the flow field is laminar or turbulent by calculating the maximum Reynolds number in the reaction zone. The boundary conditions are tabulated in Table 1. It was considered that the gas has only axial component of velocity at the inlet boundary and outlet is considered to be long enough to establish a fully developed flow so

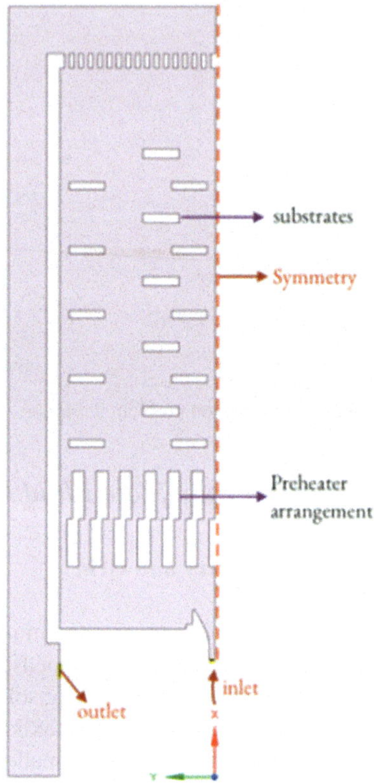

Fig. 7 Geometry and components

steady-state conditions are valid in this case. As the flow velocity considered is very low, pressure-based solver is chosen for analysis.

Based on the assumptions (2D, steady state, incompressible) the governing equations (continuity and 2D Navier–Stokes equations) solved by the numerical model are:

Continuity Equation:

$$\frac{\partial u}{\partial x} + \frac{\partial v}{\partial y} = 0$$

Conservation of Momentum:

$$x : u\frac{\partial u}{\partial x} + v\frac{\partial u}{\partial y} = -\frac{1}{\rho}\frac{\partial P}{\partial x} + \frac{\mu}{\rho}\left(\frac{\partial^2 u}{\partial x^2} + \frac{\partial^2 u}{\partial y^2}\right)$$

$$y : u\frac{\partial v}{\partial x} + v\frac{\partial v}{\partial y} = -\frac{1}{\rho}\frac{\partial P}{\partial y} + \frac{\mu}{\rho}\left(\frac{\partial^2 v}{\partial x^2} + \frac{\partial^2 v}{\partial y^2}\right)$$

Fig. 8 Structured mesh model

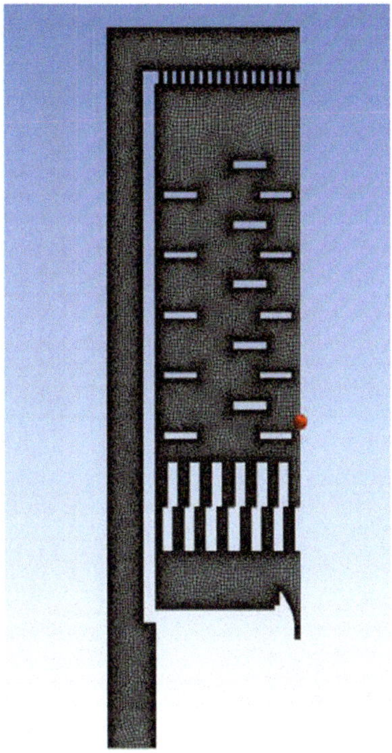

Table 1 Boundary conditions for solver

Mass flow rate at inlet	0.00029 kg/s
Pressure	20 mbar
Flow type	Laminar
Flow model	Single-phase flow with effective properties

Effect of Preheater and Substrate Arrangement

It can be observed from Figs. 9, 10, and 11 that as the flow approaches the preheater, it is diverted horizontally and then squeezed in the spacing between walls of preheater. This tends to distribute the flow from injector to reactor diameter uniformly. Comparatively, although with single preheater arrangement and axial arrangement of substrates, the flow is uniform through the length but space between substrates has very low presence of gas. In Fig. 10, flow pattern with two preheaters and criss-cross arrangement of substrates, the flow is distributed in the spacing and all the edges of substrates experience gas flow. When the gases flow from the sides of the substrates, their velocity increases as they pass through the narrow spacing. For the above-mentioned boundary conditions, a laminar flow model is used as Reynolds

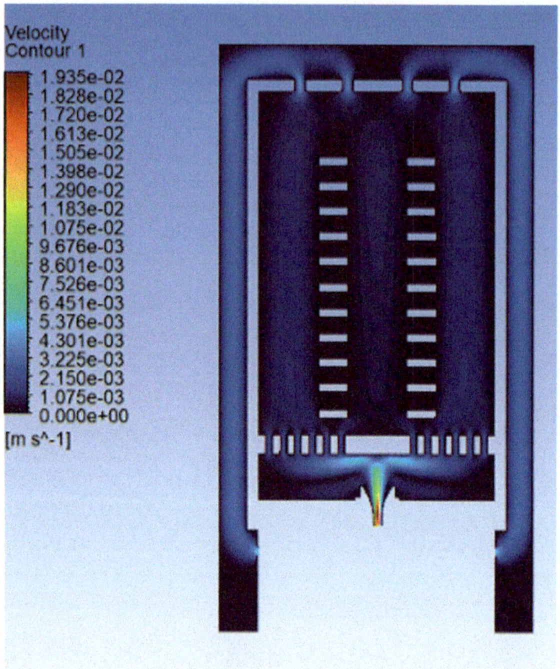

Fig. 9 Flow field with single preheater

number is in laminar regime. At higher flow rates turbulence and eddies are unavoidable due to varying flow area between components. So lower flow rates tend to drive the flow within laminar regime that has optimum residence times. Eddies lead to longer residence time of gases which can result in product species not being removed effectively.

Effect of Gas Injector Geometry and Top Lid

For the carrier and reactant gases to be injected into the reaction zone at uniform velocity, the shape of injector plays an important role. From Fig. 12, its observed that the injector diameter is small and not positioned to align with the hearth table surface which results in slightly higher velocities and uneven distribution of gases in the reaction zone. Multiple holes on top lid tend to mitigate the gas flow presence in the reaction zone. However, with the reduction of number of holes on lid, the residence time of gases is slightly higher within the reaction zone and evenly affects the flow downstream as per in Figs. 12 and 13. In comparison with complicated designs of injector, a simple cylindrical injector with sufficient diameter results in an ideal flow field as shown in Fig. 13. The stagnant flow area has also been reduced compared with the other two scenarios. These results clearly show the difference in velocity and its profile without the need for experiments.

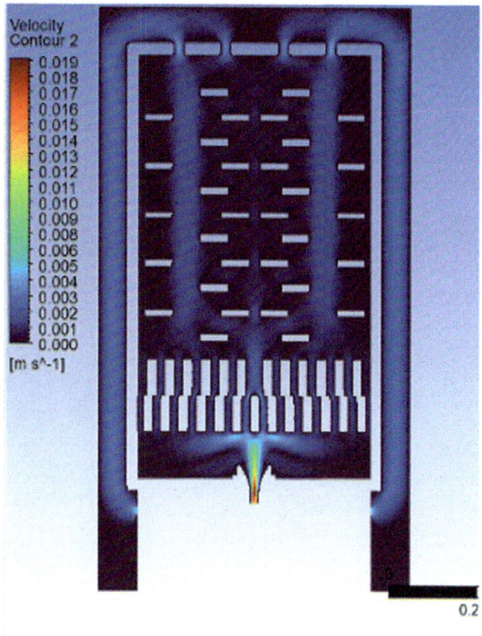

Fig. 10 Flow field with two preheaters and crisscross substrate arrangement

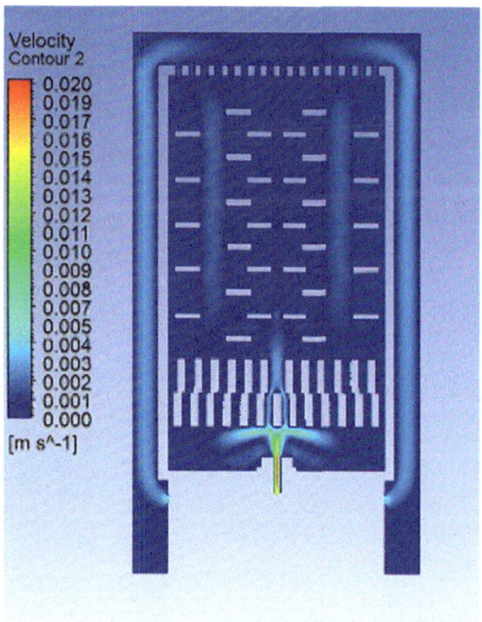

Fig. 11 Flow field with uniform diameter injector

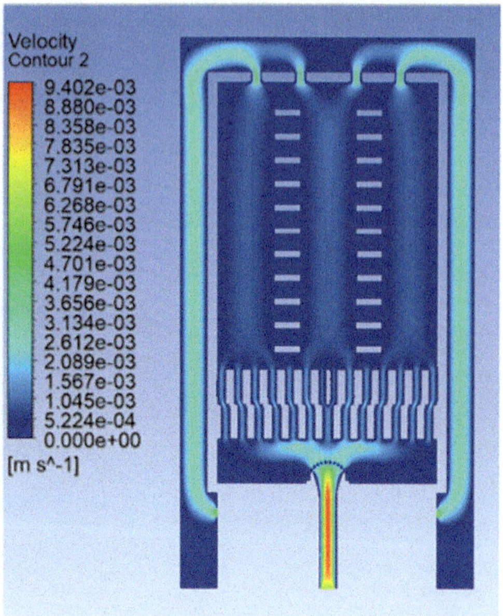

Fig. 12 Flow field with perforated injector

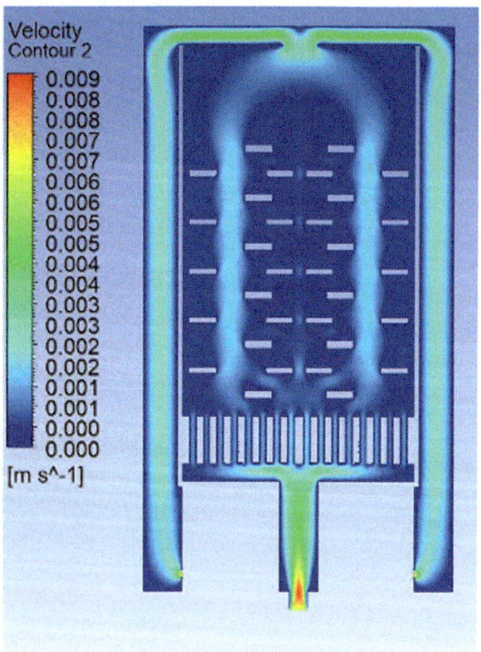

Fig. 13 Flow field with simple cylindrical injector

3 Conclusion

The component arrangement in the reactor plays a vital role in ensuring consistent gas flow to all the surfaces of the substrate. The stagnation zones especially between the substrates should be avoided as it results in poor infiltration due to the high residence time of product gases. CFD analysis can be used to study the flow patterns in order to optimize the reactor components and changes to component arrangement to avoid stagnation zones. This can help in customizing the design of tools, fixtures, and gas flow guides based on component shape and size. The actual effects of these design changes require experimental confirmation.

Reference

1. Xu Y, Yan X-T, Chemical vapor deposition, an integrated engineering design for advanced materials, 2010 edn

Lost Mandrel-Based In-Situ Skirts for Composite Rocket Motor Casing

Surjeet Kumar Sinha and Manoj Kumar Buragohain

1 Introduction

A composite rocket motor casing (CRMC) is a major structural part in a solid rocket motor that houses the propellant. It is basically a pressure vessel with associated specific load bearing attachments. It provides feature for assembly of igniter, nozzle and adjacent systems. CRMCs are primarily made up of composite material and are mostly in cylindrical shape. Any CRMCs generally consists of a composite shell, metallic polar fitting, metallic bulkhead and composite skirt [1, 2]. An elastomeric layer commonly known as shear ply is also used in Y-joint region between shell and skirt. A typical schematic composite rocket motor casing is shown in Fig. 1.

Composite shell is mainly pressure load bearing member, whereas skirt is meant for transferring structural load in the form of axial force and bending moment. Shell and skirt are primarily processed through filament winding technology, where resin-impregnated reinforcement is laid over a tool called mandrel. At times, impregnated reinforcement in fabric form is also used as composite ply. However, they are deposited by hand lay-up process.

Realization of skirt as an integral CRMC part is a quite involved task. Overhanging portion of skirt creates another problem in managing its manufacturing as an integrated part. A unique methodology has been evolved where skirt (shell with skirt) is manufactured through lost mandrel-based in-situ processing technique. This lost mandrel-based in-situ skirts for composite rocket motor casing is presented in this paper.

S. K. Sinha (✉) · M. K. Buragohain
Advanced Systems Laboratory, Hyderabad 500058, India
e-mail: surjeet_sinha@yahoo.com

M. R. Manohara Babu et al. (eds.), *Recent Advances in Composites Engineering and Research*, Springer Proceedings in Materials 39,
https://doi.org/10.1007/978-981-99-8807-5_13

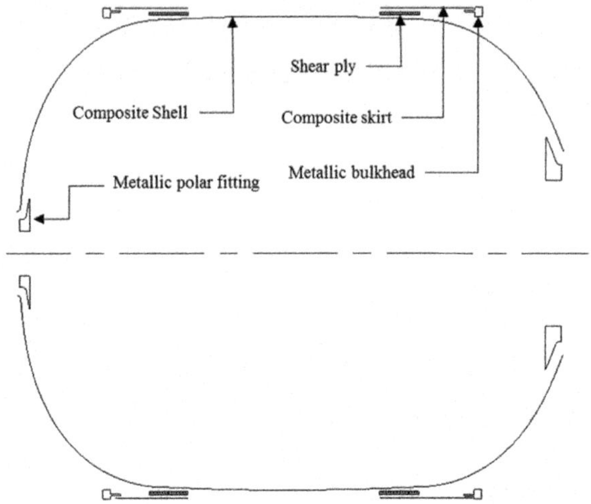

Fig. 1 Schematic composite rocket motor casing

2 Prevailing Practice

Skirt along with shear ply is used to be fabricated separately on a separate mandrel as pre-fabricated skirt [3, 4]. The processed skirt is cured and extracted from the mandrel. This pre-fabricated skirt is mounted over composite shell at its designated location/zone using fixtures. Shear ply portion of skirt is bonded to wound (processed) shell by using high-temperature curable epoxy adhesive [5]. Detail scheme is shown in Fig. 2.

Part of shell with skirt bonding zone is shown in Fig. 2. Outer boundary envelope with dimension, *D (outer diameter)*, is supposed to be bonded with shear ply zone of dimension, *d (inner diameter)*, of the pre-fabricated skirt as shown. Skirt bonding

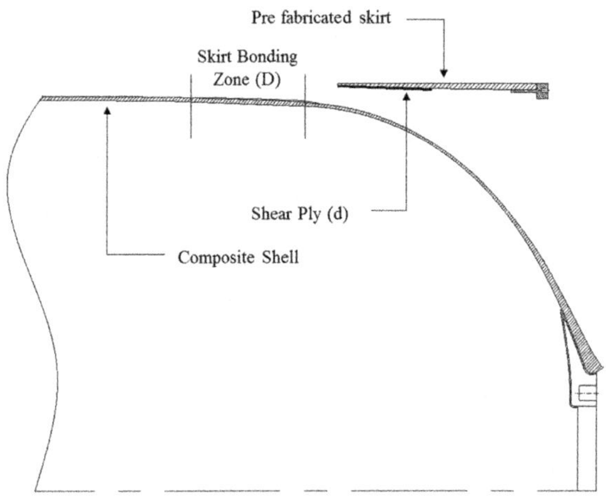

Fig. 2 Composite shell with pre-fabricated skirt scheme

zone dimension, *D*, and skirt portion dimension, *d*, are supposed to be in particular pre-finalized relation for proper processing of integrated skirt over shell in overall CRMC realization.

3 Issues

Composite shell is manufactured by filament winding process involving number of different plies. In this process, impregnated reinforcement is fed through dispensing unit and same gets deposited over rotating mandrel in pre-designated fashion. It is very difficult to achieve precise thickness of each ply especially in wet winding process, where fiber is impregnated during the process of winding itself. Accordingly, deposited ply does not have controlled thickness, and achieved thickness differs from intended thickness. This thickness variation is mainly due to variation in resin content. Since a number of plies are involved, variation in thickness keeps on getting accumulated. This accumulated thickness variation results into shift of outer boundary of the shell from intended outer boundary dimension (*D*) as referred in Fig. 2. Resultant achieved outer boundary, *D'*, as shown in Fig. 3 can either be shifted radially inward or outward compared to dimension *D*. Inner dimension of the shell is mostly accurate as this is achieved by machined mandrel over which it is wound/processed. Part of the shell zone over which pre-fabricated skirt is supposed to be bonded is shown in Fig. 3.

In a large-sized CRMC, where thickness of shell is higher and same is built through number of plies, resultant outer shell boundary/envelope shift can be in the order of 3–4 mm, which is quite significant. This shift in outer envelope/boundary results into either a clearance or an interference between shell and pre-fabricated bonding zone. Both clearance and interference cause improper matching/mating issues.

This issue has been resolved by evolving lost mandrel-based in-situ skirts for composite rocket motor casing.

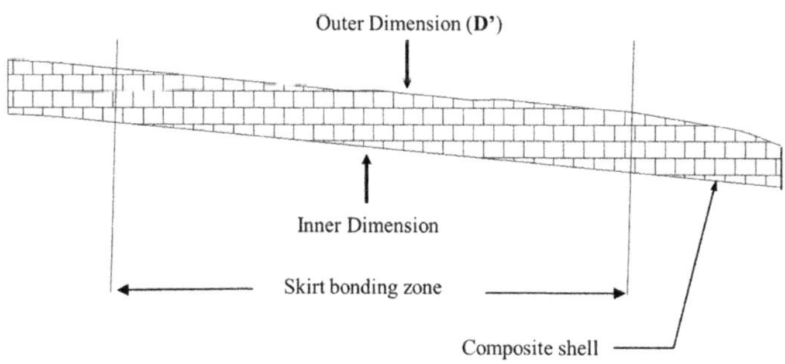

Fig. 3 Skirt bonding zone of composite shell

4 In-Situ Skirt Scheme

In this process, part of composite skirt is conceptualized separately, which is termed as 'skirt sleeve'. This skirt sleeve acts like a mandrel for major portion of composite skirt while processing (filament winding/lay-up). Subsequently, after processing and curing of composite, skirt sleeve becomes an integral part of skirt in overall integrated CRMC. In this process, skirt sleeve acts like a lost mandrel. Complete in-situ skirt scheme is shown in below Figs. 4, 5 and 6.

Skirt sleeve in overall skirt can be seen in Fig. 4. End portion of sleeve has been terminated at shell touch down point. Tip thickness of sleeve at this point has been kept low in the order of 1–2 mm to have the flexibility while mounting of sleeve during processing. Lower thickness of sleeve at this point has also been kept low to have better ply continuity for balance portion of skirt. As end portion of sleeve just requires only to touch the shell and not to go over shell for a sizeable length as in case of bonding of pre-fabricated skirt, variation in outer boundary (D) of shell does not create problem in form of clearance or interference, which requires to be managed over sizable bonding length. Minor clearance if there is any between skirt

Fig. 4 Skirt sleeve in skirt

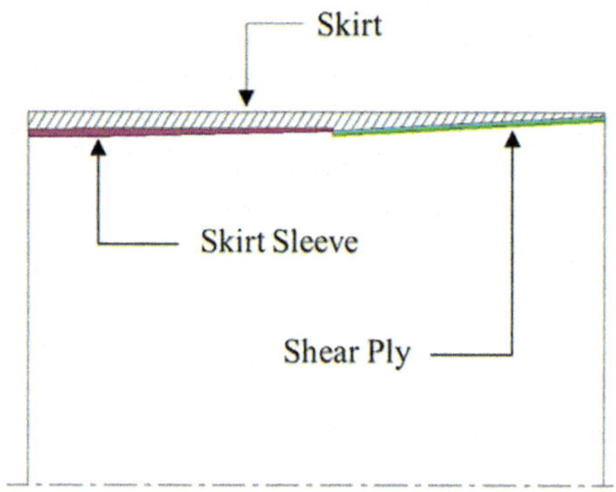

Fig. 5 Skirt sleeve mounting scheme over the shell

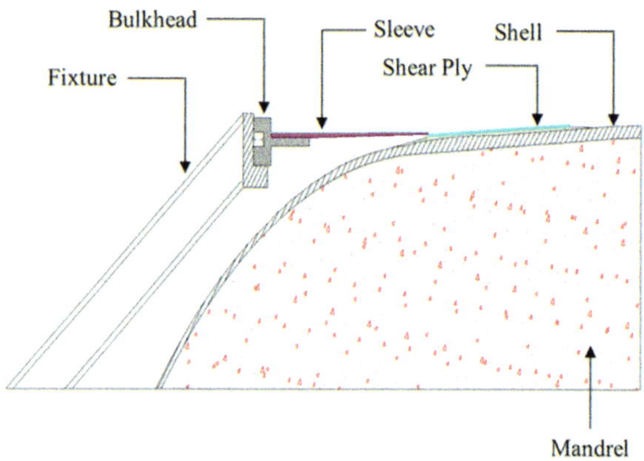

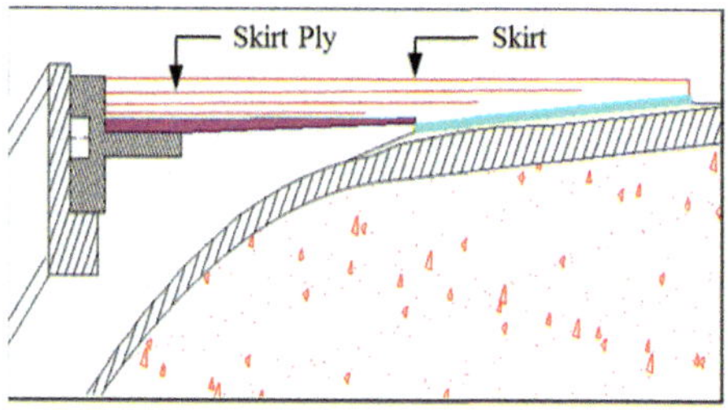

Fig. 6 Skirt ply building

sleeve end (inner diameter) and shell touch down point (outer diameter) is managed by local hoop winding or small amount of lay-up. Minor interference of this mating pair gets adjusted on its own as skirt sleeve end is flexible. Shell inherent geometry also helps in easy butting of sleeve as cylinder to dome transition starts near to sleeve touch down point over the shell.

Composite sleeve bonded separately with metallic bulkhead is brought to the identified position (skirt touch down point) of the shell with the help of mounting fixture as shown in Fig. 5. Shear ply is placed over the shell at identified location manually. Balance layers (plies) of composite skirt are deposited through filament winding as well as hand lay-up as shown in Fig. 6. This sleeve acts like a mandrel for balance plies of skirt. Subsequently, composite is cured in an oven. In this process, skirt sleeve becomes an integral part along with balance part of skirt in overall integrated CRMC, and accordingly, it loses its identity in the end.

5 In-Situ Skirt Realization

Carbon/epoxy composite has been used for realization of skirt and shell. Skirt sleeve inner diameter has been fabricated on a separate mandrel using filament winding and lay-up as per intended ply detail. Sleeve has been bonded to metallic bulkhead using epoxy adhesive. Sleeve fabrication and bonded sleeve–bulkhead (sleeve assembly) is shown in Figs. 7 and 8, respectively. This assembly with the help of fixture is brought over the shell, which is still over in processed tool (mandrel). End portion of sleeve just touches to the start of skirt bonding zone of the shell. Shear ply is bonded adjacent to skirt sleeve on shell using high-temperature curable epoxy adhesive. Skirt sleeve assembly mounting and shear ply bonding process are shown in Fig. 9. Subsequently, skirt is processed (by filament winding and lay-up if required) over shell and skirt sleeve, as per designed plies and their sequence. Skirt processing is shown in Fig. 10. Processed skirt along with shell is cured in an oven. Subsequently, skirt and bulkhead are mechanically fastened through rivets. Finally, sleeve mounting fixture is removed, and tool (mandrel used for shell processing) is extracted. Realized CRMC with integrated skirt is shown in Fig. 11.

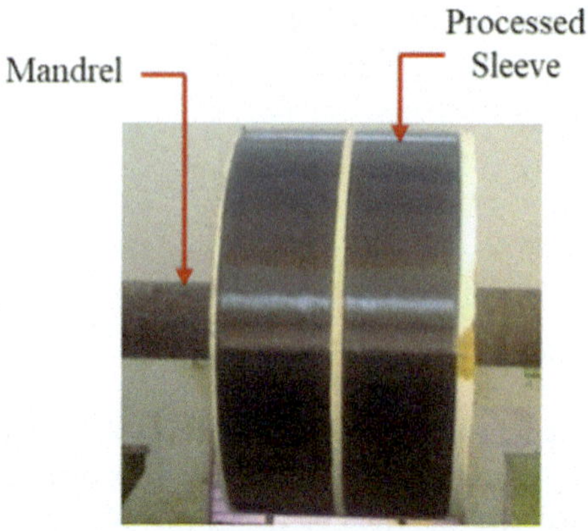

Fig. 7 Sleeve fabrication

Fig. 8 Skirt sleeve assembly

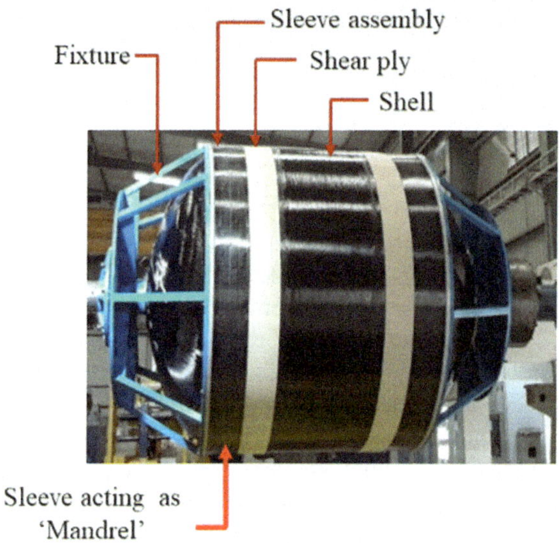

Fig. 9 Skirt sleeve assembly mounting and shear ply bonding

Fig. 10 Skirt processing

Fig. 11 CRMC with integrated skirt

Non-destructive examination (NDE) has been carried out to assess the health of skirt, interface and shell. NDE result confirmed the integrity, and no abnormal observations were reported.

6 Qualification

Skirt is configured for mainly two types of loads, namely axial force and bending moment. Skirt experiences these forces during various course of actions of the aerospace vehicles. These forces sometimes come in combination. Most of the time, axial forces are compressive in nature. However, sometimes this can be tensile in nature as well.

Integrated skirt along with shell (CRMC) has been tested for all these possible loads, which it is likely to be experiencing during its use in a vehicle. Loads have been applied through two end metallic dummy sections (one at either end) by hydraulic actuators [6]. Axial compressive force, bending moment, compressive force along with bending moment and tensile force have been applied separately. Composite rocket motor casing with assembled dummy sections under test in rig is shown in Fig. 12. Strain and displacement (axial as well as radial) at critical locations were measured during the test by strain gauge and linear variable differential transformer (LVDT), respectively.

Integrated skirt along with shell has withstood all the loads successfully without any sign of damage. Measured strain and displacement values were within acceptable limit. Subsequently, non-destructive examination (NDE) in the form of ultrasonic test and X-ray radiography test was also carried out to assess the health of the skirt as well as interface. NDE also confirmed soundness of the article. Skirt along with interface and shell were found to be intact after the test.

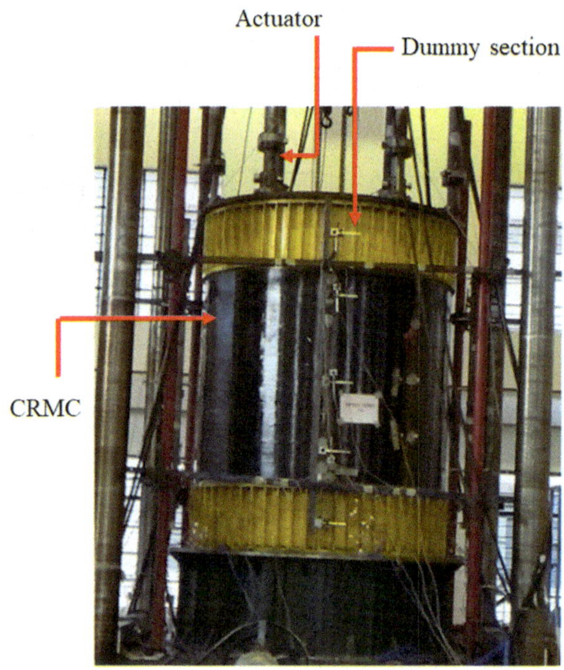

Fig. 12 CRMC under load test

7 Conclusion

A number of CRMCs have been manufactured using lost mandrel-based in-situ skirt manufacturing technique. They have been qualified successfully for all likely structural loads. Casings manufactured through this process have also been successfully proven in ground as well as flight test of the vehicles. Later, this technique has also been adopted in manufacturing of various other types of large-sized CRMCs. Lost mandrel-based in-situ skirt for CRMC has been found to be quite convenient in realization of overall integrated CRMC. Lead time for realization of CRMCs has also been found to be reasonably lower compared to pre-fabricated skirt manufacturing route.

Lost mandrel-based in-situ skirt fabrication methodology for CRMC manufacturing is fully demonstrated and adopted in realization of various large-sized CRMCs.

Acknowledgements The authors acknowledge the support provided by the test facility and NDT division of Advanced Systems Laboratory. Authors immensely acknowledge the constant support and encouragement provided by Director, Advanced Systems Laboratory, Hyderabad.

References

1. Liang C-C, Chen H-W, Wang CH, Optimum design of dome contour for filament wound composite pressure vessels based on a shape factor
2. Betti F, Perugini P, Mataloni A, Tessitore N, Leone A, Di Cosmo A, Design and development of VEGA solid rocket motors composite cases
3. Liang C-C, Chen H-W (2003) Optimum design of fiber reinforced cylindrical skirts for solid rocket motor cases subjected to buckling and overstressing constraints. Compos Part B 34:273–284
4. Addanki R, Gangadhar Sinha Y, Ramaswamy K, Sankara Rao B (2006) Pressure and structural testing of filament wound motor cases. In: ISAMPE National conference on composites (INCCOM-5), pp 472–479
5. Rajput R, Sinha YG, Ramaswamy K, Buragohain MK, Chatterji SK, Sankara Rao B (2004) Development of large diameter composite rocket motor casings. In: ISAMPE National conference on composites (INCOM-3)
6. Addanki R, Buragohain M, Aneesh KV, Sinha SK, Chakravarty VK, Thakar PJ, Jayaraman K, Manufacture of composite skirts for conical rocket motor casing

Contributions of Non-Destructive Testing in the Development of Dimensionally Stable Space Composite Camera Structures

S. Hari Krishna, M. B. Sobhanan, Mathew Sebastian, K. M. Usha, C. R. Bijudas, H. Priyadarshan, and R. Kalyanaraman

1 Introduction

In the domain of aerospace, structures made of CFRP materials [1, 2] occupy a distinct place because of their advantages such as specific strength, specific stiffness, low thermal expansion, tailorability, ability to provide reduced part count, etc. The realization process of a composite component is such that the component and its constituent material are created at the same time. When it comes to the space camera structure [3–5], the spatial alignment of the optical elements has a great role in its performance. Appropriate choice of fiber orientations is important to achieve very low or even negative in-plane thermal expansion, whereas choice of matrix is important to provide minimum moisture expansion. Initial assembly accuracies can be perturbed during the vibration, acoustic loads during launch, microgravity in space along with the effect of thermal, vacuum and atomic oxygen on the composite material of the camera structure. These effects need to be properly accounted for, in the design and if required, appropriate active correction mechanism needs to be implemented [6].

The onboard cameras of the Earth Observation Satellites of ISRO that operate from the low earth orbit are designed with less than 1 m spatial resolution [7]. Such a spatial resolution is achievable by highly dimensionally stable structures. Figure 1 shows a typical schematic diagram of a composite camera structure of 1 m height. The cylindrical shell has a removable flat base plate of 0.1 m height on one end and a secondary mirror on the other end. The primary mirror and detector are centrally positioned on either side of the base plate. The secondary mirror (SM) is mounted on

S. Hari Krishna (✉) · M. B. Sobhanan · M. Sebastian · K. M. Usha · R. Kalyanaraman
VSSC, Trivandrum, Kerala, India
e-mail: shkris@rediffmail.com

C. R. Bijudas · H. Priyadarshan
IIST, Trivandrum, Kerala, India

M. R. Manohara Babu et al. (eds.), *Recent Advances in Composites Engineering and Research*, Springer Proceedings in Materials 39,
https://doi.org/10.1007/978-981-99-8807-5_14

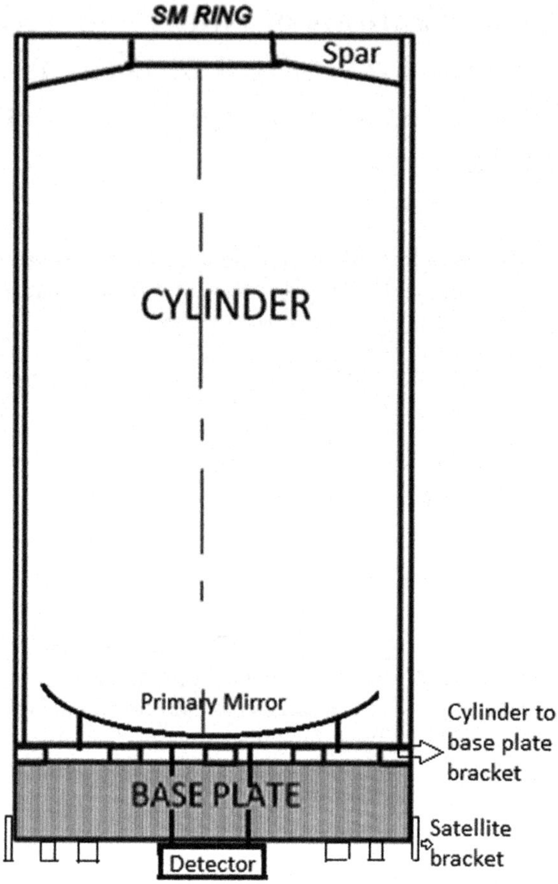

Fig. 1 Schematic of camera structure

a low thermal expansion alloy, which in turn is connected to the cylinder using 3 Nos of components called spars. Figure 2 shows the schematic diagram of the SM ring and spars. The primary and secondary mirrors are separated by a distance of 1 m and need to be maintained within a very stringent tolerance of 5 μm for a temperature variation of 10 °C. For comparison purposes, for the temperature variation of 10 °C, the slip gauges made of steel used in metrology applications will have a length variation of 105–125 μm.

One of the material options for the cylinder structure could be carbon–carbon [8–10]. Considering inheritance and process cycle advantages, CFRP material has been chosen for the construction. Figure 3 shows the smaller-sized camera structures flown in earlier missions of ISRO. In view of future large camera structures, a new composition of CFRP has been chosen as the material of construction, for the 1 m class camera structure.

The design, development and qualification of the new composite camera structure required the support of NDT at each stage: evolution of processing parameters using

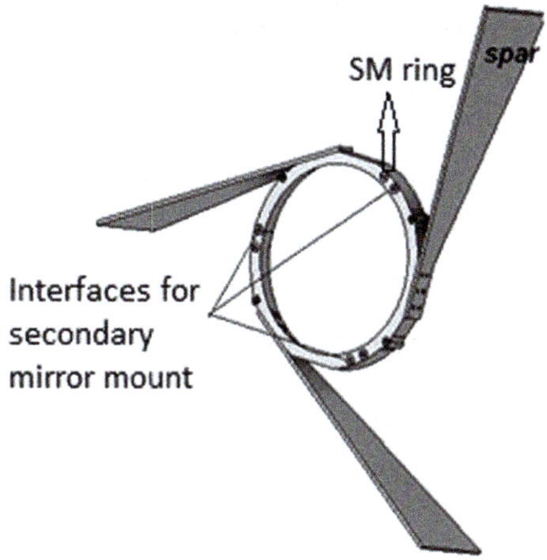

Fig. 2 Schematic of the secondary mirror (SM) ring and spars

Fig. 3 Camera structures flown in earlier missions of ISRO

the new composition joint configuration studies at preliminary design level; component, sub-assembly, assembly stages and at different environmental (such as thermo vacuum, vibration, acoustic) qualification stages. NDT requirements at different stages of product realization included mapping of discontinuities and differences in material characteristics without destroying the serviceability of the part or system; to understand the reasons for variations and deviations from the expected behavior.

Primary NDT methods for aerospace composite structures [11–15] include: visual, tap, ultrasonic, radiography, infrared thermography and laser shearography. Visual and tap tests provide the first-level product assessment. Alcohol wipe test, where the industrial alcohol applied on the surface acts as a liquid penetrant for detection of surface opening defects, can also be considered as a part visual test. Further information about the material/product can be obtained by the other test methods.

Different ultrasonic techniques use the sound wave propagation characteristics for the material/defect detection. However, the physical coupling, i.e., proper contact of the ultrasonic transducer to the test component, is a major requirement for ultrasonic testing techniques, in general. Advanced ultrasonic techniques using air-coupled and laser excitation require appropriate automation support for the coverage of the test component. Radiography techniques can provide visualization of the internal details of a product, however are limited by the safety and simultaneous accessibility requirement of both sides of test component.

Infrared thermography can be used for defect detection by using different thermal excitation methods based on time dependent surface temperature variations. With appropriate loading/stressing methods, the optical interferometry-based laser shearography can be used for the detection of defects that exhibit differential out of the plane movement with respect to the surrounding areas. While infrared thermography and laser shearography are non-contact, area coverage techniques, their detectability comes down as the defect depth from the surface increases. Also, they require line of sight while testing, which may not be possible at assembly stages.

The choice of NDT method(s) is dependent on the applicability, availability and requirement. A particular NDT method can provide only limited information about a flaw or a defect because of its ability to measure a particular aspect of physical properties. Based on requirement, a combination of NDT methods may be necessary, for better defect characterization as well as cross reference purposes. This paper provides the adaptation of different NDT methods, techniques, which are chosen based on the applicability and traceability for cross-verification purposes from component to assembly.

2 Contributions of NDT in Product Development of Composite Camera Structure

2.1 NDT Inputs During Initial Design

Achieving the stringent dimensional accuracies required for the camera structure demanded high levels of uniformity in component and assembly levels. During initial design, NDT assessment through a combination of visual, tap tests and automated ultrasonic air-coupled C-scan imaging has been used to provide the inputs about the processing quality of components. Initial processing trials revealed poor laminate consolidation and a high degree of non-uniformity. Figure 4 shows the ultrasonic through transmission C-scan images (the mapping of ultrasonic signal transmission between the ultrasonic transducers that are placed on either side of the laminate) showing the variability in trial laminates.

Based on the NDT feedback, processing was improved to obtain better uniformity levels as can be seen in the C-scan image shown in Fig. 5. NDT assessment was also provided on the laminate coupons after different environmental life cycle tests and the specimens for mechanical property evaluation. Some of the laminates with minor

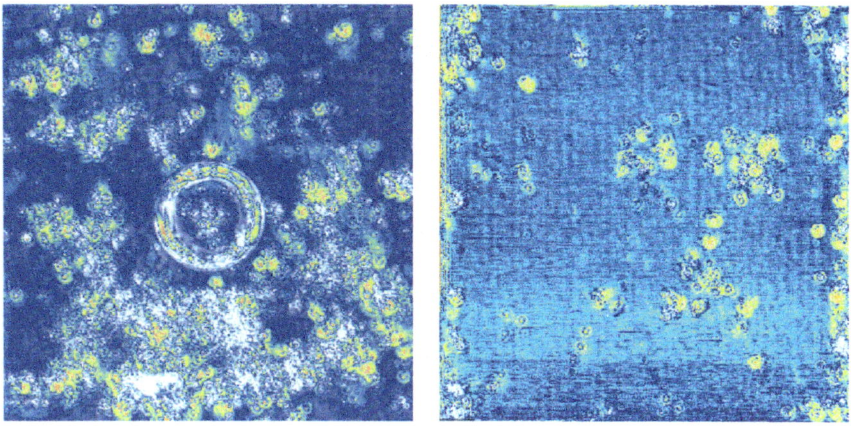

Fig. 4 C-scan images of defective laminates during initial process trials

Fig. 5 C-scan image of a good laminate

porosity were subjected to mechanical tests. Based on the satisfactory failure loads and modes, acceptance criterion was developed.

For the evaluation of joint configurations, manual ultrasonic pulse-echo using phased array technology with industrial alcohol as couplant and ultrasonic through transmission with silicone rubber as dry couplant were used. NDT was carried out before and after different load levels to understand the structural performance or material behavior. This exercise has provided the inputs for process and design modifications/improvements.

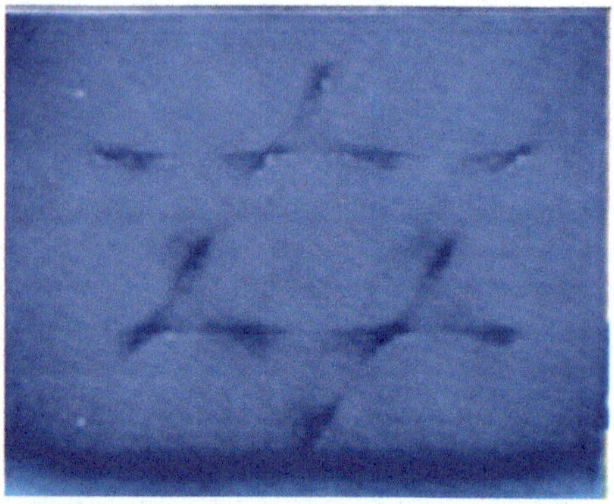

Fig. 6 C-scan image of a base plate laminate

2.2 NDT Inputs After Final Design

In one of the skin laminates of the base plate, a star pattern was noticed in the C-scan image as shown in Fig. 6. Measurement showed that the separation distance between the parallel lines is 0.3 m. Further exploration showed that this 0.3 m corresponds to the width of the prepreg roll. Physical verification of the remaining prepreg roll showed that the edges of the roll had lost the tackiness, which could be due to the edge damage or deviation in the prepreg roll. In subsequent usage, prepreg roll was used after trimming off the edges, resulting in a good-quality laminate.

2.3 NDT Support During Environmental Qualification Tests

After assembly realization, NDT was also carried out after thermo vacuum cycling. Subsequently, the structure was assembled with the dummy masses of optical elements for the vibration qualification test to assess the integrity of the structure at different load conditions. After the full level vibration test along the axial direction, the vibration signature showed stiffness reduction of the structure compared with that of the initial vibration signature. The reason for the deviation could be revealed by NDT by the identification of disbonds between the spar and cylinder joint regions.

Subsequently, the sizes and locations of disbonds were updated in the finite element (FE) model. The analysis confirmed the quantified stiffness drop due to the disbonds. Based on the match between the NDT observations and the updated FE analysis results, the structure was tested again and no further stiffness drop was observed. Subsequently, other vibration qualification tests were also completed successfully.

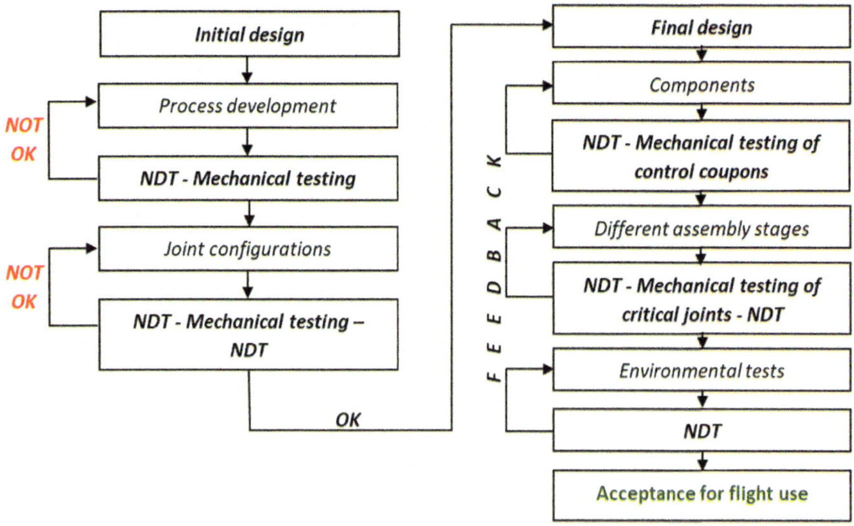

Fig. 7 Summary of the contributions of NDT in the development of camera structure

3 Summary and Conclusions

Figure 7 shows the summary of contribution of NDT at different stages in the development cycle of the 1 m class camera structure from the process feasibility studies, screening of specimens for mechanical property evaluation, feature-based joint qualification studies, individual component to the assembly-level quality assessment, after different stages of environmental qualification tests, till the acceptance for flight use.

NDT at each stage provided the necessary feedback for improvement and product quality assurance. The composite spacecraft camera structure has been successfully qualified, flown and is performing satisfactorily in orbit. All these learnings have been used in the realization of subsequent camera structures.

Acknowledgements The authors thank Shri Sanjeev Kumar and colleagues of CMSE for their valuable support in carrying out these studies.

References

1. Campbell FC (2010) Structural composite materials. ASM International
2. Composites in the race to space. http://www.compositesworld.com. Last accessed 30 Oct 2022
3. Composites stabilize space-based telescope. http://www.compositesworld.com. Last accessed 30 Oct 2022
4. Kim Y-S, Lee E-S, Woo S-H (2003) System trade-off study and opto-thermo analysis of a sunshield on the MSC of the KOMPSAT-2. J Astron Space Sci 20(4):393–402

5. Planche G, Massol C, Maggiori L (2004) HRS camera: a development and in-orbit success. In: ICSO 2004 International conference on space optics. Toulouse, France
6. Ling J, Chen B, Liu C, Hsueh E (2008) Innovative design concepts for the low cost remote sensing satellites. In: 22nd Annual AIAA/USU conference on small satellites SSC08-IV-11
7. Hari Krishna S, Sebastian M, Premdas M, Usha KM, Bijudas CR, Priyadarshan H (2021) Role of non-destructive testing in the life cycle of space composite structures—current and future perspectives. In: International conference on product life cycle modelling, simulation and synthesis (PLMSS). Trivandrum, India
8. Alain Gleyzes M, Perret L, Kubik P (2012) PLEIADES system architecture and main performances. In: International archives of the photogrammetry, remote sensing and spatial information sciences, vol XXXIX-B1. 2012 XXII ISPRS Congress, Melbourne, Australia
9. Meftaha M, Leeb S, Irbaha A, Ostergren S (2011) Carbon/Carbon for satellite applications. In: Proceedings of SPIE. The International Society for Optical Engineering
10. Bailly B, Cornu J-L, Capdepuy B, Leleu F, Kornmann M, Pradier A (1997) High stability carbon/carbon telescope structure. In: International conference on space optics—ICSO 1997. Toulouse, France
11. Kapadia A (2007) Non-destructive testing of composite materials. National composites network (NCN) best practice guides
12. Wang B, Zhong S, Lee T-L, Fancey KS, Mi J (2020) Non-destructive testing and evaluation of composite materials/structures: a state-of-the-art review. Adv Mech Eng 12(4):1–28
13. Bossi RH, Georgeson GE (2020) Nondestructive testing of aerospace composites. Woodhead Publishing
14. Meola C, Boccardi S, Carlomagno GM (2017) Infrared thermography in the evaluation of aerospace composite materials. Woodhead Publishing
15. Newman JW (2018) Shearography nondestructive testing of composites. In: Comprehensive composite materials II, vol 7. Elsevier, pp 270–290

Finite Element Simulation and Full Scale Component Testing to Evaluate the Static Strength of a Composite Wing

M. Mohan Kumar, Kumari Asha, N. Lohith, Byji Varughese, C. Surendra, and C. M. Manjunatha

1 Introduction

Validation of aircraft structures is critical to the design process so that it provides assurance that the structure will perform as intended in all respects. The validation and certification approach is primarily through numerical studies, supported by test evidence component levels limit or ultimate load test. Static strength tests are required to demonstrate the validity of both internal loads and design allowable. In addition, because of the more and more extensive application of composite materials in the field of aviation the design and manufacture of light-weight and high-strength composite structures has become one of the challenges faced by the development of aircraft [1]. Zhang et al. [2] carried out lightweight design for the structure of small ultra-light solar UAV according to the mechanical characteristics of the structure, and the rationality of the structure was verified by the finite element analysis and static testing method. Chunzhi et al. [3] designed the composite structure of the wing of a low-speed aircraft in the way of layering and geometric modeling and analyzed the strength and stiffness of the wing in different layering by ANSYS finite element software. The results showed that the composite wing had lighter weight, higher strength and greater stiffness than the traditional wing under the same working condition.

In the present study, composite wing of a two-seater trainer aircraft has been designed and developed using Just-in-time Prepreg (JIPREG) fabrication process. As part of weight optimization exercise, thicknesses of wing sub-parts are optimized based on strength and stiffness criteria. The optimized design is evaluated through

M. Mohan Kumar (✉) · C. Surendra · C. M. Manjunatha
Structural Integrity Division, CSIR-National Aerospace Laboratories, Bangalore 560017, India
e-mail: mmk@nal.res.in

K. Asha · N. Lohith · B. Varughese
Advanced Composites Division, CSIR-National Aerospace Laboratories, Bangalore 560017, India

M. R. Manohara Babu et al. (eds.), *Recent Advances in Composites Engineering and Research*, Springer Proceedings in Materials 39,
https://doi.org/10.1007/978-981-99-8807-5_15

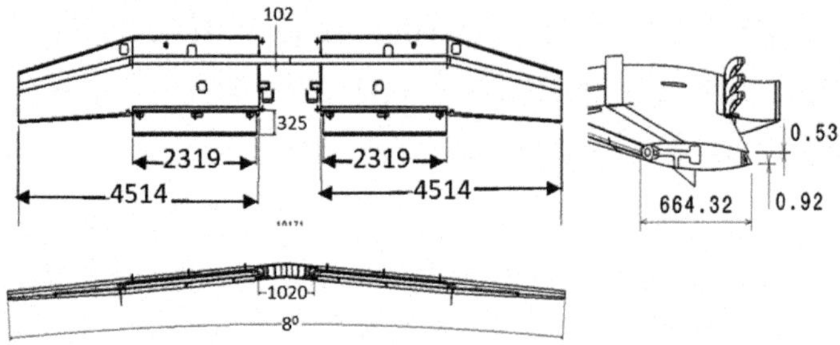

Fig. 1 Outline dimension diagram of composite trainer aircraft wing (mm)

finite element simulation, for various critical flight and landing load cases to estimate the strains and deflections, followed by Static Strength test, carried out on the full-scale wing to meet the requirements of JAR VLA.

2 Structural Design of the Composite Wing

2.1 Wing Structure

The composite wing consists of skins, a main load carrying spar, two shear attachment spars and transverse ribs. Skins, spar webs and ribs are made of sandwich construction with foam as core and glass bi-directional fiber composite as face sheet material. Main spar flange is made of monolithic carbon composite. The geometrical details of the composite wing are shown in Fig. 1.

3 Finite Element Analysis of Composite Wing

3.1 Finite Element Analysis Model

In the finite element analysis, Hyper Mesh software was used for mesh generation, boundary condition setting, other pre-processing and post-processing. MSC Nastran software was used as solver. The finite element model used the triangle/quadrilateral shell elements. Idealization of the structure for creating the FE model simulating the reality of structural elements was critical. Spar flange was modeled using PCOMP card to represent the monolithic UD CFRP material. Sandwich construction components, viz. skins, rib webs and spar webs are modeled using PCOMP card with foam as separate layer in the laminate (Fig. 2).

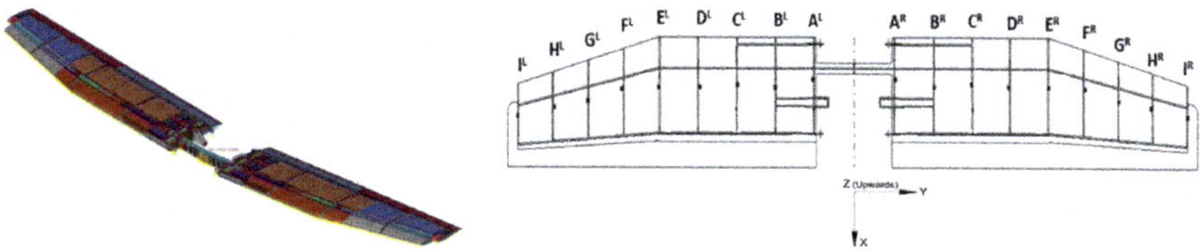

Fig. 2 FE model of composite aircraft wing with control surfaces

3.2 *Material Details of Wing*

Wing comprises top skin, bottom skin, 9 ribs each at LH & RH and three spars, viz. leading edge (LE) spar, front spar and rear spar. The skin, spar and ribs are made of sandwich construction with PVC foam core and GFRP bi-directional face sheets. Spar flange comprises monolithic UD CFRP. At all attachment locations, foam core is replaced with plywood core and aluminum reinforcement. PVC foam used has E = 34.7 MPa, v = 0.13, tensile strength = 1 MPa and compressive strength = 0.6 MPa.

3.3 *Loads and Boundary Condition Details*

The wing was tested for the critical case of maximum up-bending load case which is considered for Design Limit Load (DLL) as well as Design Ultimate Load (DUL) test. Only X-direction and Z-direction loads (Global X- and Z-direction) are considered for testing which derives most of the design parameters. Y-direction loads will not be applied on structure during the test as their effect is negligible. Test loads are computed with inertia relief. Statically equivalent discrete point loads were derived from design loads shown in Table 1. The details of boundary conditions are depicted in Fig. 3.

3.4 *Optimization of Wing Structure*

FE-based weight optimization of wing structure was carried out on the spar flange which is of monolithic CFRP construction. The optimization problem was formulated as below:

Design Variables: Ply thickness of 0^0 carbon composite ply of spar flange

Design Constraints: Failure-index (Tsai-Wu failure theory) ≤ 1

Eigenvalue for buckling ≥ 1

Design Objective: Minimize the mass

Weight reduction of 4.6 kg was obtained in the spar flange optimization of wing structure.

Table 1 Design ultimate load details of maximum up-bending load case

Rib stations		Global X (mm)	Global Y (mm)		Global Z (mm)	X- distance from leading edge (mm)	Load (N)	
			Left	Right			F_X	F_Z
A^L	A^R	1117	−615	615	−362	417	0	2227
B^L	B^R	1194	−1189	1189	−322	494	0	2668
C^L	C^R	1169	−1752	1752	−282	467	−1197	3659
D^L	D^R	1194	−2318	2318	−243	494	0	3021
E^L	E^R	1194	−2934	2934	−197	494	0	2982
F^L	F^R	1215	−3497	3497	−176	393	0	3110
G^L	G^R	1332	−4009	4009	−155	398	0	2825
H^L	H^R	1393	−4521	4521	−135	348	0	2276
I^L	I^R	1489	−5083	5083	−113	322	0	1010

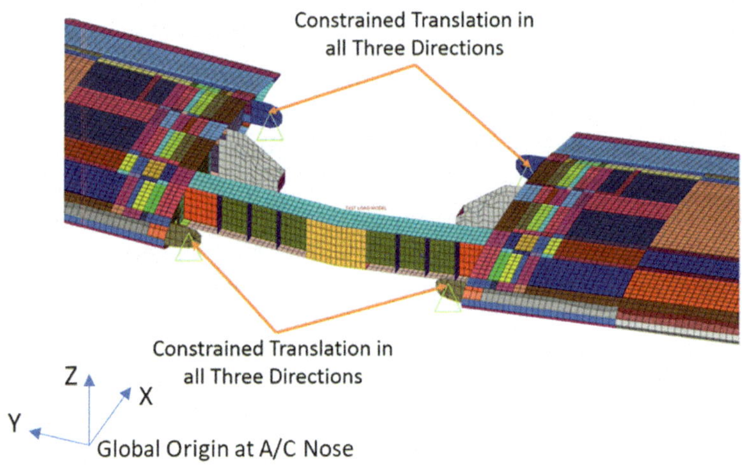

Fig. 3 Simulation of boundary conditions in the wing FE model

4 Full Scale Static Testing of the Composite Wing

4.1 Test Rig and Loading Details

The static test setup shown in Fig. 4 comprises a main test rig consisting of combination of circular and square steel cross-sectional vertical columns. The vertical box sections are interconnected by box beam to form a framework. Four steel attachment brackets simulating the fuselage end of the trainer aircraft are mounted on the support test rig which in turn connects the wing attachment lugs in the form of lap joint by four attachment bolts of 9.53 mm diameter fastened with 11 N-m torque.

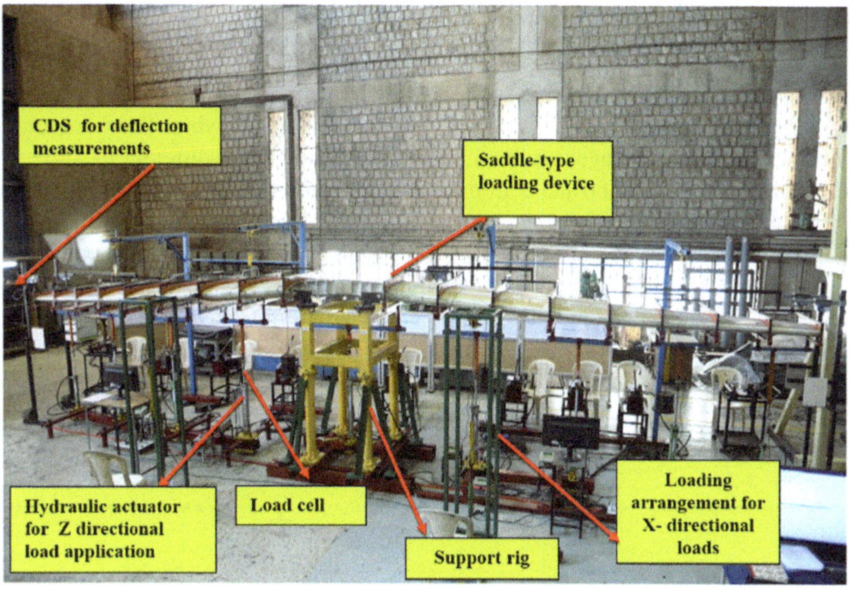

Fig. 4 Static strength test setup of composite wing for maximum up-bending load case

Wing is mounted with bottom surface facing upward to simplify the loading arrangement and the loads on the wing are introduced at each rib stations by means of wooden saddle with metal framework loading device. In this load case (max Up-bending case), 'C' rib station (on both LHS and RHS) are having two loading points, i.e., X- and Z-direction, for which a special type of loading fixture is designed to facilitate the same. For X-direction loading on the rib station C, turn-buckle type mechanical arrangement is employed to apply the low magnitude loads along its direction on both the sides of the wing. The number of loading stations is reduced using the industry standard of whiffle-tree loading arrangement. The wooden saddles-metal framework loading device at each rib station has been designed to carry more than twice the maximum load expected under DUL. Hydraulic actuators which are mounted on a rigid base frame are connected to saddle framework attachments through calibrated load cells which monitors the test load applied on Wing. The base frame consists of two C channel beams connected by C channels which are in turn fixed to the reaction floor by 30 mm reaction bolts. The actuators are mounted on the base frame through connecting flange. The rig is designed to support the vertical loading actuators and to sustain the reaction loads applied from them. The complete test rig along loading fixtures is designed, analyzed and fabricated in-house. Monotonically increasing loads will be applied on test article by 09 hydraulic actuators mounted at the load application points. Since the load magnitudes applied at each rib station are low, 'S' type load cells of different load capacities (ranging from 1 to 20 KN) with load indicators are used for the applied load measurement.

4.2 Test Instrumentation

During the static strength test, structural response of the wing such as strains is monitored by strain gauges bonded at various critical locations of wing. There are totally nine stacked rosette gauges and two linear gauges bonded at different locations, at the outer skins on both top and bottom side, main spar flanges, web and root rib webs of wing. The locations of strain gauges bonded on wing are as shown in Fig. 5.

For the displacement measurements, four calibrated Cable Displacement Sensor (CDS) at the wing tips two each at LHS and RHS front and trailing edges of the wing shown in Fig. 6 for max up-bending load case (DLL and DUL).

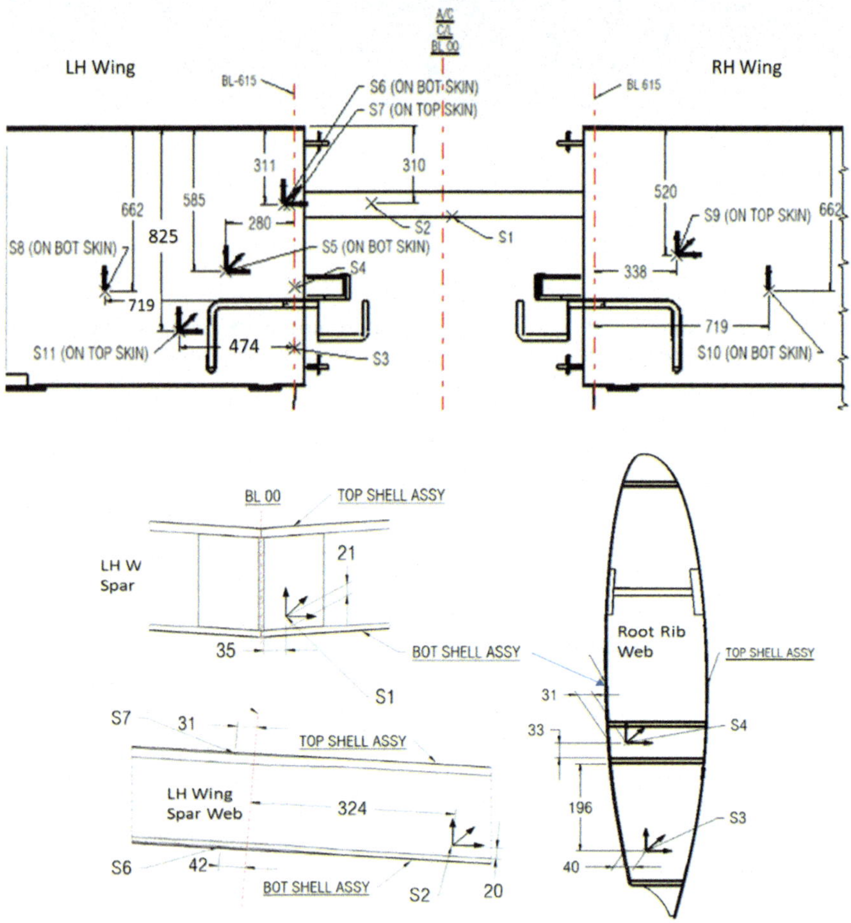

Fig. 5 Strain gauge installation on wing top, bottom and web skins

Fig. 6 CDS used in the displacement measurements

5 Static Testing and Finite Element Analysis Results

During the DLL and DUL test, the wing was tested up to 100% of DLL and up to 225% DLL (DUL) for maximum up-bending load case. The wing component was initially loaded to 20% limit load and then released to remove the slackness in the loading arrangement. The wing is loaded to 10% of (DLL), and deflection and strain reading at 10% of load are recorded.

Further, the load is increased in steps of 10% of DLL up to 100% DLL and recording the strains and deflections at every load step. The loading was stopped after reaching 100% DLL; strain and deflection readings were noted. The wing was unloaded in the steps of 10% DLL to zero load. After unloading to zero, a check is made for permanent set (from the deflection measurements).

For the testing of wing for DUL test, the wing is loaded up to 100% DLL and recording the strains and deflections at every load step. The loading is further increased up to 225% DLL. Beyond 220% and up to 225%, DLL the load increment are at 5%.at every load step. Loading is stopped after reaching 225% DLL and held at 225% DLL for 3 s. The wing is unloaded in the steps of 20% DUL to zero load. The test results monitored during the test were in terms of *deflections and strains* measurements recorded at the specified locations and were compared with FE Analysis results.

In the FE and test strain comparison at 225% DUL, a maximum compressive strains of −2690 $\mu\varepsilon$ was observed compared to FE strains of −2204 $\mu\varepsilon$ at the rosette gauge S9_0° which was located on RHS of the outer surface of top skin of wing. A maximum tensile strains of 3144 $\mu\varepsilon$ were observed compared to FE strains of 3140 $\mu\varepsilon$ at the rosette gauge location S6_0° which was located on LHS of the outer surface of bottom skin of wing.

Strain variation at the rosette strain gauge locations S6, S9, S7 which is located on LHS of the outer surface of top skin of wing and S2 which is bonded on the spar web toward bottom side of the wing are plotted v/s % load as shown below in Fig. 7. Linearity was observed in strain values measured during the test.

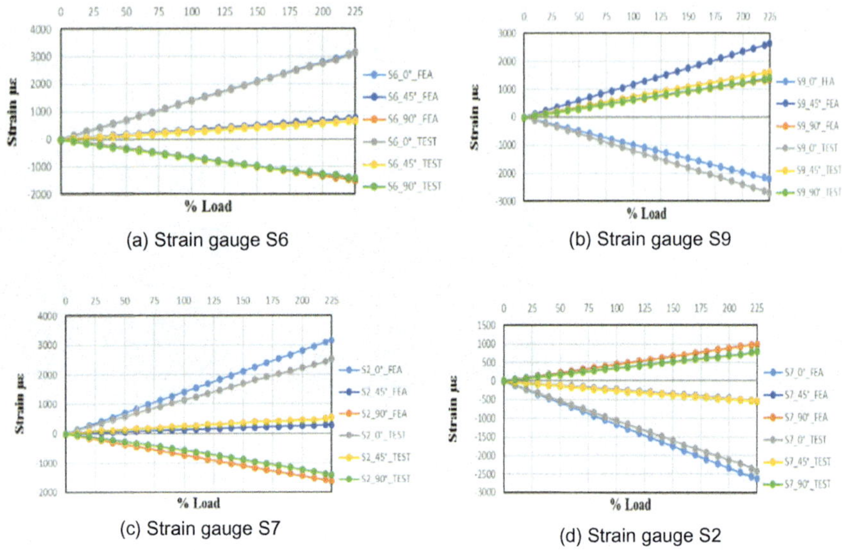

Fig. 7 Test and FEA strain results comparison in DUL case (up to 225% DLL)

In the design ultimate load case, a maximum deflection of 389 mm and 400 mm in Z-direction was measured during the test by displacement sensors (CDS), respectively, which was mounted at the wing tips at trailing edges on both LHS and RHS of wing respectively in comparison to predicted FE deflections of 454 mm and 422 mm and a maximum deflection of 369 mm and 380 mm in Z-direction was measured at the wing tips at leading edges on both LHS and RHS of wing, respectively. The deflection readings monitored during test compared with FEA results are plotted as shown in Fig. 8. Linearity was observed in deflection values measured during the test.

Maximum compressive stress observed in foam is 0.18 MPa in the wing structure (allowable foam compressive stress is 0.6 MPa).

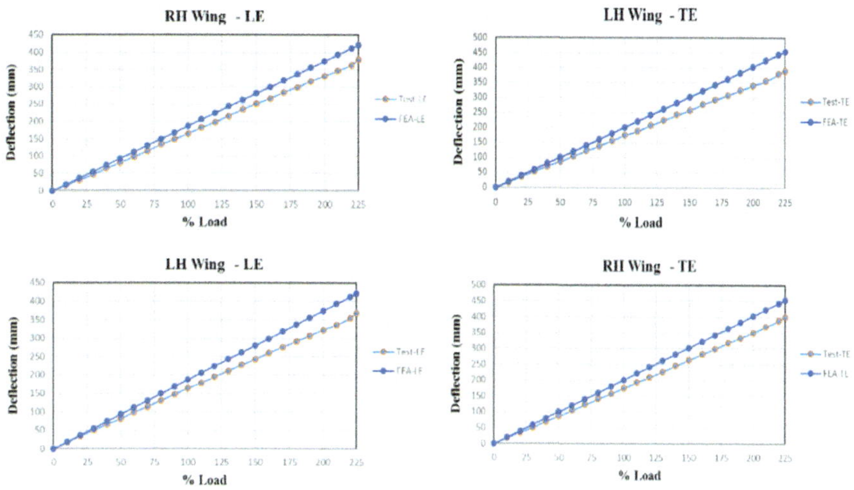

Fig. 8 Test and FEA deflections results

6 Conclusion

The static strength test of composite wing for the 100% DLL and 225% DLL (DUL) for max up-bending case was successfully completed. The wing was able to support the limit loads without detrimental/permanent deformation up to 100% DLL (indicated by the deflection values returning back to near zero values during unloading to zero load). The wing was able to support the ultimate loads without failure for 3 s for max up-bending load case. There is a good agreement between the FE and test strains except for few locations which may be due to finite element modeling (absence of joining layers, joggle in the skin/foam) which is conservative assumption. CFRP and GFRP strain values measured during the test were well within the DLL and DUL strain limits.

References

1. Chunzhi DU, Boyu FU, Zhihao QIU (2019) Design and finite element analysis of composite wing for a low speed aircraft. J Mach Des 36(S2):55–58
2. Wang T, Wu Z (2011) Finite element analysis for simulate composite wing of X model aircraft. J Shenyang Aerospace Univ 28(05):48–51
3. Smith HW (1988) Static test of an ultralight airplane. J Aircraft 25(1):37–40. https://doi.org/10.2514/6.1986-2600

Thermo-structural Qualification of a Composite Control Surface

J. Justina Geetha, S. Narendar, R. Gopikrishna, M. Varadanam, and Atul R. Bhagat

1 Introduction

Aerospace vehicles, such as missile systems and launch vehicles, operating at elevated Mach numbers, encounter the phenomenon of aerodynamic, or kinetic, heating [1]. Due to the phenomenon of kinetic heating, the airframe undergoes thermal elevation to varying degrees along the entirety of the missile's structure. The thermal conditions experienced by an airframe during the duration of its flight are contingent upon the prevailing flow dynamics in the vicinity. At higher temperatures, the materials employed in aerospace structures exhibit a discernible decline in their mechanical strength [2]. The structural framework of these aerospace systems is meticulously engineered to ensure that it remains impervious to failure resulting from diminished strength under heightened thermal conditions. Ensuring the preservation of structural integrity in an aircraft's airframe throughout the duration of a flight, particularly in the face of heightened thermal conditions, holds paramount significance. Furthermore, the airframe serves as a housing structure for a multitude of electronic systems and subsystems that are indispensable for the proper functioning of the control surface. The aforementioned electronic systems possess the capability to function within a restricted spectrum of ambient temperatures and necessitate safeguarding against elevated temperatures through the utilization of thermal coatings and protection systems [3].

Thompson and Scott [4] conducted a comprehensive examination of a co-cured carbon fiber composite control surface through static testing, wherein the obtained outcomes were subsequently juxtaposed with finite element analysis. The control

J. Justina Geetha · A. R. Bhagat
Advanced Systems Laboratory, Hyderabad, India

S. Narendar (✉) · R. Gopikrishna · M. Varadanam
Defence Research and Development Laboratory, Hyderabad, India
e-mail: narendar.drdl@gov.in

M. R. Manohara Babu et al. (eds.), *Recent Advances in Composites Engineering and Research*, Springer Proceedings in Materials 39,
https://doi.org/10.1007/978-981-99-8807-5_16

surface was effectively loaded to the utmost extent of its design capacity, encompassing the displacement caused by wing bending, without any discernible indications of structural failure. Doepke et al. [5] introduced a sophisticated actuation mechanism that demonstrates exceptional performance in the context of controlling morphing spoilers. This mechanism relies on the utilization of pressurized flexible matrix composite actuators. A diverse array of actuators were manufactured and subjected to rigorous experimentation involving elevated pressures and substantial loadings in order to assess their performance capabilities. A conclusive iteration of a spoiler prototype was devised to assess the efficacy within the context of a comprehensive design. The ultimate iteration of the prototype was executed in a comprehensive manner, adhering to the full scale in the chord direction. It was meticulously crafted to perfectly align with the spoiler profile of the actual design scenario. The ultimate iteration underwent initial testing to ascertain its utmost displacement and performance when subjected to pseudoaerodynamic loading.

The scholarly work conducted by Guo et al. [6] delved into the realm of neural network technology, specifically focusing on the development of an adaptive composite dynamic surface control strategy. This strategy was designed to facilitate high-accuracy tracking control of electro-hydraulic systems operating at extremely low velocities. The analysis of the stability of the proposed control strategy is conducted using the Lyapunov method. The comparative simulation and experimental findings substantiate the commendable control performance exhibited by the novel control strategy employed in valve-controlled cylinder systems operating at extremely low velocities. Min et al. [7] have undertaken a study that delves into the intricacies of composite neural tracking control as it pertains to the longitudinal dynamics of hypersonic flight. The system's dynamics have been effectively partitioned into distinct subsystems, namely the velocity subsystem, the altitude subsystem, and the attitude subsystem. The reference command for the flight path angle in the altitude subsystem is derived from the attitude subsystem. Through the implementation of dynamic surface control utilizing a groundbreaking neural design, the neural system achieves a heightened level of convergence, resulting in an accelerated rate of performance improvement and enhanced tracking capabilities. The scholarly work conducted by Shrivastava and Mohite [8] delved into an intricate investigation concerning the formulation of a systematic approach aimed at optimizing the first-ply failure index in unidirectional fibrous laminates. This endeavor was accomplished through the utilization of genetic algorithms (GA) in the context of quasi-static loading conditions. The application of the genetic algorithm yielded ply orientations combinations that were reasonably well-optimized, achieving a faster rate of convergence. Nevertheless, the ultimate optimized sequence of ply orientations is achieved through the fine-tuning of the sequences provided by genetic algorithms, which are rooted in industrial practices and informed by valuable experience, whenever necessary. The ongoing investigation pertaining to the transformation of an entirely metallic configuration into a partially carbon fiber composite (CFC) structure has yielded a notable reduction in weight, amounting to approximately 12%.

The study conducted by Iyas et al. [9] sought to examine the potential impact of control surface deflection on both flutter speed and flutter frequency. An experimental investigation was conducted utilizing a wind tunnel to assess the aerodynamic characteristics of a flat plate wing composed of composite material. The researchers arrived at the determination that the excitation of control surfaces would yield the following outcomes: (a) Induce wing flutter at diminished velocities. (b) Diminish the frequency of flutter. (c) Diminish the extent of the flight envelope for an aircraft. Na and Shin [10] have devised a highly effective equivalent-plate analysis technique for a composite wing with a control surface, characterized by a low aspect ratio. This analysis is rooted in the first-order shear-deformation theory of a plate, which enhances its intellectual rigor and analytical precision. It has been demonstrated that the natural frequencies of the lowest 10 modes exhibit a discrepancy of no more than 5% when compared to the corresponding values obtained from a meticulously constructed three-dimensional model using Nastran. Additionally, it was discovered that the torsional spring coefficient exerts a substantial impact on the dynamic characteristics within the context of their parametric investigation. Furthermore, it is worth noting that this particular influence can manifest itself in the presence of structural concentrated nonlinearities, such as the phenomenon of free-play nonlinearity.

Phelan [11] expounded upon the conceptualization, examination, construction, and experimentation of a composite control surface tailored for a diminutive submersible. The author provided a comprehensive exposition on the selection of construction materials and the rationale behind their usage, as well as an in-depth exploration of design requirements and the analytical approach employed, culminating in the presentation of the obtained results. The intricate procedures encompassing the fabrication processes of diverse materials are meticulously expounded upon in a comprehensive manner, delineating each step in a sequential format. The comprehensive exposition of a laminated and fully scaled prototype is provided in meticulous detail.

In order to evaluate the structural soundness of aerospace vehicles' airframes under high temperatures prior to flight testing, it is imperative to conduct a series of thermal and thermo-structural examinations through rigorous ground testing procedures [12–15]. The successful execution of both the functional and thermo-structural validation of control surfaces in hypersonic scenarios has been accomplished. The test in question possesses a remarkable degree of singularity, with its methodology and execution having undergone a successful evolution. The verification of the shaft's behavior under thermal and structural loading conditions has been conducted, thereby ensuring the functionality of the system. The assessment of the structure's well-being is further substantiated by the examination of its inherent frequency both prior to and subsequent to the conducted test.

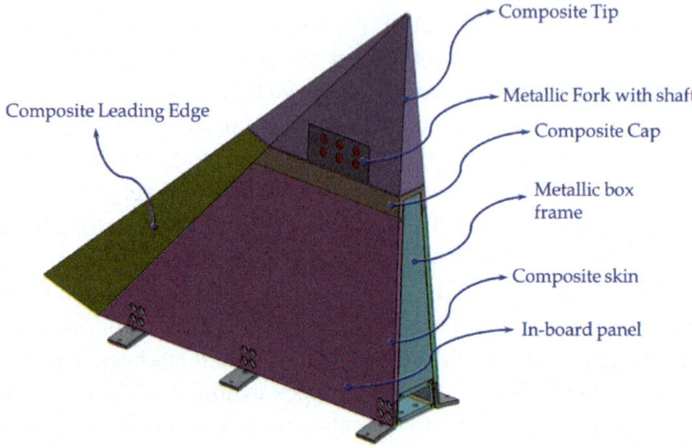

Fig. 1 Configuration of the control surface

2 Configuration of Control Surface

The aerospace vehicle has four aerodynamic control surfaces situated to the rear end of the vehicle. The control is required only during the initial phase of the vehicle and therefore, separated along with the first-stage propulsion system during flight. In certain aerospace vehicles, these control surfaces experience very high loads and temperature, during their operation at high Mach numbers, within the atmosphere. Therefore, the control surface is designed with a composite tip portion and a composite leading edge. The inboard panel is a metallic box frame, with a composite layer skin to ensure lightweight construction. It houses the control actuator with a shaft connected to the tip of the control surface. The material of the tip and leading edge are chosen to withstand very high temperatures. Figure 1 shows the details of the control surface configuration. The control surface construction is a very unique and tailor made one to meet the specific flight requirements. It is first of its kind configuration, involving challenges in design and fabrication. The control surface has been successfully realized and before application in the mission; it is ready to undergo ground qualification test for thermo-structural loads.

3 Test Setup with Structural Loads

The structural loads experienced by the control surface are experienced due to the aerodynamic loads on the vehicle at supersonic and hypersonic Mach numbers. Simulation of loads and boundary condition is very critical in experimental evaluation of the control surface [16]. The total aerodynamic load acting as a resultant force at the center of pressure of the control surface is resolved to be applied as 6 concentrated loads as shown in Fig. 2. The loads are distributed on the leading edge, tip as well

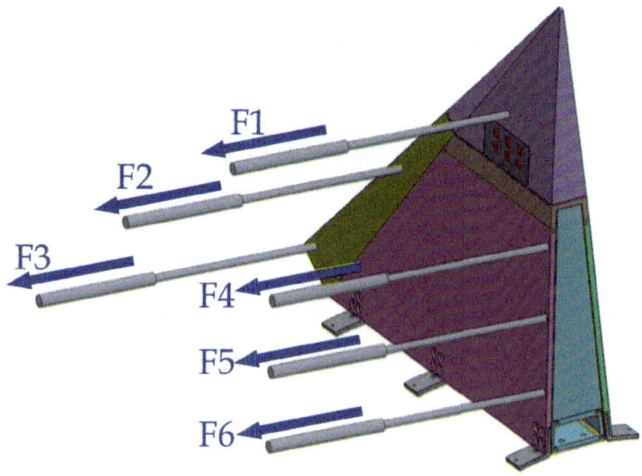

Fig. 2 Loads acting on the control surface

as the inboard panel as shown. In order to apply these concentrated loads, holes of diameter 6.5 are drilled at the required locations apriori, without causing damage to the test specimen. High-strength material linkage rods connect the control surface with the double-acting hydraulic actuators. In order to ensure load application normal to the surface of contact of the test specimen, metallic wedges are used to distribute the load locally. Figure 3 shows the photograph of the wedges and linkage rod used to apply the structural loads. The control surface is attached to a cylindrical shell during its actual application. This boundary condition is simulated using a fixture with the contour matching to that of the shell. The fixture is designed with necessary safety margins for the loads and moment it experiences during the test. A critical part of the structural set-up is the delinking of the tip load F1 at the end of the thermal cycle, to verify functionality of the tip of the control surface. In order to ensure this, a mechanism was devised to connect the linkage with loading actuator, through a bolt attached to another vertical actuator. During loading, this bolt will ensure the connectivity; however, it will be retracted at the end of the thermal cycle to facilitate the tip rotation. A very simple and effective mechanism has ensured achieving the main objective of the test as human intervention during the test is unsafe due to the thermal environment nearby the test specimen. Figure 4 shows the photograph of the delinking actuator for F1 load.

4 Open-Loop Heat Flux Control

To achieve the desired heat flux, an open loop control scheme was used. To arrive at the correct values of the input command voltages to be given to the power controller to produce the desired heat flux, a set of computational and experimental steps are taken. This procedure is shown in Fig. 5, in the form of a flowchart.

Fig. 3 Wedges with linkage rod for load application

Fig. 4 Photograph of delinking mechanism for F1 load

5 Thermo-structural Test Setup

Once the hardware is mounted on to the bottom fixture, which in turn is fixed to a test rig base plate, the necessary set-up to simulate the thermal loads is carried out using the heater cassette assembly as shown in Fig. 6. The heating system comprises heater cassettes containing IR lamps which generate the necessary heat through radiation. The distance between the cassettes and the hardware and also the power input are decided from the open-loop heat flux calibration procedure mentioned above. Different heat flux profiles are to be simulated at different zones of the control surface, as well as on the windward and leeward sides of the control surface. Accordingly,

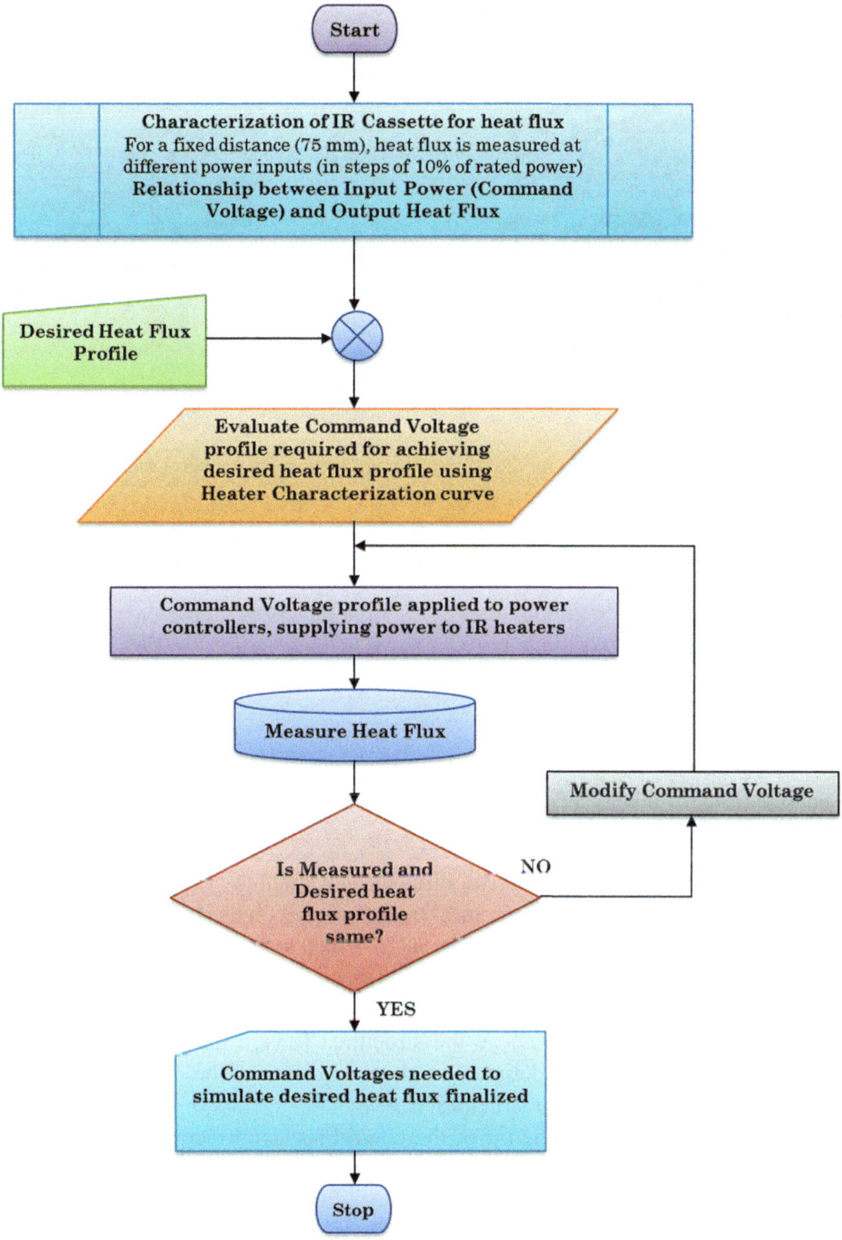

Fig. 5 A flowchart of open-loop heat flux control system

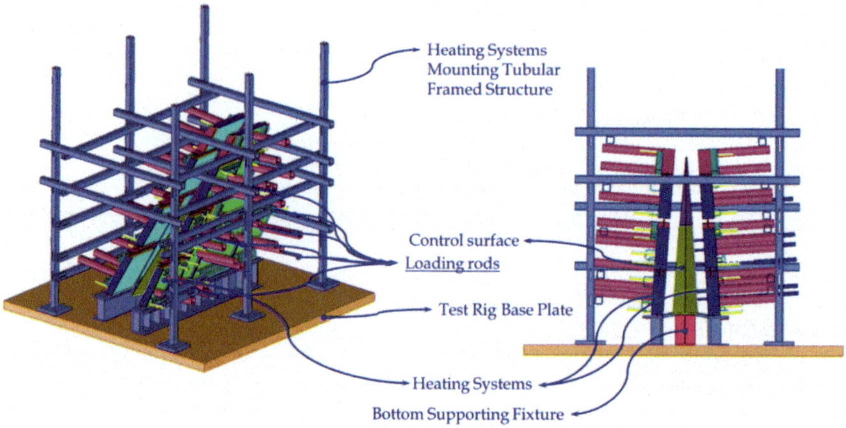

Fig. 6 Heater cassette assembly for thermo-structural qualification of control surface

the heater cassette input is defined during the test. Four inlets and four outlets are used for cooling of the reflector back surface of heating systems. Heating systems are grouped for water circulation using ½" braided flexible pipes.

6 Test Sequence

The functionality check of the control surface under thermo-structural loads is executed in the following steps.

i. An impact hammer test on the control surface assembly prior to the actual test is carried out to measure the natural frequency and mode shape of the structure.
ii. Application of only structural loads in 10% load steps on the control surface to measure the response of strains and tip deflection.
iii. Application of structural loads followed by application of thermal loads (five different heat flux profiles) on the structure. Toward the end of the thermal cycle, the tip actuator is delinked to allow actuation of the tip. Figure 7 shows photograph of the thermo-structural test.
iv. Functionality check of the tip of the control surface by powering the control actuator and ensuring $\pm$ 25° deflection of the tip of the control surface.
v. Impact hammer test on the control surface after all the test cycles to verify the health of the structure.

Fig. 7 A snapshot taken during the thermo-structural test

7 Results and Discussion

The control surface withstood the applied loads without any visual damage. Also, the functionality of the tip was ensured by smooth operation at the end of the thermal cycle. The natural frequency of the first mode of the control surface before and after test turned out to be the same (124 Hz) inferring that the structure has not degraded during the test. Figure 8 shows the mode shape of the first mode of the control surface.

A maximum of 20.5 mm tip deflection was observed due to the structural loads applied. The maximum strain measured during the structural cycle was of the order of 350 $\mu\varepsilon$ on the composite surfaces, which is less than the failure strains of the material. The metallic actuator shaft experienced strains about 2680 $\mu\varepsilon$; however, the maximum value was less than the yield strength of the material. Linearity of the strain and deflection data confirmed the health of the structure as well as the correctness of loading procedure. The thermocouple on the shaft measured room temperature values, despite the skin temperature being in the order of 800 °C. This confirms the effectiveness of the skin construction to prevent the shaft from getting heated up during flight. This has in turn ensured the functionality of the tip during the test. Figure 9 shows photograph of the structure after the completion of the thermo-structural cycle.

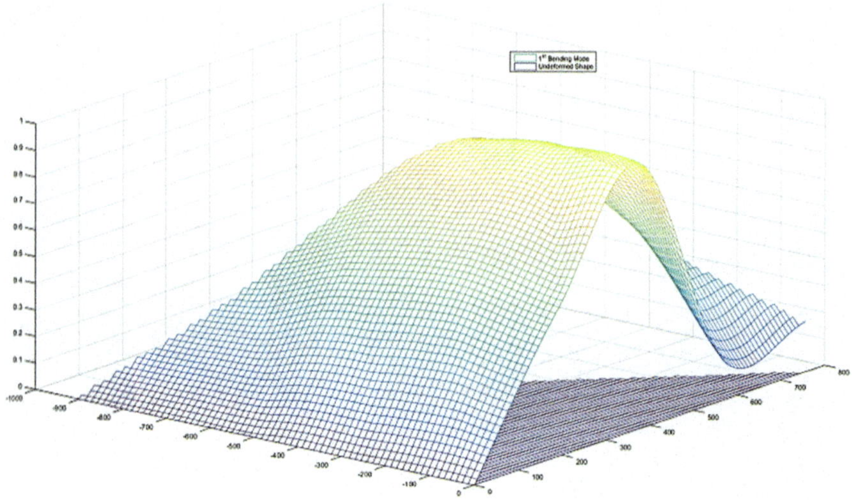

Fig. 8 1st mode shape of the control surface

Fig. 9 Photograph of the control surface after the test

8 Conclusions

The functionality and thermo-structural qualification of control surfaces in hypersonic applications have been carried out successfully. It is a very unique test, and the methodology and procedure to execute it has been successfully evolved. The behavior of the shaft under thermal and structural loading environment is verified and the functionality ensured. The natural frequency checks before and after the test give additional validation of the health of the structure.

Acknowledgements The authors are thankful to Project Hypersonics, Composite fabrication team, Actuation control division and the structural testing team for their support throughout the test. The authors also extend their gratitude to Shri RK Gupta and Shri. Y Krishna, retired scientists for their guidance throughout the execution of the test.

References

1. Maykapar GI (1968) Aerodynamic heating of the lifting bodies. NASA-CR-97576, ST-AD-HT-10770
2. Wesley Truitt R (1960) Fundamentals of aerodynamic heating. The Ronald Press, New York
3. Thomas HJ, Lance RW, Leslie G (1977) A technique for transient thermal testing of thick structure. NASA-TM-4803
4. Thomson RS, Scott ML (2000) Experience with the finite element modelling of a full-scale test of a composite aircraft control surface. Compos Struct 50:331–345
5. Doepke B, Kim D, Philen M (2007) Design and demonstration of a flexible matrix composite morphing control surface for air gap control in a Fowler flap. J Intell Mater Syst Struct 28(20):3139–3151
6. Guo Q, Shi G, Wang D, He C (2017) Neural network–based adaptive composite dynamic surface control for electro-hydraulic system with very low velocity. Proc Inst Mech Eng Part I: J Syst Control Eng 231(10):867–880
7. Shang Min Z, Chun Wen L, Ji Hong Z (2015) Composite dynamic surface control of hypersonic flight dynamics using neural networks. Sci China Inf Sci 58:1–9
8. Shrivastava S, Mohite PM (2015) Design and optimization of a composite canard control surface of an advanced fighter aircraft under static loading. Curved Layered Struct 2(1). https://doi.org/10.1515/cls-2015-0006
9. Iyas MM, Sallehuddin M, Sukri MA, Shuhaimi MM (2013) Wind tunnel testing of composite wing flutter speed due to control surface excitation. Appl Mech Mater 315:359–363
10. Na YH, Shin S (2013) Equivalent-plate analysis for a composite wing with a control surface. J Aircraft 50(3):853–862
11. Phelan R (1995) Design, analysis, fabrication and testing of a composite control surface. Nav Eng J 107(2):41–55
12. Vijayakumar G, Narendar S, Justina Geetha J (2020) Thermal design and testing of external protuberance of hypersonic carrier vehicle airframe. In: Voruganti H, Kumar K, Krishna P, Jin X (eds) Advances in applied mechanical engineering. Lecture notes in mechanical engineering, pp 287–295. Springer, Singapore. https://doi.org/10.1007/978-981-15-1201-8_33
13. Narendar S, Gopikrishna R (2019) Evaluation of structural integrity of tactical missile ceramic radomes under combined thermal and structural loads. Procedia Struct Integr 14:89–95
14. S. Narendar, R. Gopikrishna, T. Srinivasa Kumar, M. Varadanam, Experimental Evaluation of Circular Cylindrical Shells with Cutouts Under the Combined Effect of Structural and Thermal Loads. In: Singh, S., Raj, P., Tambe, S. (eds) Proceedings of the International Conference on Modern Research in Aerospace Engineering. Lecture Notes in Mechanical Engineering. Springer, Singapore, (2018) https://doi.org/10.1007/978-981-10-5849-3_27
15. Narendar S, Srinivasa Kumar T (2022) Design and development of a short-wave electric infrared heater of 215 kW capacity. In: Kumar R, Chauhan VS, Talha M, Pathak H (eds) Machines, mechanism and robotics. Lecture notes in mechanical engineering. Springer, Singapore. https://doi.org/10.1007/978-981-16-0550-5_35
16. Thermo-structural test report of LV fin of HSTDV, DRDL internal test report

Load Monitoring of Composite Aircraft Wing Using Fibre Bragg Grating Sensors

Saransh Jain, M. J. Augustin, S. Venkatesh, S. R. Viswamurthy, Nitesh Gupta, Kundan Kumar Verma, and M. G. Kotresh

1 Introduction

There is substantial increase in use of composite material in aerospace industry. Structural health monitoring systems [1] for composite structures can be extremely useful where conventional NDE methods are difficult to implement. CSIR-NAL and IIT Kanpur together are working towards developing and demonstrating a reliable and rugged fibre-optic sensor-based aircraft load and usage monitoring system. This system is designed for online strain monitoring, which will provide continuous information on the loading experienced by the structure. Hansa-3 aircraft designed and built by National Aerospace Laboratories, Bangalore available at the Flight Laboratory of IIT Kanpur was chosen as the platform for demonstration of this technology. The aircraft wing is made of composite materials, and the primary load carrying member is the main spar. The Hansa wing spar was identified as the structural component to be monitored using this technology. Several studies have shown that Fibre Bragg grating (FBG) sensors are well suited for strain measurements in composite aircraft components due to their low weight, multiplexing capability and resistance to EMI [2]. FBG sensors will be installed at suitable locations on the aircraft wing spar. Usually, the fibres containing FBGs are surface bonded on the host structure.

The purpose of this research activity was to develop algorithms for load estimations for aircraft wing. Towards this, a ground test article closely representing HANSA-3 spar was designed and fabricated. The ground test article was tested in four-point bending configuration. The HANSA-3 wing is designed for several symmetrical and unsymmetrical load cases. Shear force and bending moment generated in the spar due to these load cases were evaluated. The loads in four-point bending

S. Jain (✉) · M. J. Augustin · S. Venkatesh · S. R. Viswamurthy · N. Gupta · K. K. Verma · M. G. Kotresh
Advanced Composites Division, National Aerospace Laboratories, Old Airport Road, Bengaluru, Karnataka 560017, India
e-mail: jainsaransh@nal.res.in

M. R. Manohara Babu et al. (eds.), *Recent Advances in Composites Engineering and Research*, Springer Proceedings in Materials 39,
https://doi.org/10.1007/978-981-99-8807-5_17

tests was applied in such a manner that equivalent shear force and bending moment were generated in the proposed region to be monitored in the spar. This support condition closely simulated the spar supporting condition in the wing root region of the aircraft. Fibre Bragg Grating sensor patches and strain gages were bonded at critical locations identified based on finite element analysis and accessibility to acquire strains during tests. The acquired strains were then plotted. Comparison were made between test results and finite element results. Good agreement was found between tests and numerical results.

The strains generated in ground test article due to applied loads were indicative of the applied bending moment. Strains measured during tests and applied bending moments were used to develop mathematical models (algorithms) which can then be used to estimate bending moments in Hansa-3 wing spar from flight test strain data. Several data driven techniques have been used [3, 4] over the years for development of algorithms in SHM systems. In this work, the bending moments were computed at loading ends and then interpolated in the gage region. The known bending moments and measured strains were used to obtain a highly reliable linear fit. Curve Fitting Toolbox available in MATLAB (R2022a) [5] was used.

The mathematical model estimated bending moment in the spar with good accuracy. The average error in estimations was about 5%. Such accuracy in estimation provides us with confidence in implementing these algorithms in estimating bending moments during flight trials.

2 Test Specimen

The front spar or main spar is continuous spar throughout the wing. This carries major bending and shear and has an 'I' shaped cross section. Based on physical inspection of aircraft it was found that only certain portion of main spar root section was accessible for bonding sensors and monitoring during flight trials. The intention of this activity was to replicate sensor locations and loads during flight trials; followed by comparison of the ground tests results with flight trials. Therefore, ground test article was idealized as main spar (shown in Fig. 1); where-in it is proposed to bond FBG sensors on the locations identical to sensor locations in aircraft during flight trials. The region of interest in main spar is 1.23 m long. Hence, ground test article was 1.23 m long and had 1 m overhang on both sides to instal supports (Fig. 1).

The ground test article had four major components: (a) Top flange, (b) Bottom flange, (c) Web, (d) Gusset plates. The ground test article was manufactured by hand layup using UD Carbon Tape (T-300), glass cloth 92,125 (8 mil), glass cloth 92,110 (4 mil) as reinforcement, LY5052 as resin and HY5052 as hardener. The materials and processes used for fabrication of ground test article and fabrication of HANSA-3 aircraft were same.

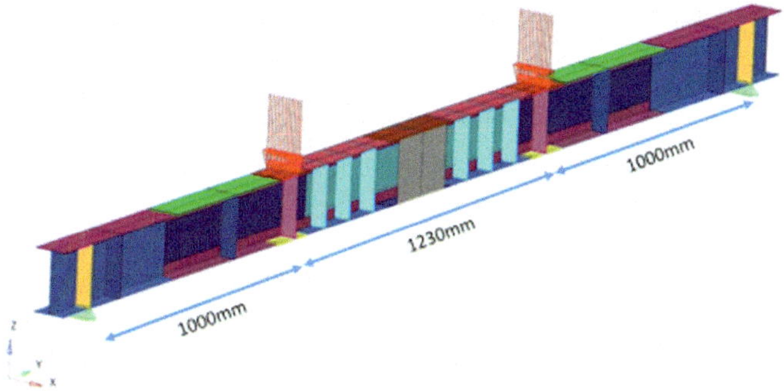

Fig. 1 Proposed ground test article

3 Loads and Test Scheme

The ground test article essentially carries only bending and shear loads, torsion is not carried by main spar by design. Therefore, bending moment and shear loads carried by the main spar in all the above critical load cases was calculated. Bending moments and shear forces for all the load cases were derived from Hansa wing design document using hand calculations. Critical load cases for design and testing were identified and subsequently applied loads were computed for those cases only.

The ground test article was proposed to be loaded in four-point bending conditions (Fig. 2). Loads to be applied are presented in Table 1. Loads P1 and P2 were applied using hydraulic actuators. Two actuators were used to apply desired loads according to the load case being simulated in the tests. The sequence of loading was such that the tests started with least severe case to most severe case. Therefore, the test started with load case F, subsequently the ground test article was inverted and loads were applied on top flange starting from load case Aa-u and ending with load case D. In Table 1 the loads are presented in order of this sequence.

The test set-up for these tests is shown in Fig. 3. Ground test article was supported at two extreme ends with the help of cross beams as shown in Fig. 3. The ground test

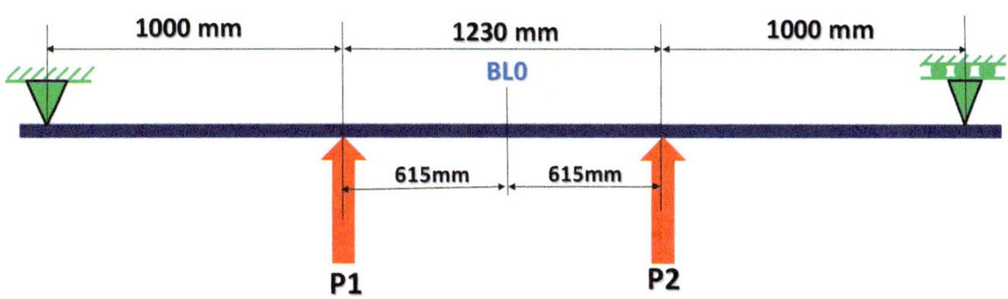

Fig. 2 Proposed four-point bending test for ground test article

Table 1 Actuator forces to be applied for achieving desired bending moment & shear forces in the test article

S. No.	Load Case	Bending Moment (M) @ P1 (kg-m)	Bending Moment (M) @ P2 (kg-m)	Shear force (S) @ BL-615 (kg)	Load, P1 (kg)	Load, P2 (kg)
1	Load case F	945	945	0	−945	−945
2	Load case Aa-u	−1579	−1311	218	1797	1092
3	Load case Ca-u	−1609	−1306	247	1856	1059
4	Load case R-ib	−1730	−1841	−90	1640	1932
5	Load case D	−2285	−2285	0	2285	2285

article was supported in such a manner that simply supported boundary conditions were simulated during test. The test article was supported on rollers to simulate these boundary conditions (Fig. 4). Actual test set-up is shown in Fig. 5.

For application of load on to the test article, two hydraulic actuators with 5 T capacities were used. The hydraulic actuators were connected with load cell used to monitor the loads. Displacement on spars were measured using dial gauge. The dial gauge location is shown in Fig. 6.

Strains were measured in the ground test article under four-point bending loading. Locations of FBG sensors and strain gauges are shown in Fig. 6. These locations were kept identical to location proposed in wing spar to be monitored during flight trials. This sensor layout was suited only for estimation of bending moment in the main spar between BL-615 and BL + 615. Due to limited accessibility in flight article as well as difficulty in installation, sensors were not bonded to spar web towards measurement of shear strains. The top flange of spar had three FBG fibres bonded, (1) on the side face with 4 FBG sensors, (2) on the bottom face with 2 FBG sensors, and (3) on top face with 2 FBG sensors. The bottom flange had one FBG fibre bonded to the bottom face with 4 FBG sensors. Apart from these FBG sensors, strain gauges were bonded as show in Fig. 6. Totally 12 FBGs and 12 strain gauges are used for measurements. All the strain gauges are linear gauges. Figure 6 shows details of FBG sensors and strain gauges. Figure 7 shows sensor IDs on top and bottom flange of the ground test article.

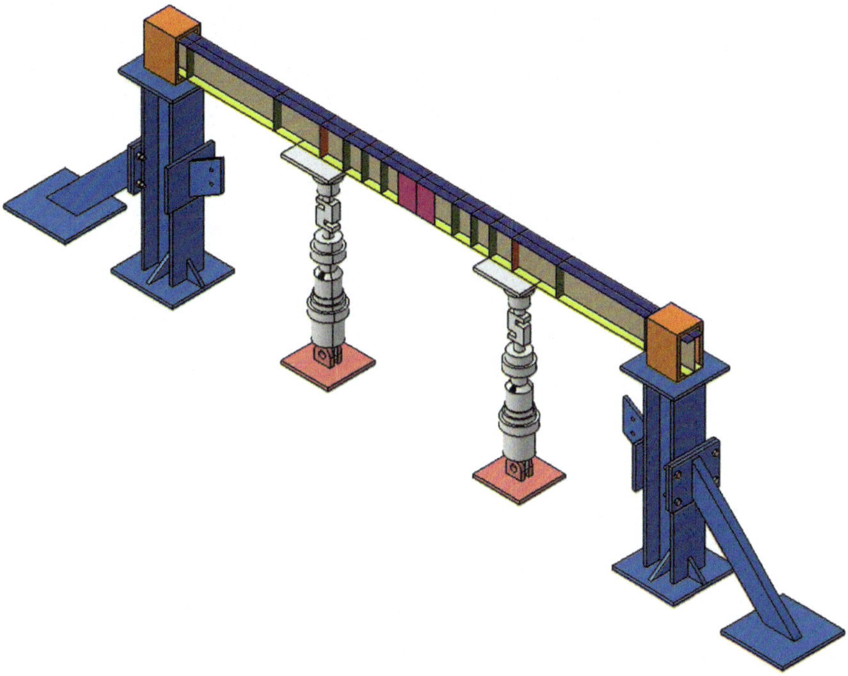

Fig. 3 Set-up for four-point bending tests on ground test Article

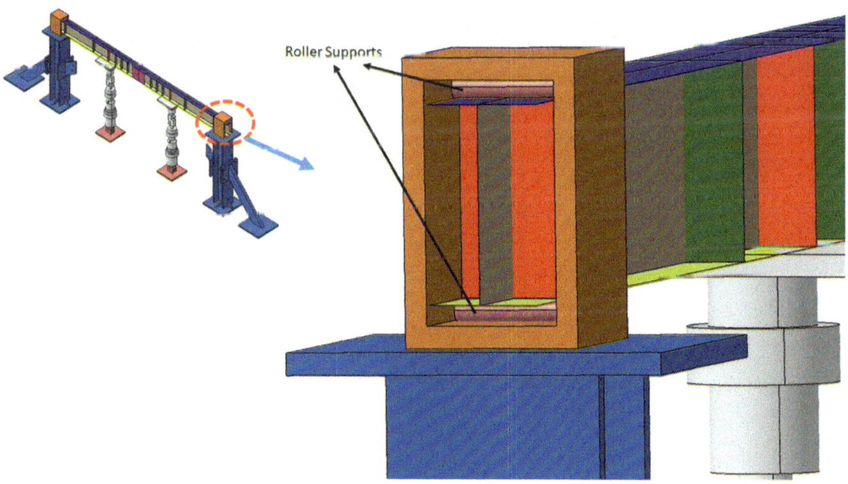

Fig. 4 Roller support at the ends of ground test article

(a) Side view of ground test article with dial gauges, actuators, and routing of sensors

(b) Complete test set-up

Fig. 5 Test set-up for ground test article

Strain data from the FBG fibres were measured using FBG interrogator SmartScan Aero Mini having 4 channels. At zero load, a baseline (wavelength values) file was recorded, which was used for calculating strain for different loads. Load was applied in steps of 10% of maximum load for all load cases. For a particular load case, at each of the load step, the wavelength shift was measured in all the FBG sensors. The shift in wavelengths is a measure of strain experienced by spar at that location (temperature is assumed to be constant during ground test).

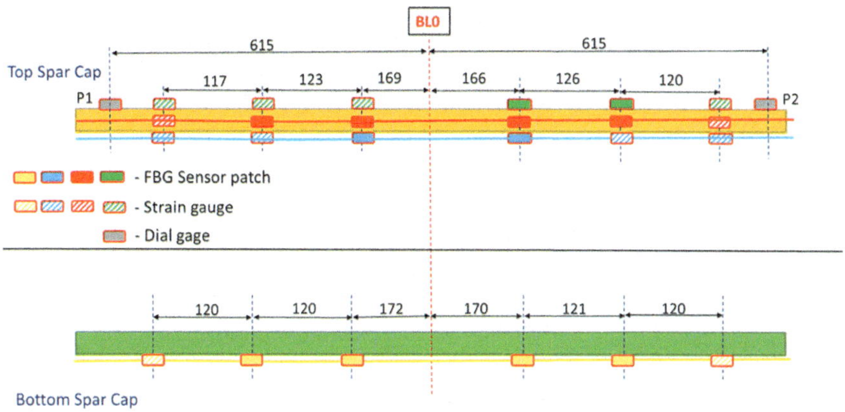

Fig. 6 Location of FBG sensors and strain gages in ground test article

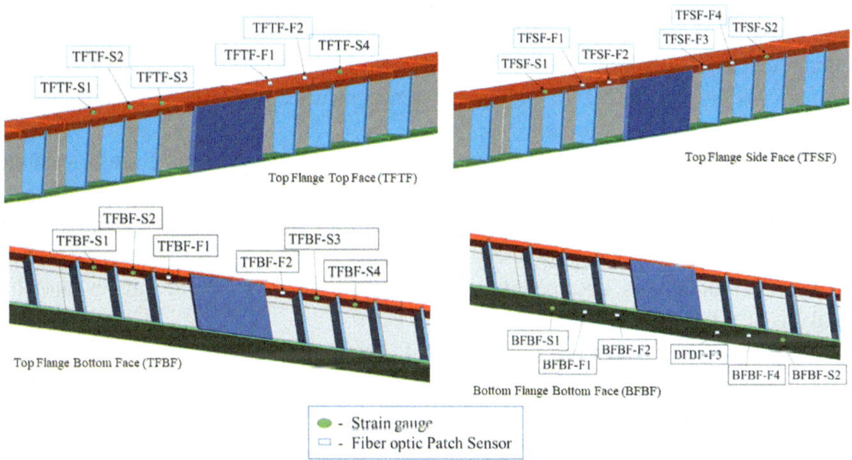

Fig. 7 Sensor IDs on top and bottom flange of ground test article

4 Test Results

As mentioned in previous section, quasi-static load tests were conducted for five load cases. During these tests, strains were measured using FBGs and strain gauges. Displacements were also recoded at loading ends during the tests. This section presents strain and displacement data acquired during the tests for one of the critical load cases, i.e. load case D. Also, comparison of test data is presented with finite element simulation results.

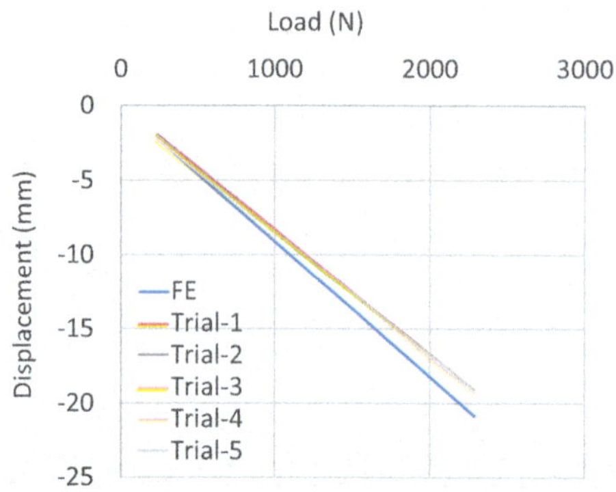

Fig. 8 Load versus displacement for load case D at P2

4.1 Load Case D

Load case D was most critical load case. In this load case total five trials were conducted. Figure 8 show load vs displacement plots at location P2 (Fig. 6). Figure 8 also show comparison between displacements measured in tests and displacement obtained from FE simulations.

Figures 9 and 10 show comparison plots for test vs FE strains for strains measured by FBG sensors and strains measured by strain gauges, respectively. These comparisons were made at 100% load levels. Another plot was generated in which strain obtained from all the sensors were plotted along with FE strains. Figure 11 shows comparison of strains obtained from FBG patch sensors, strain gauges and FE analysis at 100% load.

Good agreement between displacements measured in tests and displacements obtained from FE simulation was seen. The agreement between displacements from tests and FE in all the load cases provides confidence that overall stiffness of ground test article was captured well in FE simulations. Also Figs. 9 and 10 indicate very good agreement between measured strains and strains obtained from FE simulations. Figure 11 shows that strains captured by all the different types of sensors at the same locations were in agreement with each other and with FE strains as well. This agreement enhances confidence of using FBG patch sensors.

5 Bending Moment Estimation Using Measured Strains

Strains measured during tests were used to develop mathematical models which can then be used to estimate bending moments in Hansa wing spar from flight test strain data.

During computation of loads for testing, bending moments for all load cases were generated (refer Sect. 4—LOADS). Bending moments for all critical load cases at

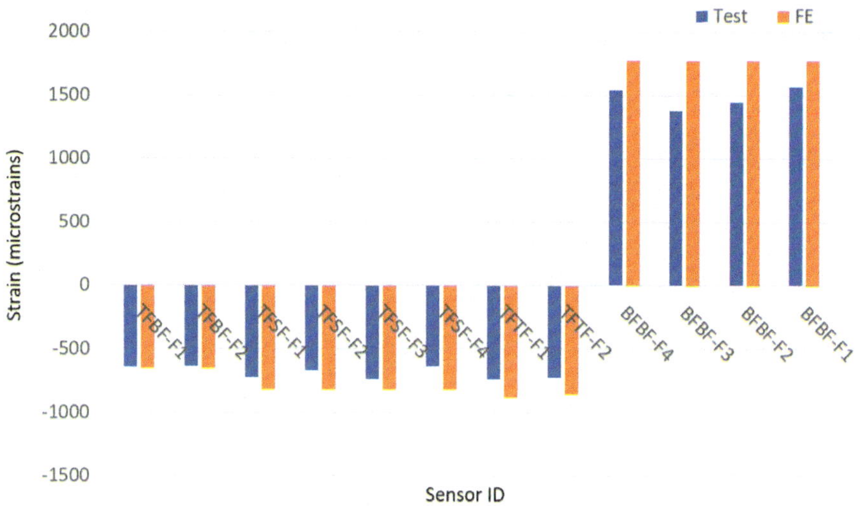

Fig. 9 Comparison of test versus FE strains for FBG sensors at 100% load (Load case D)

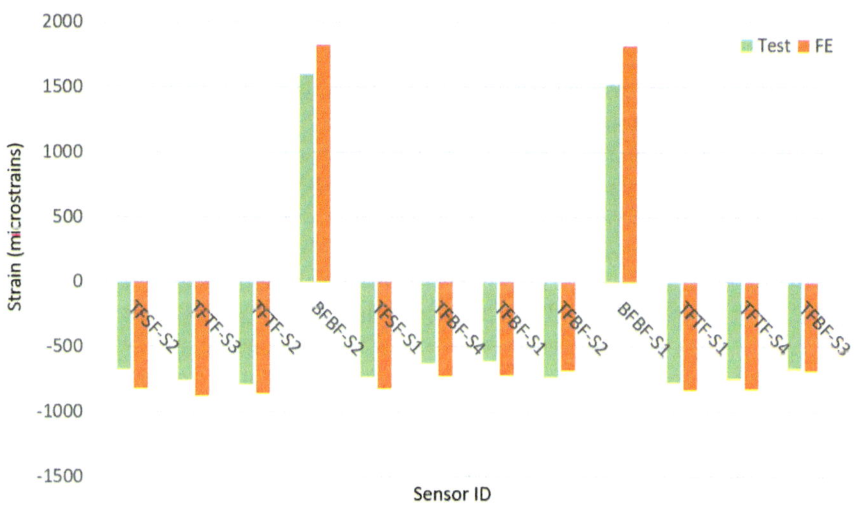

Fig. 10 Comparison of test versus FE strains for strain gauges at 100% load (Load case D)

100% load level are shown in Table 1. The bending moments for load case D and load case F were equal at P1 and P2, because they were symmetric load cases. Rest of the load cases (i.e. load case Aa-u, Ca-u, and Ri-b) were unsymmetrical, therefore their bending moments at P1 and P2 were not equal.

Using these known bending moments and their corresponding strains (measured during tests), data set was prepared to generate the mathematical model. Table 2 shows data set for load case D. Only load levels at 60% and above were considered

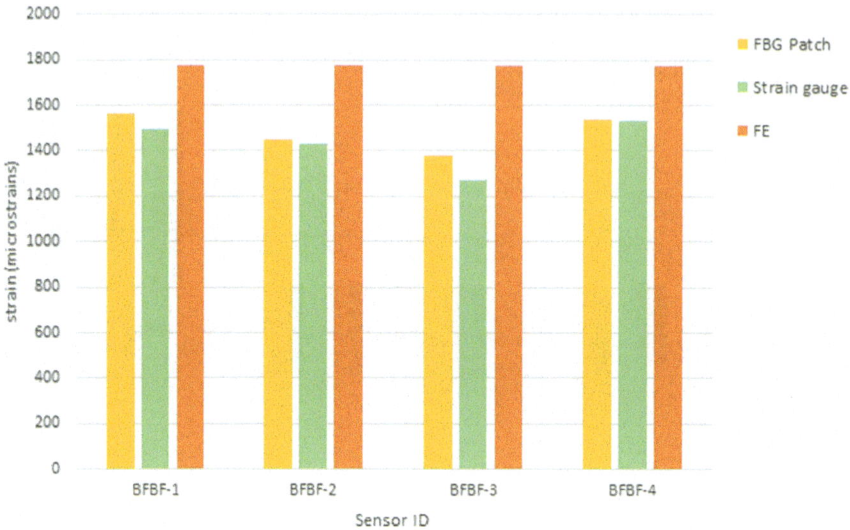

Fig. 11 Comparison of FBG Patch, Strain gauge with FE data at 100% load (Load case D)

for data generation, because strains below these load levels were not significant. Such table data set was generated for all the other critical load cases (Annexure 1). Only strains measured on top flange were considered to generate data. Figure 6 shows locations of sensors on top flange of the ground test article.

Strains and bending moments corresponding to 70% load level were not used for generating mathematical model. Data at 70% load was used to test the developed model. Data shown in Annexure-1 was then plotted as strains vs bending moment (Fig. 12). Figure 12 contains measured strains vs bending moment for all the load cases. As shown previously (Table 2), each of the load case had 24 data points to be plotted. The tests were conducted for five load cases. Therefore, totally 120 (i.e. 24 (data points for each load case) X 5 (no. of load cases)) data points were plotted in Fig. 12.

A linear fit was then obtained between strains and bending moment using Curve Fitting Toolbox MATLAB (R2022a) [5].

The linear equation between bending moment and strain obtained using curve fitting was

$$BM = P1 * S + P2,$$

where

BM Bending Moment.
S Measured Strains.
P1, P2 Coefficients of linear fit.

Table 2 Data Set for Load Case D

Load Case	Load %	Location	Strain	Bending Moment (kg-m)
Case D	60%	BL -409	-453.9	-1371.2
		BL -292	-458.8	-1371.2
		BL -169	-427.2	-1371.2
		BL 166	-395.5	-1371.2
		BL 292	-387.8	-1371.2
		BL 412	-438.1	-1371.2
	70%	BL -409	-527.9	-1599.8
		BL -292	-536.9	-1599.8
		BL -169	-500.0	-1599.8
		BL 166	-465.4	-1599.8
		BL 292	-451.0	-1599.8
		BL 412	-511.9	-1599.8
	80%	BL -409	-607.7	-1828.3
		BL -292	-616.2	-1828.3
		BL -169	-573.5	-1828.3
		BL 166	-534.0	-1828.3
		BL 292	-517.2	-1828.3
		BL 412	-587.0	-1828.3
	90%	BL -409	-686.0	-2056.9
		BL -292	-694.3	-2056.9
		BL -169	-649.3	-2056.9
		BL 166	-597.7	-2056.9
		BL 292	-584.9	-2056.9
		BL 412	-660.1	-2056.9
	100%	BL -409	-765.0	-2285.4
		BL -292	-771.5	-2285.4
		BL -169	-723.5	-2285.4
		BL 166	-666.4	-2285.4
		BL 292	-647.5	-2285.4
		BL 412	-733.0	-2285.4

For the above fit, following values for coefficients are obtained:

P1 = 3.043 kg-m.

P2 = 1.261 kg-m.

This linear fit served as the mathematical model showing relation between strains and bending moment. As mentioned earlier data corresponding to load levels of 70% were used for testing. Strains from 70% load levels were inserted in the linear fit and

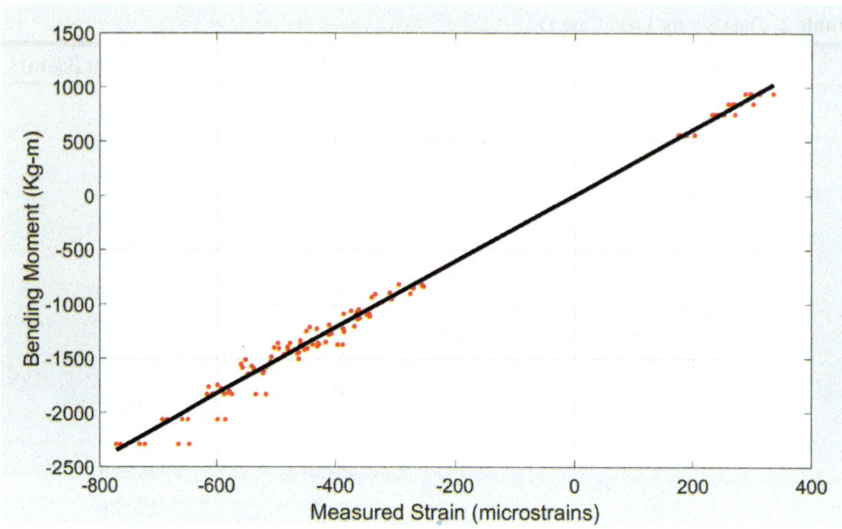

Fig. 12 Strains vs Bending Moments

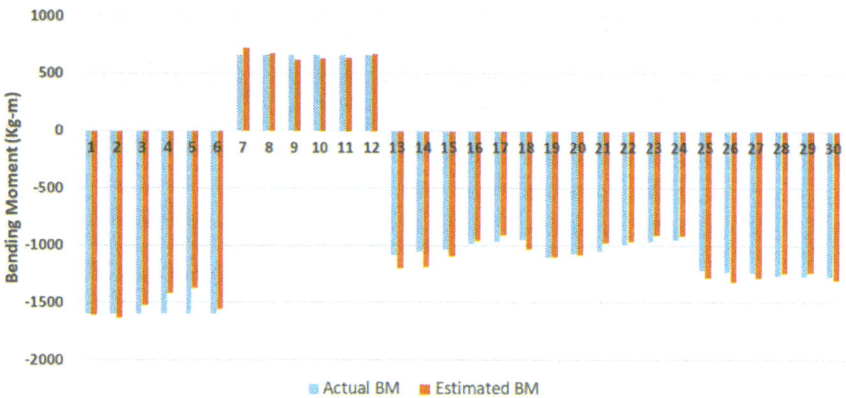

Fig. 13 Percentage error in bending moment estimations

bending moment was estimated for all the load cases. Figure 13 shows actual and estimated bending moments. Table 3 shows actual and estimated bending moments along with error in the estimations.

Table 3 and Fig. 13 show that the average percentage error in estimations was about 5% and maximum percentage error was about 14%. Therefore, it can be concluded that a very reliable mathematical model was obtained based on measured strains and known bending moments, to estimate bending moments. Such good fits provide confidence in implementation similar approach to estimate bending moments during

Table 3 Estimated bending moments using generated mathematical model

S. NO	Load Case	Strain	Actual BM (kg-m)	Estimated BM (kg-m)	Absolute Error	Percentage Error
1	D	-527.9	-1599.8	-1605.2	5.4	0.3
2		-536.9	-1599.8	-1632.7	32.9	2.1
3		-500.0	-1599.8	-1520.2	-79.6	5.0
4		-465.4	-1599.8	-1414.9	-184.9	11.6
5		-451.0	-1599.8	-1371.2	-228.6	14.3
6		-511.9	-1599.8	-1556.4	-43.4	2.7
7	F	237.3	661.3	723.2	-61.9	9.4
8		220.8	661.3	673.2	-11.9	1.8
9		203.2	661.3	619.5	41.8	6.3
10		206.9	661.3	630.8	30.4	4.6
11		209.9	661.3	640.1	21.2	3.2
12		218.3	661.3	665.5	-4.3	0.6
13	Aa-u	-393.8	-1086.4	-1197.1	110.7	10.2
14		-388.2	-1056.1	-1180.0	123.9	11.7
15		-359.5	-1037.3	-1092.7	55.3	5.3
16		-313.7	-985.7	-953.4	-32.3	3.3
17		-299.3	-966.9	-909.6	-57.4	5.9
18		-339.5	-949.1	-1031.8	82.7	8.7
19	Ca-u	-360.5	-1104.9	-1095.6	-9.3	0.8
20		-355.1	-1070.7	-1079.3	8.6	0.8
21		-321.3	-1049.4	-976.4	-73.0	7.0
22		-315.9	-991.0	-960.1	-30.9	3.1
23		-299.3	-969.7	-909.5	-60.2	6.2
24		-300.3	-949.5	-912.7	-36.9	3.9
25	R-ib	-420.3	-1219.4	-1277.8	58.5	4.8
26		-434.1	-1231.8	-1319.6	87.8	7.1
27		-420.6	-1239.6	-1278.6	39.1	3.2
28		-406.5	-1260.9	-1235.6	-25.2	2.0
29		-405.7	-1268.6	-1233.3	-35.3	2.8
30		-426.3	-1276.0	-1295.9	20.0	1.6

flight trails, where-in only measured strains will be available at the end of flight and we will have to use such mathematical model to estimate bending moments.

6 Conclusions

A load and usage monitoring scheme proposed to be implemented on HANSA-3 aircraft for aircraft loads and usage monitoring was designed and developed. As part of this scheme, aspects related to sensor preparation, routing, sensor bonding, data acquisition, primary processing, and algorithm development of this data, were evaluated on a ground-based tests specimen representing the structure to be monitored during flight trails. Towards this, a ground test article was designed and fabricated. The ground test article closely represents HANSA-3 spar (essentially structure to be monitored during flight trials). Several load cases were evaluated for ground tests based on wing design document. The ground test article was then tested under all the critical load cases under four-point bending condition. Fibre Bragg grating sensors and strain gauges were bonded at critical locations to acquire strain data during tests. The acquired strains were compared with finite element results. Good agreement was found between tests and numerical results. Also measured strains and known bending moment data was used to establish a mathematical model to subsequently estimate bending moments in ground test article. A data driven model based on linear fit using Curve Fitting Toolbox in MATLAB was developed. A highly reliable mathematical model with good accuracy in estimation was obtained.

This work demonstrated load monitoring approach on a wing like composite structure. Responses measured through strain sensors (such as FBGs) depend on damage and load acting on the structure. In an SHM scheme, it is imperative to separate these two aspects. Current research work dealt with one key aspect that of load monitoring. This will pave way for health and usage monitoring of Hansa-3 composite wing in future.

Annexure 1

Data used for generating mathematical model is as follows

Note: Data for load case D was already shown in Table 2.

Load %	Loca-tion	Strain	BM (kg-m)	Strain	BM (kg-m)	Strain	BM (kg-m)	Strain	BM (kg-m)
		Case Aa-u		Case F		Case Ri-b		Case Ca-u	
60%	BL -409	-339.4	-931.2	203.1	566.8	-360.7	-1045.2	-306.1	-947.1
	BL -292	-333.5	-905.3	188.8	566.8	-374	-1055.9	-301.7	-917.7
	BL -169	-309.6	-889.2	175.6	566.8	-363.6	-1062.5	-274.7	-899.5
	BL 166	-269.4	-844.9	177.4	566.8	-344.1	-1080.7	-270.4	-849.4
	BL 292	-253.7	-828.8	180.1	566.8	-347.4	-1087.4	-252	-831.2
	BL 412	-293.9	-813.5	187.5	566.8	-365.8	-1093.7	-255.3	-813.9
70%	BL -409	-393.8	-1086.4	237.3	661.3	-420.3	-1219.4	-360.5	-1104.9
	BL -292	-388.2	-1056.1	220.8	661.3	-434.1	-1231.8	-355.1	-1070.7
	BL -169	-359.5	-1037.3	203.2	661.3	-420.6	-1239.6	-321.3	-1049.4
	BL 166	-313.7	-985.7	206.9	661.3	-406.5	-1260.9	-315.9	-991
	BL 292	-299.3	-966.9	209.9	661.3	-405.7	-1268.6	-299.3	-969.7
	BL 412	-339.5	-949.1	218.3	661.3	-426.3	-1276	-300.3	-949.5
80%	BL -409	-449.2	-1241.6	270.5	755.7	-481.1	-1393.6	-412.4	-1262.7
	BL -292	-442.8	-1207	251.7	755.7	-495.9	-1407.8	-406	-1223.6
	BL -169	-410	-1185.5	232	755.7	-480.4	-1416.7	-368.2	-1199.3
	BL 166	-359.5	-1126.5	236.1	755.7	-461.5	-1441	-362.2	-1132.6
	BL 292	-342.9	-1105.1	241.9	755.7	-464.4	-1449.8	-342	-1108.3
	BL 412	-385.8	-1084.7	250.7	755.7	-486.4	-1458.3	-343.5	-1085.2
90%	BL -409	-503.5	-1396.8	301.7	850.2	-540.6	-1567.8	-464.9	-1420.6
	BL -292	-496.8	-1357.9	280.5	850.2	-556.1	-1583.8	-458.3	-1376.6
	BL -169	-459.6	-1333.7	258.9	850.2	-537.9	-1593.7	-415.4	-1349.3
	BL 166	-406.1	-1267.4	265.8	850.2	-521.5	-1621.1	-409.8	-1274.1
	BL 292	-385.1	-1243.2	269.4	850.2	-522.4	-1631.1	-385.2	-1246.8
	BL 412	-430.8	-1220.2	280	850.2	-546.9	-1640.5	-387	-1220.8
100%	BL -409	-558.2	-1552	336.1	944.7	-598.2	-1741.9	-516.2	-1578.4
	BL -292	-551.3	-1508.8	312.2	944.7	-613.1	-1759.8	-508.9	-1529.5
	BL -169	-508.9	-1481.9	288.1	944.7	-591.5	-1770.8	-461.3	-1499.2
	BL 166	-447.2	-1408.2	296.5	944.7	-577.7	-1801.2	-451.7	-1415.7
	BL 292	-430.2	-1381.3	297.7	944.7	-579.5	-1812.3	-430.4	-1385.3
	BL 412	-477.2	-1355.8	313.1	944.7	-603.4	-1822.8	-429.7	-1356.5

References

1. Staszewski WJ, Boller C, Tomlinson GR (2004) Health monitoring of aerospace structures. Wiley
2. Sundaram R, Kamath GM, Gupta N (2012) Structural health monitoring of composite structures—issues and challenges. Int J Veh Struct Syst 4(3):74–85
3. Cao X, Sugiyama Y, Mitsui Y (1998) Application of artificial neural networks to load Identification. Comput Struct 69(1):63–78

4. Li ZX, Yang XM (2008) Damage identification for beams using ANN based on statistical property of structural responses. Comput Struct 86(1–2):64–71
5. https://in.mathworks.com/products/matlab.html, last accessed 20 Oct 2022.

Characterisation of Composite Materials

Characterization of Quartz–Cyanate Ester Composites Fabricated by RTM

Anu Anna Abraham, B. Padma, K. V. Aneesh, and S. Seetharaman

1 Introduction

Cyanate ester resins are high-performance resins with excellent mechanical, thermal and dielectric properties. They are replacements to epoxy resin, bismaleimide resin which are generally used in high-performance applications [1, 2].

Cyanate ester resins are processable by different methods, like RTM, filament winding, prepregging technique, sheet moulding compound, etc. The processing versatility of this resin is due to the tailorable properties that can be achieved by using various monomers and their partially polymerized prepolymers [2].

RTM is a processing technique by which components of near net shapes can be obtained easily, by using resins of low viscosity which ensures proper flow and wetting of fabric [3, 4].

The combination of exceptionally low dielectric constants, dissipation factors coupled with good mechanical and thermal properties makes cyanate ester as an excellent choice of matrix material, for composites for radome applications, when used along with quartz fabric reinforcement which has very low dielectric properties [5, 6].

The objective of the present study is to evaluate physical, mechanical, thermal and dielectric properties of quartz–cyanate ester composite laminates fabricated by RTM and to find the effect of post curing on the above-mentioned properties.

A. A. Abraham (✉) · B. Padma · K. V. Aneesh · S. Seetharaman
Advanced Systems Laboratory, Dr. Abdul Kalam Missile Complex, Kanchanbagh, Hyderabad, India
e-mail: anuanna6@gmail.com

M. R. Manohara Babu et al. (eds.), *Recent Advances in Composites Engineering and Research*, Springer Proceedings in Materials 39,
https://doi.org/10.1007/978-981-99-8807-5_18

2 Materials

The material selected is a toughened two-part cyanate ester resin which is mixed in 1:1 ratio. Quartz fabric with 99.95% silica content, having twill weave, was used in the study.

3 Experimental

3.1 Fabric Testing

The fabric was checked for its areal density, break strength on a universal testing machine as per ASTM 5035 standard and thickness by dial gauge.

3.2 Resin Studies

Viscosity. The viscosity of the uncured resin mix was tested by a Brookfield viscometer at 25 °C.

Differential Scanning Calorimetry (DSC). Cure behaviour of the resin mixed in 1:1 ratio was checked by a Perkin Elmer 8000 DSC, in dynamic mode at a heating rate of 10 °C/min. The laminate after curing was checked for cure completion.

Rheology. Rheology studies of the resin, i.e., gelation temperature was studied using rheometer, Aton Par in dynamic condition at heating rate of 3 °C/min and frequency of 1 Hz.

3.3 Fabrication of Composite Laminate by RTM

Laminates were fabricated by RTM by injecting the resin into a closed mould. The fabric was cut into size of 320 mm × 320 mm using a template and stacked into the mould as per thickness requirement as in Fig. 1.

The mould was closed with the top plate after laying up of fabric and was preheated to a temperature of 80 °C, and the resin was heated to a temperature 40 °C and injected into the mould under vacuum of 640 mm Hg till the mould was filled with the resin as shown in Fig. 2.

The mould was then shifted to an oven for curing. Curing was carried out for a period of 10 h at 175 °C as recommended by the manufacturer. The demoulded laminate after cure completion is as in Fig. 3.

Fig. 1 RTM mould bottom plate with fabric laid-up

Fig. 2 RTM mould bottom half with fabric laid-up

3.4 Laminate Testing

Mechanical Testing. Cure behaviour of the resin mixed in 1:1 ratio was checked by a differential scanning calorimeter (DSC) Perkin Elmer 8000 in dynamic mode at a heating rate of 10 °C/min. The laminate after curing was checked for cure completion by DSC where the ΔH was measured.

Fig. 3 Cured quartz–cyanate ester laminate

Dynamic Mechanical Analysis (DMA). Dynamic mechanical analysis was done on a DMA, Q800 TA instruments to find out the glass transition temperature (Tg) of the cured laminate. Testing was carried in three-point bend mode at 5 °C/min from RT to 300 °C at a frequency of 1 Hz. The drop in the storage modulus was used for calculation of Tg.

Differential Scanning Calorimetry (DSC). The laminate after curing was checked for cure completion and Tg by DSC.

Dielectric Testing. Dielectric testing was carried out by wave guide method as per ASTM D5568-14 on PNA (Keysight N5224A), to find the dielectric constant and loss tangent of the cured composite material. Samples were fitted in a waveguide flange, and measurements were done in X band in the frequency range 9–10 GHz.

Resin Content and Density Test. The resin content was tested as per ASTM D3171 by burn-off method. Density was tested as per ASTM D792.

4 Results and Discussion

The quartz fabric was twill weave and was measured to have a 300 gsm and thickness 0.3 mm. The break strength of fabric was 220 kg/inch.

DSC curve of the uncured resin mix shows a peak and a shoulder, exhibiting a complex reaction as in Fig. 4. The heat of polymerization ΔH during cure was around 690 J/g, with a peak temperature of 211 °C by DSC. The gelation temperature was around 162 °C in dynamic run done on rheometer.

The viscosity of the resin at 25 °C is 380cP. The resin was heated to 40 °C to bring down the viscosity to help in the impregnation of fabric with resin by RTM technique.

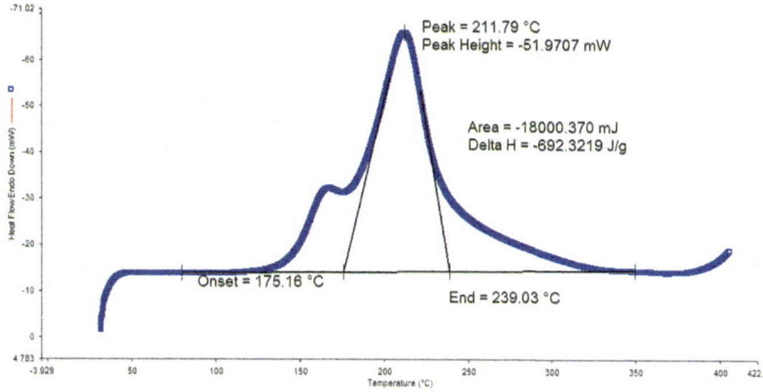

Fig. 4 DSC run of uncured resin

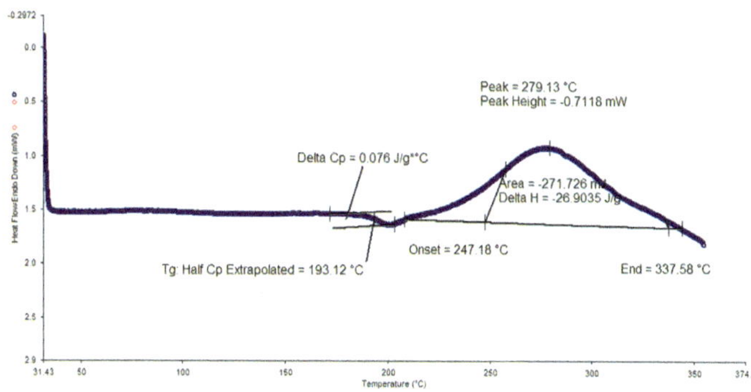

Fig. 5 DSC of laminate cured at 175 °C

The laminate cured at 175 °C for 10 h was found to have a small single-peak exotherm as shown in DSC Fig. 5. The density of the laminate was around 1.64 g/cc and resin content 37% by weight.

The degree of conversion, calculated from DSC after 175 °C for 10 h, corresponded to around 89.4%. The glass transition temperature Tg, tested by DMA, was found to be around 187 °C for the cured laminate (Fig. 6).

The mechanical properties tested include flexural strength and ILSS. ILSS of 68 MPa indicates a good bonding between quartz fabric and the cyanate ester resin. Flexural strength of 545 MPa was reported for the cured samples. The strength values for the composite are as per reported in literature [4]. The dielectric constant of cured laminate is 3.298 and loss tangent 0.02 @10 GHz. The dielectric values, especially the loss tangent value, were found to be high, and hence, post curing was essential to see if it affects the properties especially dielectric properties of the composite.

As per literature, the physical and mechanical properties of cyanate ester resins are directly affected by the degree of cure [7]. Mackezine et al. [5] mentions that post cure is optional and used either to maximize used either to maximize Tg for primary structural applications or to minimize loss tangent for radome applications.

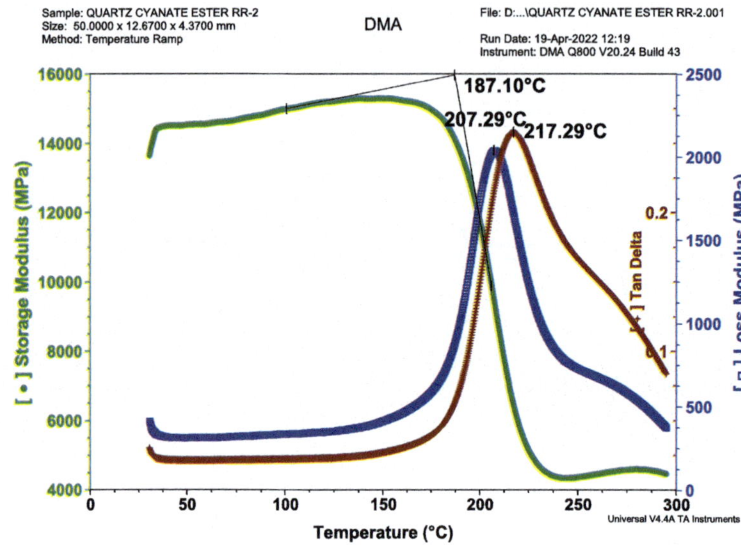

Fig. 6 DMA of laminate cured at 175 °C

The cured quartz–cyanate ester laminates were post cured at 200 °C for 4 h and evaluated for all properties. Post curing resulted in improving the degree of conversion is > 99% as observed in DSC Fig. 7.

Dielectric properties of the laminate were lowered by post curing. Dielectric constant and loss values were of 3.26 and 0.005@ 10 GHz after post curing. The drop in the dielectric values, as degree of conversion increased is due to decrease in polarity with conversion of cyanate functionality to symmetrical triazine structures and as a result of a network free volume expansion [4].

After post curing, the ILSS was 64 MPa and flexural strength 527 MPa. These mechanical properties were not significantly changed with post curing.

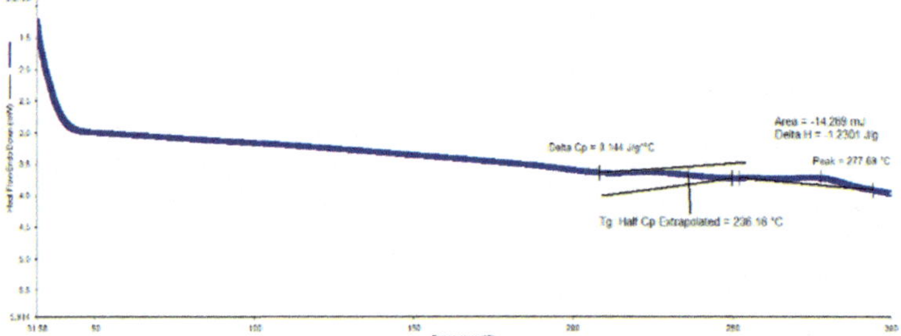

Fig. 7 DSC of laminate post cured laminate

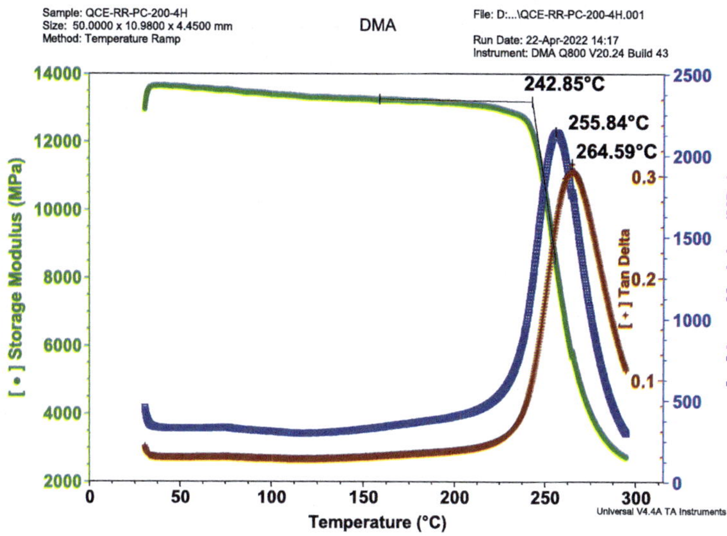

Fig. 8 DMA of post cured laminate

The dynamical mechanical analysis was used to evaluate the glass transition temperature of the material. The glass transition temperature of the material was significantly improved with post curing to 242 °C (Fig. 8).

5 Conclusion

The post cured samples exhibited excellent dielectric properties along with good thermal and mechanical properties.

References

1. Reghunadhan Nair CP, Mathew D, Ninan KN (2001) Cyanate ester resins, recent developments. In: Advances in polymer science, vol 155. Springer-Verlag Berlin Heidelberg, pp 1–99
2. Hamerton I, Hay JN (1998) Recent technological developments in cyanate ester resins. High Perform Polym 10:163–174
3. Kevin P (1997) Resin transfer moulding. Chapman & Hall, London
4. Guo Y, Zhang X, Cui H, Liao Y, Gao K, Zhang S, Jing K, Deng J, Cheng M (2017) Properties of quartz fabrics reinforced cyanate ester resin, composites produced by RTM processing. In: 21st International conference on composite materials. Chinese Society for Composite Materials, Xi'an
5. Hamerton I (ed.) (1994) Chemistry and technology of cyanate ester resins. Chapman & Hall
6. Russo O, Colasante A, Bellaveglia G, Maggio F, Marcellini L, Scialino L, Rolo L, Angevain JC, Midthassel R (2012) State-of-the-art materials for KU and KA band mobile satellite antenna

radomes. In: Telecommunications ground segments workshop, ESA/ESTEC, Noordwijk, The Netherlands
7. Ising SJ, Shimp DA, Christenson JR (1989) 3rd International SAMPE electronics conference, pp 360–370

Shock Response Behaviour of E-Glass/Epoxy Composites

Challa Jayarami Reddy, B. Venkataramudu, K. Gopinadha Reddy, Ranjith Kumar Singh, and Ashish A. Naukarkar

1 Introduction

Glass fibre reinforced polymer (GFRP) composites are being widely used in various industries including aerospace, defence, and other manufacturing industries due to their light weight, good mechanical properties, high specific strength, high-energy absorption and corrosion resistance properties. Depending on the applications, GFRP composites are being subjected to various types of mechanical and thermal loads. Many studies were reported on the mechanical and thermal behaviour of GFRP composites. However, to use these GFRP materials effectively in airframe structures, combat vehicles and hypersonic vehicles applications, it is very essential to understand their behaviour against different shock loads. Shock tubes have been extensively used for studying the dynamic behaviour of various materials under shock loads [1]. Shock tubes also offer the advantage of generating plane wave fronts and also provide experimental parameters that can be easily controlled and measured [2, 3]. Few studies on the response of GFRP under shock and blast loads using shock tube have been reported. Puneet Kumar and co-workers studied the effect of plate curvature on blast response of carbon composite panels using 3D digital image correlation. The blast loads were generated using shock tube and found that curved sample showed good blast mitigation efficiency. The failure mechanisms reported were fibre breakage and inter-layer delamination [4]. Response of E-glass reinforced vinyl ester and urethane panels of varying structures subjected to shock loading and drop weight impact loading has been studied by Micheal Hebart et al. [5]. Shock waves were created using a shock tube with a testing range of 3.08–7 MPa peak incident pressure and found that finer glass fibre structures perform better under shock

C. Jayarami Reddy (✉) · B. Venkataramudu · K. Gopinadha Reddy · R. K. Singh · A. A. Naukarkar
Defence Metallurgical Research Laboratory, Hyderabad 500058, India
e-mail: cjrreddy78@gmail.com

M. R. Manohara Babu et al. (eds.), *Recent Advances in Composites Engineering and Research*, Springer Proceedings in Materials 39,
https://doi.org/10.1007/978-981-99-8807-5_19

wave loading. Le Blanc and Arun Shukla studied the response of E-Glass/vinyl ester curved composite panels subjected to underwater explosive loading using ultrahigh speed system and recorded the full field shape and displacement profiles in real time [6]. Pankow et al. have studied the shock response of 3D woven composites by using shock tube and measured the deformation of composites using digital image correlation [7]. Reddy et al. have studied the behaviour of E-glass and H-glass composites using explosive blast tests [8]. Despite the vast studies [9, 10] on the response of composite materials using shock tube, the understanding on the failure behaviour of composites when subjected to different shock wave pressures is very limited.

Therefore, in the present paper, an attempt was made to understand the failure behaviour of E-glass/epoxy composites, by subjecting them to various shock wave profiles generated in the shock tube. Different types of shock wave pressure profiles are realized by changing the diaphragm thickness. The effect of diaphragm thickness on incident and reflected pressures is also studied. The real time deformation of E-glass/epoxy composites subjected to different shock pressures was studied using 3D DIC. The real time displacements and velocities of the composites were determined and studied. Various types of deformation modes exhibited by the composite plates are also discussed.

2 Experimental

2.1 *Materials*

The plain weave woven rovings of E-glass supplied by M/s. Ovens Cornings, India were used as reinforcements for fabrication of composite laminates. The areal weight of E-glass woven rovings was 360 gsm. The Fineset®—1556 epoxy resin and Fineset®—1972 amine-based hardeners were supplied by M/s. Fine Finish Organic chemicals, Mumbai.

2.2 *Equipment and Diagnostics*

Hydraulic Press of 250 ton capacity which was manufactured and supplied by M/s. Orion Hydraulics, Belgaum, India was used for fabrication of composites. Universal testing machine from M/s. DAK Systems, India was used for evaluation of mechanical properties of composites. The Shock tube system designed and developed by DMRL, Hyderabad was used in the experiments. Pressure sensors (Model No. 113 B24 and 113B22) and the oscilloscope Model DPO5104B supplied by M/s. PCB-Piezotronics Ltd., USA and M/s. Tektronix, USA, respectively, were used to record the pressure profiles. Two Phantom v1612 high speed video monochrome cameras

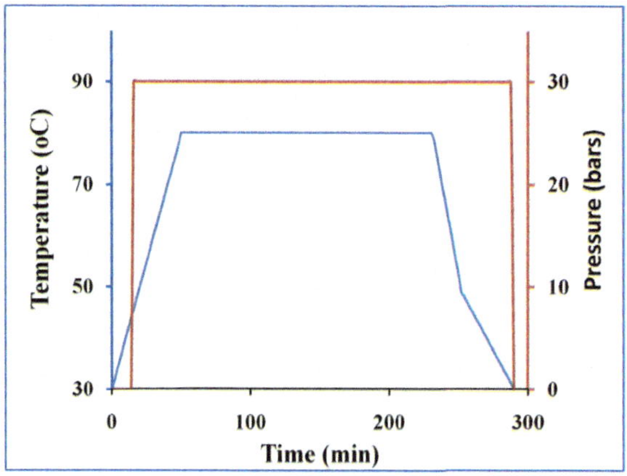

Fig. 1 Curing cycle of E-glass/epoxy composite

supplied by M/s. Vision Research, USA and 3D DIC developed by M/s. Correlated solutions USA were used to capture the real time deformations of the targets.

2.3 *Fabrication of Composite Panels*

The prepregs of E-glass plain woven rovings were prepared by applying the mixture of epoxy resin and amine hardener. The resin and hardener were taken in 100:60 weight ratio and was thoroughly mixed and degassed before applying on to the rovings. The prepregs were slightly air dried for 1 h at room temperature and cut into pieces of 300 × 300 mm. Suitable number of prepreg plies were stacked to build 3 mm thick composite laminates. The stack was kept between the Teflon coated glass release fabrics. The stack of plies along with the release fabrics were placed in the steel mould. The spacer rods were kept around the stack to arrest the flow of resin and to maintain the thickness of the composites. The steel mould was mounted in 250 ton hydraulic press and the temperature and pressure were applied as per the curing cycle shown in Fig. 1. After curing the mould was demounted from the hydraulic press and the cured panel was removed from the mould. The panel was trimmed to required dimensions using diamond wheel cutting machine.

2.4 *Physical and Mechanical Properties of Composites*

The physical properties such as resin content and the density of E-glass/epoxy composites were evaluated using burn off test method and density measurement kit, respectively. The composites were also evaluated for their mechanical properties such as tensile strength, flexural strength, and inter laminar shear strength. All the tests were performed as per the ASTM standards. The specimens were cut using

Table 1 Physical and mechanical properties of E-glass/epoxy composites

Sl. No.	Property	Standard/method	E-glass/epoxy composite
01	Density (g/cc)	Density measurement kit	1.7
02	Resin content (wt.%)	Burn-off test	21 ± 1
03	Hardness	Barcol hardness	50–55
04	Tensile strength (MPa)	ASTM D 3039	360
05	Flexural strength (MPa)	ASTM D 790	120
06	Interlaminar shear strength (ILSS) (MPa)	ASTM D 2344	8

diamond wheel cutting machine. Five test specimens were used for every measurement and the average of five tests was considered as the final result. The mechanical and physical properties of the composites are shown in Table 1.

2.5 Shock Tube Set Up

In the present study, the composite materials were tested against shock waves using shock tube consisting of 1 m length driver section and 3 m length driven section. The shock tube used for the experiments is shown in Fig. 2. The 50 mm gap was maintained between the muzzle end of the tube and composite target plate. The gap was used to avoid repeated loading of the shock wave on the target panel. The target plate of 240 × 240 mm was kept between FRP spacer as shown in Fig. 3a. The target plate surrounded by FRP spacer was sandwiched between two steel rigid flanges. The target plate along with flanges was bolted tightly to the end flange of the shock tube. The gap of 50 mm between the shock tube and the target was maintained by using aluminium hollow cylinders. The photographs of the spacer, gap by using aluminium hollow cylinders, target set up are shown in Fig. 3a–c respectively. 2, 3, and 4 mm thick aluminium diaphragms were used in the experiments. The 'X' score was made on the diaphragm to facilitate proper opening of the diaphragms. The depth of the groove was maintained as half of diaphragm thickness. The helium gas was used as driver gas for bursting of diaphragms. Air at atmosphere pressure was used as driven or test gas in all the experiments. The bursting pressure was monitored by static pressure gauge located at the driver section. Two pressure sensors (S1 and S2) were deployed on the end section of the driven section to measure the incident shock wave profiles and shock velocities. The photograph of the sensor location on tube is shown in Fig. 4a. Pressure sensors need to be deployed on the target plates in order to determine the loading pressure profiles on target plates corresponding to burst pressures. However, due to the complication of deploying pressure sensors on composite plates, pressure sensors were deployed on the thick steel plate assuming that the data provided by them would be the representative pressure profiles experienced by the target. The sensors were placed at centre (S3) and at the edge (S4) of the flange as shown in Fig. 4b.

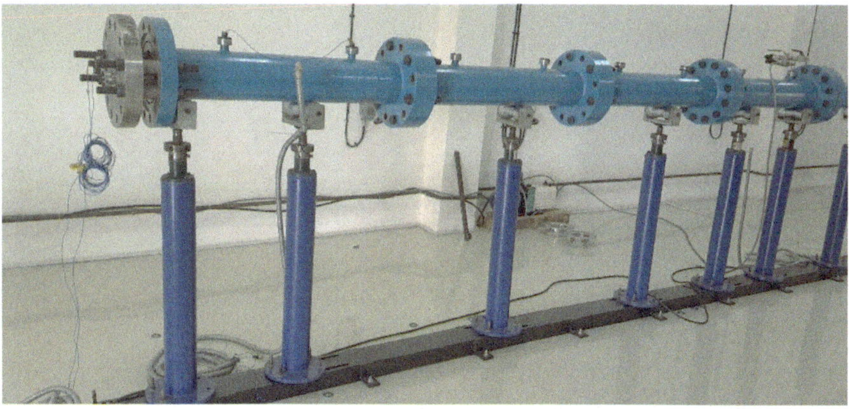

Fig. 2 Shock tube system at DMRL, Hyderabad

Fig. 3 Experimental test set up **a** spacer with target **b** gap using cylinders **c** target set up

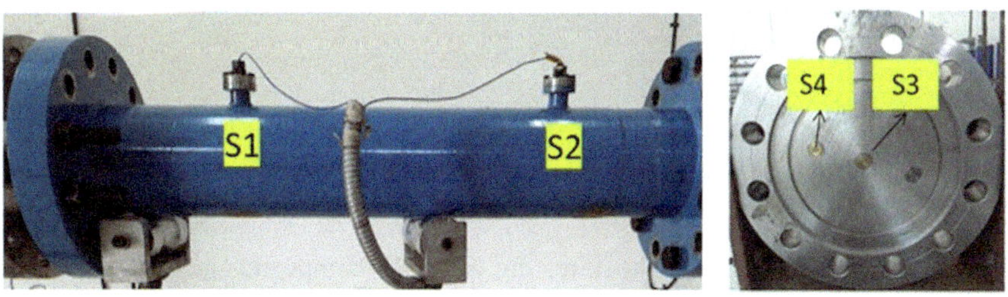

Fig. 4 Sensor location **a** on tube **b** on end steel plate

Table 2 Experimental details of shock tube testing

Sl. No.	Target	Diaphragm thickness (mm)	Groove depth (mm)	Driver gas pressure (P_4) (bar)	Driven gas pressure (P_1) (bar)
01	Sensor plate	2	1	14	1
02	Sensor plate	3	1.5	22	1
03	Sensor plate	4	2	37	1
04	E-glass/epoxy	2	1	14	1
05	E-glass/epoxy	3	1.5	22	1
06	E-glass/epoxy	4	2	37	1

2.6 Shock Tube Testing of GFRP Plates

The shock tube experiments were carried out on thick steel plates with sensors and the burst pressures (driver gas pressure P_4), incident and reflected pressure profiles were recorded. The details of experimental conditions including the diaphragm thickness, details of driver and driven gases are given in Table 2. The pressure profiles recorded by pressure sensors on steel plate can be treated as reference or calibration curves. For generating calibration curves, the experiments were carried out on sensor plate (thick steel plate) using 2, 3, and 4 mm thick Al diaphragms. Groove depth was maintained as half of diaphragm thickness.

The thick steel plate was replaced with the sample holding flange along with target samples and tests were carried out at the same experimental conditions which were used for generating calibration curves and hence the same pressure profiles can be treated as loading pressure profiles on the composite plates. 3 mm thick E-glass/epoxy composites were used as target plates. 2, 3, and 4 mm thick Al diaphragms (with groove depth as ½ of diaphragm thickness) were used to generate various shock pressures on the GFRP plates. Two tests were performed for similar set up conditions to ensure the repeatability of the results. The real time deformations of the target plates subjected to the shock pressures generated by bursting the above-mentioned diaphragms were recorded using 3D DIC interfaced with two high speed video cameras.

2.7 3D DIC Set up

The 3D DIC set up used in shock testing of GFRP composites is shown in Fig. 5. Target sample for 3D DIC analysis was prepared by applying speckle pattern on one side of the E-glass/epoxy composite. The composite sample was fixed in shock tube target holder and was illuminated using LED lights. The two high speed video cameras were fixed on the crowbar which was supported by tripod. Suitable lens were used in camera to capture the images. Cameras were interfaced with 3D DIC

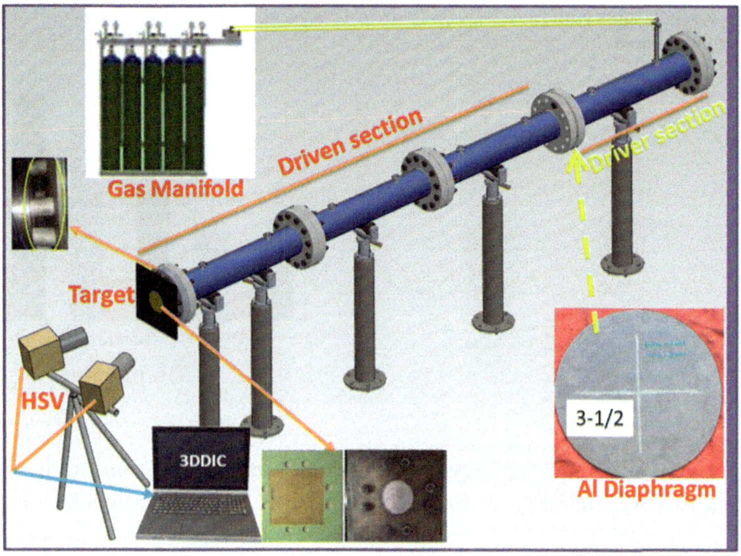

Fig. 5 3D DIC set up

software using suitable cables. The cameras were adjusted to stereo angle to capture the out of plane displacement. Cameras were switched on and the focus, brightness and depth of field of images were optimized by adjusting the distance between the cameras, stereo angle, light positions, distance between the sample and cameras. The camera settings were fixed. The calibration of the camera set up with respect to the target was done using suitable calibration grids at higher resolution and low frame rate. Then speckle images of the target during the shock tube event were captured using VIC-Snap software at higher frame rate of 40,000 fps by triggering the cameras just before the shock wave reaches the target plate. The acquired images were processed through VIC-3D to get various deformation parameters such as out of plane displacement and velocity of the plate at different time intervals.

3 Results and Discussion

3.1 Pressure Measurements

The shock tube experiments were carried out as per the experimental conditions mentioned in Table 2. The burst pressures (P_4) of aluminium diaphragms when bursted by using Helium gas were recorded using static pressure gauges and the corresponding incident and reflected pressures were recorded using piezoelectric pressure sensors. The burst pressures of each diaphragm and their corresponding

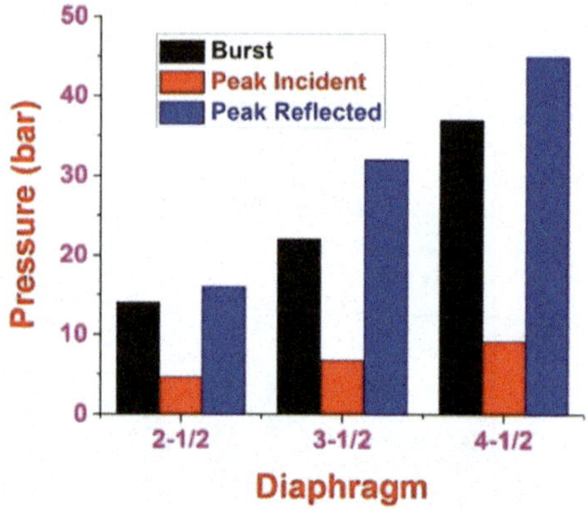

Fig. 6 Peak pressures with diaphragm thickness

peak incident and reflected pressures are shown in Fig. 6. From the figure, it is observed that with increase in diaphragm thickness, burst, peak incident, and peak reflected pressures are also increasing. This is due to the dependence of shock strength on burst or driver gas pressure (P_4) required to burst the diaphragm.

The incident pressure profiles recorded by the pressure sensor S1 when 2, 3, and 4 mm diaphragms are bursted using helium gas are shown Fig. 7. A sharp rise in the pressure observed in all the pressure profiles can be attributed to the arrival of shock front at the sensor location. The shock pressure after reaching its maximum, starts decaying for a small duration, due to the passage of shock wave from S1 towards the end plate. Once the wave reaches the end plate, it got reflected and travels back towards S1 in shock tube as well as dissipates outside through 50 mm gap. Due to the arrival of reflected shock waves at S1, a small secondary peak at around 4 ms is observed in all the pressure profiles. The final decay of pressure observed is due to the meeting of low-pressure expansion wave with the reflected shock front. This can also be mainly attributed to the diffusion of wave through 50 mm gap. It is also observed that the incident pressure is increasing with increase in thickness of the diaphragm from 2–1/2 to 4–1/2. The velocity of the shock wave was calculated by using the shock arrival time at S1 and S2 sensor locations and the distance between the S1 and S2 locations. The velocities of incident shock waves generated by bursting the 2, 3, and 4 mm diaphragms are found to be 764, 895, and 983 m/s, respectively.

The shock wave at sensor location S1 move forward and traverse through the 50 mm gap to reach the end plate and subsequently got reflected. The reflected pressure profiles recorded by S3, when 2, 3, and 4 mm thick diaphragms are bursted using helium driver gas are shown in Fig. 8. The trend observed in the incident pressure is also observed in the case of reflected pressure profiles where the reflected pressures are also increased with increase in diaphragm thickness.

In Fig. 8, the pattern of the pressure profiles recorded by S3, are similar for all the three diaphragms. However, the difference lies in the magnitude of the pressures and time scales. The sharp rise in the pressure profiles observed, is due to the reflection of the incoming shock wave from the end plate. The pressure dropped slowly due to the

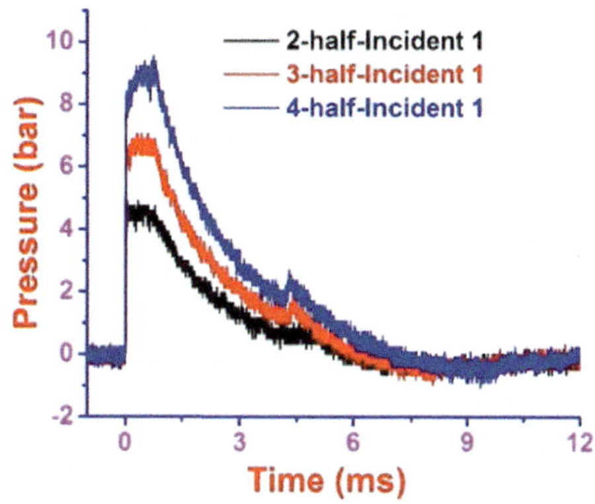

Fig. 7 Incident pressure profiles

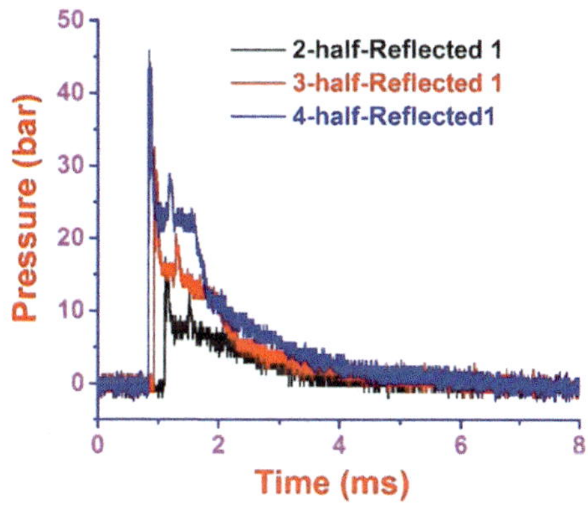

Fig. 8 Reflected pressure profiles

meeting of expansion fan with the reflected shock wave and the dissipation of the gas through 50 mm gap. However, there is also a small portion of pressure stagnation in each pressure profile, which may be due complex interaction mechanisms of contact surface and reflected shock wave and even expansion fan. The maximum reflected shock pressures experienced by plate, when 2, 3, and 4 mm diaphragms are bursted using helium gas are 16, 32, and 45 bars, respectively. As the composites were also subjected to same experimental conditions that were used on steel plates, it is assumed that they also would have experienced the same incident and reflected pressures.

3.2 *Deformation of E-glass/Epoxy Composites*

Post-test Visual Observations. E-glass/epoxy composites were tested as per the experimental conditions mentioned in Table 2. The samples were collected after the tests and were visually observed for various deformation mechanisms. The photographs of the front and back sides of E-glass/epoxy composites which were subjected to the reflected shock pressures produced by bursting the 2, 3, and 4 mm diaphragm using helium driver gases are shown in Figs. 9, 10, and 11, respectively. The maximum permanent midpoint displacements showed by E-glass/epoxy composites are found to be around 1, 26, and 33 mm when subjected to the 16, 32, and 45 bars shock pressures, respectively. As shown in the figures when E-glass composites are exposed to the reflected shock pressures, the samples got deflected at the exposed area. However, due to the boundary constraints and gradients in pressures at the centre and edge, they showed maximum deflection at the centre. Though permanent deflections of composites were measured after tests, 3D DIC was also deployed to determine the deformation of composites at different time intervals.

Deformation Measurements Using 3D DIC. The real time deformation of the E-glass/epoxy composites subjected to different shock pressures was captured using high speed video cameras and 3D DIC. The images captured through high speed video cameras were processed through VIC-3D software and determined various parameters such as displacements in x, y, and z directions. The out of plane displacements of E-glass/epoxy composites, subjected to different shock pressures were

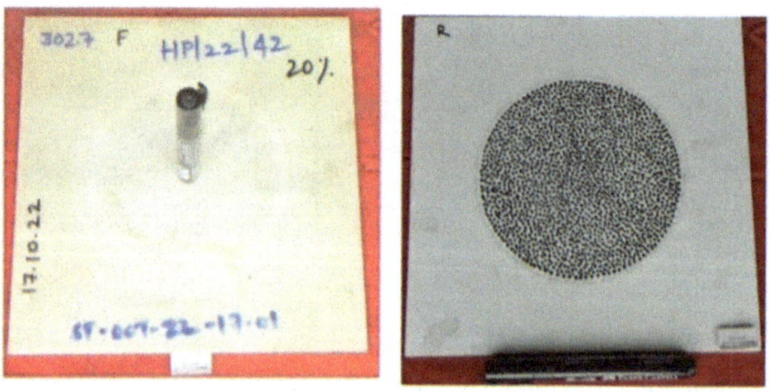

Fig. 9 Deformation patterns of E-glass/epoxy composites subjected to 16 bars shock pressure

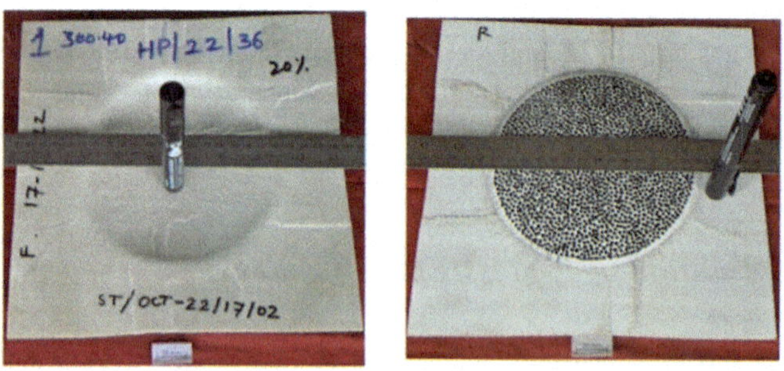

Fig. 10 Deformation patterns of E-glass/epoxy composites subjected to 32 bars shock pressure

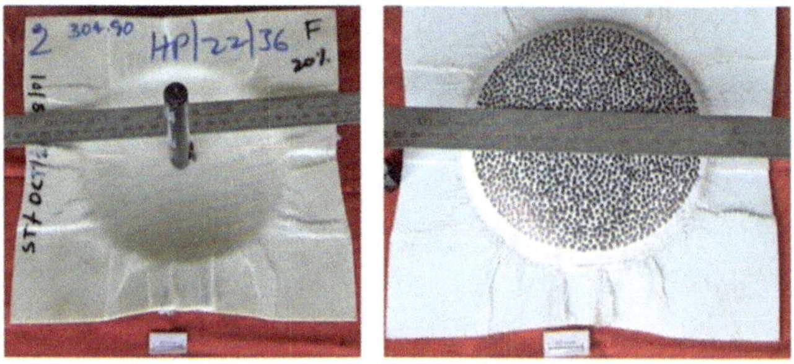

Fig. 11 Deformation patterns of E-glass/epoxy composites subjected to 45 bars shock pressure

recorded at various time intervals and are shown in Fig. 12. The out of plane velocities of the composites at different shock pressures were also determined and shown in Fig. 13. Though 3D DIC images of the deformation of composites, subjected to different shock pressures are captured at different time intervals, only the images corresponding to peak (transient) displacement and permanent displacement are shown in Figs. 14 and 15.

From Fig. 12 it is observed that the sample reached its maximum displacement in loading direction immediately after the shock impact in all the cases of shock pressures. However, samples exhibited different recovery patterns from peak displacement point. The sample subjected to lower pressure of 16 bars recovered completely to less than 1 mm deflection from its peak deflection of 12 mm, which shows that the shock loading is in the elastic range of the composite. The maximum displacement of the composites in the form of smooth bulge or dome and complete recovery can also be observed in 3D DIC image shown in Fig. 14. The sample took 200 μs to reach the peak displacement immediately after the impact. However, in other two cases where the shock pressures are higher such as 32 and 45 bars, the recovery of the samples from peak deflection point is very less. The sample subjected to 32 bars of shock pressure showed peak displacement of 30 mm and permanent displacement of 26 mm

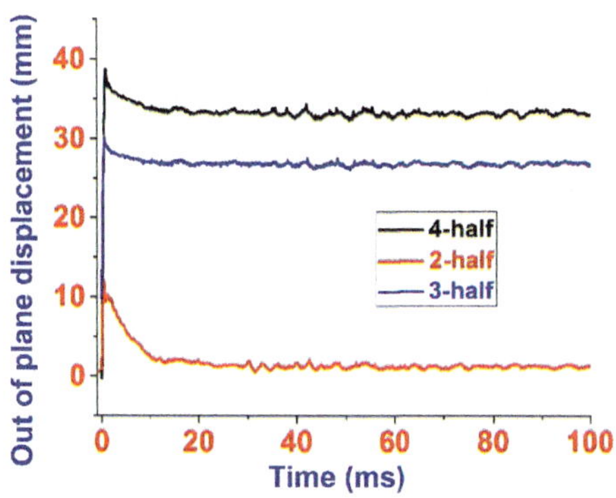

Fig. 12 Real time out of plane displacements of E-glass/epoxy at different shock pressures

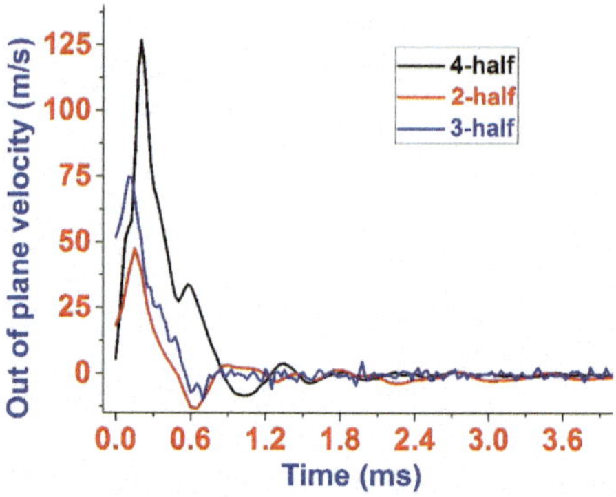

Fig. 13 Real time out of plane velocity of E-glass/epoxy at different shock pressures

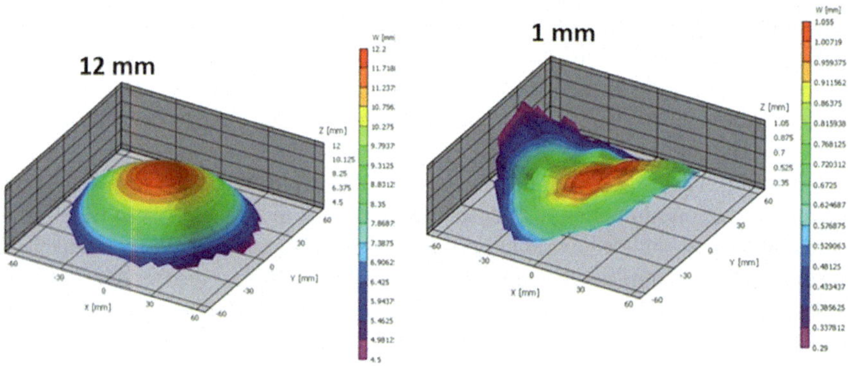

Fig. 14 3D DIC images of maximum transient displacement and permanent displacement of E-glass/epoxy subjected to 16 bars of shock pressures

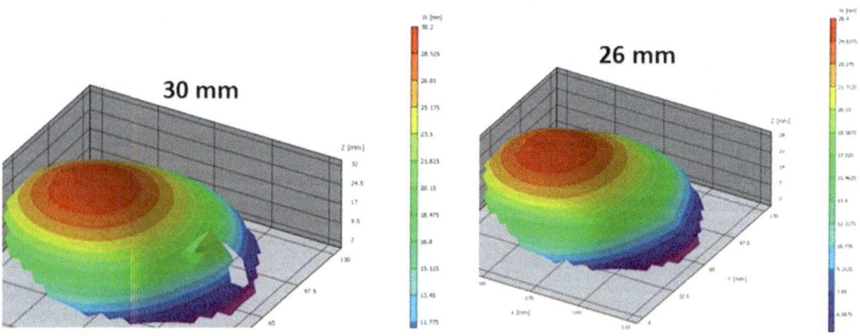

Fig. 15 3D DIC images of maximum transient displacement and permanent displacement of E-glass/epoxy subjected to 32 bars of shock pressures

indicating very small recovery. The other composite sample which is subjected to 45 bars showed the peak displacement of 39 mm and permanent displacement of 33 mm, which also indicating very small elastic recovery of the sample. So, the results indicate that at higher shock pressures, the composites exhibit more plastic deformation compared to the lower pressure. The permanent out of plane displacement determined using 3D DIC are also closely matching with the permanent displacement values measured after the tests. The maximum displacement in the form of large inelastic deformation or bulge of the composite subjected to 32 bars and 45 bars can be visualized from Figs. 15 and 16, respectively. The maximum displacements of both the samples subjected to 32 and 45 bars, respectively, were observed at 450 and 825 μs, respectively. The increase in the time for composite to reach maximum displacement is due to the large displacement. From Fig. 13, the velocity of the composite panel was found to be increased with increase in the loading shock pressure. The velocities of the composite panels subjected to 16, 32 and 45 bars of reflected pressures were found to be 45, 75, and 125 m/s respectively. A small negative velocity observed in all the graphs is due to the vibration of sample.

Damage Patterns of E-glass/Epoxy Composites. From the Figs. 9, 10, and 11 the E-glass/epoxy composites exhibited various types of damage patterns when subjected to shock pressures. The damage modes include large inelastic deformation (mode I) and partial tearing at the boundary or at clamped edge which indicates the inception of mode II failure. The pulling of plate from the edges to the centre is also observed. The matrix cracking, fibre breakage, and minor delamination at the clamped edges are observed in composites. Stretching of the composite around the shock exposed area is also observed. Matrix cracking is due to the brittle nature of the epoxy resin and poor load transfer between the matrix and reinforcement at the shock pressures. The large inelastic deformation is attributed to the low flexural strength of the composite. Pulling of the composite from the edges and slight delamination observed in the composite are due to the weak interface properties of the composite. The less fibre

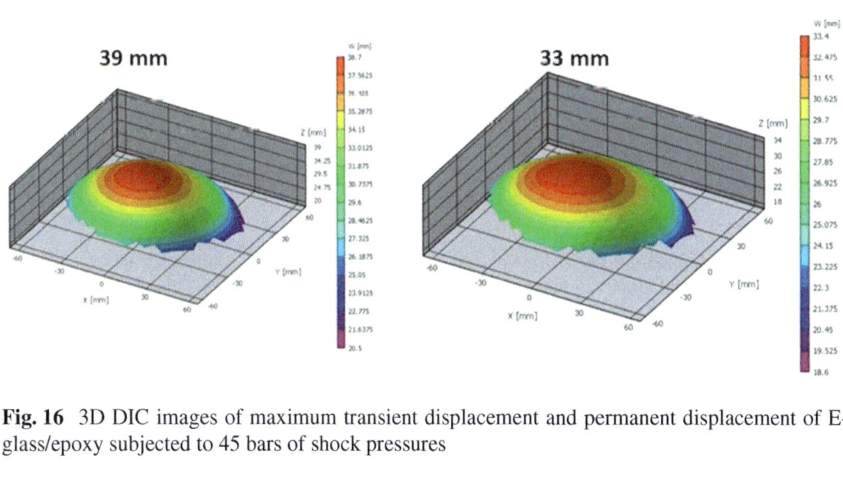

Fig. 16 3D DIC images of maximum transient displacement and permanent displacement of E-glass/epoxy subjected to 45 bars of shock pressures

damage is the result of high tensile strength of the composites. The composites also absorb the shock impact energy through these various types of damage modes.

As shown in Fig. 9, at low loading pressures of 16 bars, no matrix cracking, fibre breakage, and inelastic deformations are observed visually. However, when observed in 3D DIC, large elastic deformation was observed. The permanent displacement was found to be less than 1 mm. As observed in Fig. 10 with the increase in shock pressures to 32 bars, composite showed only minor elastic deformation and large inelastic deformation (Mode I failure). The matrix cracking and the initiation of fibre breakage, tensile tearing (Mode II) at the clamped edges and pulling of the fibre near the clamping edges are also observed. The increase in the magnitude of the damage modes is observed in Fig. 11, when the shock pressures increased to 45 bars. The pulling of plate from the edges to the centre has become prominent at higher pressures indicating shear failure of the composite materials at higher pressures.

4 Conclusions

Response of the E-glass/epoxy composites was studied using compressive gas driven shock tube and 3D DIC. Various shock pressures were generated by changing the diaphragm thickness and helium driver gas. The incident and reflected pressures were recorded using shock pressure sensors. The incident, reflected shock pressures and shock wave velocities were found to increase with burst pressures. The real time deformation of the composites at different time intervals was captured using 3D DIC. The composites showed maximum out of plane deflection of 39 mm and 125 m/s velocity when subjected to 45 bars of shock pressures. The effect of various shock pressure profiles on the failure behaviour of E-glass/epoxy composite was also studied. E-glass composites showed only elastic deformation when subjected to 16 bars of shock pressures. However, composites showed large inelastic deformation (mode I failure) and tensile tearing (mode II failure) at higher shock pressures. The other damage modes at higher shock pressures observed were matrix cracking, fibre breakage, and shear pulling of the composite.

References

1. Holmberg AD (2010) Development and characterization of shock tubes for laboratory scale blast wave simulation. Engineering Mechanics Dissertations & Theses, Paper 26
2. Bruce L, David H (2010) Blast/impact on engineered (nano) composite materials. Comp Part B Eng 40:413–4441
3. Srinivasan AT, Kunigal S, Arun S (2008) Mechanical behavior and damage evolution in E-glass vinyl ester and carbon composites subjected to static and blast loads. Comp Part B Eng 39:57–65
4. Puneet K, David SS, Arun S (2013) Effect of plate curvature on blast response of carbon composite panels. Comp Struct 99:19–30

5. Michael H, Carl-Ernst R, Arun S (2008) Shock loading and drop weight impact response of glass reinforced polymer composites. Comp Struct 84:199–208
6. James LB, Arun S (2007) Shock loading of 3D Shock loading of three-dimensional woven composite materials. Comp Struct 79:344–355
7. Pankow M, Justusson B, Salvi A, Waas AM, Chian-Fong Y, Seth G (2011) Shock response of 3D woven composites: an experimental investigation. Comp Struct 93:1337–1346
8. Reddy CJ, Venkataramudu B (2019) Blast response of Hollow glass fibre reinforced epoxy matrix composites. Procedia Struct Integr 14:634–641
9. Fengyuan Y, Zhijie L, Zhanli L (2021) Shock loading mitigation performance and mechanism of the PE/Wood/PU/foam structures. Int J Impact Eng Procedia Struct Integr 155:103904
10. Mouritz AP (2019) Advances in understanding the response of fibre-based polymer composites to shock waves and explosive blasts. Comp Part A 125:105502

Effect of Fibre Volume Fraction and Ply Angle on Performance of Glass Composite Laminates Under Low Velocity Impact

T. Sreekantha Reddy, K. Mogulanna, P. Prasanna, P. Rama Subba Reddy, and B. Ramakrishna

1 Introduction

Armour is a material or system which will provide protection to either personnel or combat system against ballistic threats based on kinetic or chemical energy. Traditionally, steel is used as armour to protect against these threats and the standard nomenclature for steel armour is Rolled Homogenous Armour (RHA). In recent years, polymer composites and ceramic materials have gained importance as a light weight armour materials due to their high specific strength and their effectiveness in absorbing the impact energy. To some extent these two materials have been used to replace RHA wherever it is possible. Globally, attempts are going on for making completely composite armoured vehicle.

Keeping in view of above mentioned advantages of composites, various researchers have started working on development of composites for armour applications. It is advantageous to carry out the low velocity impact for screening of materials during development stage because it not only save the material cost but also reduce the developmental cycle. Hence, quite a few studies are reported the low velocity impact behaviour of glass fibre-based composite laminates [1, 2]. For instance, Belingardi et al. examined the effect of fabric type on low velocity (0.7–2.42 m/s) impact response of glass/epoxy composite laminates. Their prime focus was to find out the saturation impact energy and degree of damage. They concluded that mechanical characteristics of glass/epoxy composite laminates were independent of strain rate in the selected impact velocity range [3]. The effect of impact force and impact

T. Sreekantha Reddy (✉) · K. Mogulanna · P. Rama Subba Reddy · B. Ramakrishna
Armour Technology Centre, DMRL, Hyderabad 500058, India
e-mail: sreeku2020@gmail.com

K. Mogulanna · P. Prasanna
Jawaharlal Nehru Technological University, Hyderabad 500085, India

M. R. Manohara Babu et al. (eds.), *Recent Advances in Composites Engineering and Research*, Springer Proceedings in Materials 39,
https://doi.org/10.1007/978-981-99-8807-5_20

energy (2–28 J) on energy absorption capability of E-glass/epoxy and carbon/epoxy composites was examined by Kersys et al. [4]. They found that elastic deformation for E-glass/epoxy laminate was one and half times higher than that of carbon/epoxy laminate. Yang and Cantwell also carried out similar study on glass/epoxy composites to examine the effect of different experimental variables like impactor size and shape, test specimen dimensions and testing temperature on damage initiation energy [5]. They observed that damage initiation energy is proportional to impact force with a factor of $t^{3/2}$, where t is the thickness of test specimen. Amaro et al. compared the response of glass/epoxy composite laminates when subjected to multiple impacts of lower magnitude and single impact energy of higher magnitude. They reported that sole impact energy induced more damage than that of the cumulative damage from multi impacts of lower magnitude [6]. In an another study, effects of the stacking sequence (namely (0/90)2 s and (±45/90/0) s) on the behaviour of composite laminates was carried out at different impact energies by Riccio et al. through numerical simulations [7].

Reddy et al. [8] studied the influence of specimen thickness and test temperature on E-glass/epoxy composite laminates under low velocity impact and assessed their performance in terms of peak force, energy absorption and failure behaviour. They concluded that maximum force and minimum energy required for perforation have increased linearly with increase in laminate thickness. In one more study, Kallagunta and Tate [9] conducted a comparison study of low velocity impact behaviour of E-glass epoxy composites that are modified with 10 wt% nanosilica and 2.5 wt% alumina nanofibres. These composites were manufactured by using vacuum-assisted resin transfer moulding. The effect of laminate thickness of Z-pinned laminates on low velocity impact behaviour was studied by Liao et al. [10]. The authors showed that Z-pins could not influence the delamination initiation but those Z-pins helped for reduction in degree of delamination. Impact damage suppression effectiveness depends on failure mode of composites and Z-pin damage. Durdu et al. [11] studied the effect of delaminations on the low velocity impact and mechanical behaviour of the composite materials. From the test results, they found that the maximum contact force of the specimens increases with the increment of the delamination diameter.

Substantial research has been carried out on the behaviour of glass composites under low velocity impact. However, very limited information is available on low velocity impact behaviour of glass composites with varying fibre volume fraction and architecture. Low velocity impact test is useful to study the effect of debris, tool drops, bird impact, hail stones, etc., on performance and failure behaviour of composite laminates. Hence, the objective of the present study is to understand the effect of fibre volume fraction, laminate ply angle and impact energy on E-glass/epoxy composite laminate's performance under low velocity impact.

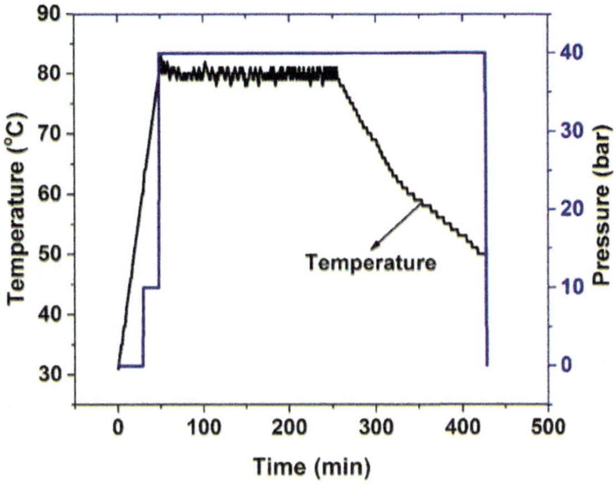

Fig. 1 Cure cycle of E-glass/epoxy composite laminate

2 Experimental

2.1 *Preparation of Laminates*

E-glass/epoxy composite laminates of sizes 350 × 350 mm were made through hand lay-up technique followed by compression moulding using laminating hydraulic press. The curing of E-glass/epoxy laminates was carried out at 80 °C for 3 h under 40 bar pressure as per the curing cycle shown in Fig. 1. Three composites with varying fibre volume fractions, viz. 0.6, 0.53 and 0.45 (labelled as C1, C2 and C3, respectively) and three composites with different ply angles, viz. 15°, 30° and 45° (named as HC_15, HC_30 and HC_45, respectively) were fabricated.

2.2 *Mechanical and Low Velocity Impact Tests*

Mechanical properties like tensile, flexural and interlaminar shear stress (ILSS) tests were carried out using Universal Testing Machine (Make: Dak Systems Inc, Model: 7200). Instrumented drop weight impact tester shown in Fig. 2 (CEAST- 9350) was used to carry out low velocity impact tests according to ASTM D7136 [12]. Low velocity impact tests were performed using instrumented drop weight impact tester which is equipped with 12.7 mm dia hemispherical impactor. The specimens were cut into the dimension of 100 × 150 mm using diamond wheel cutting machine. Resistance offered by the laminate during the impact test was recorded by using force transducer with capacity of 45 kN. Data acquisition system was used to record the force–time, force–displacement history. The sampling rate was kept at 500 kHz. The required impact energy was achieved by adjusting the drop height of the impactor. Parameters like absorbed energy, displacement and impactor velocity were calculated

Fig. 2 Instrumented drop weight impact tester and impactor

from the force–time data with the help of application software integrated with the equipment.

Impact experiments were carried out at three different impact energies, viz. 100, 150 and 200 J for all the six categories of laminates to cover the range of perforation and non-perforation. The corresponding impact velocities were 6.08, 7.51 and 8.68 m/s, respectively. Minimum of three samples was tested at each energy level for each laminate.

3 Results and Discussion

This section is divided in to two parts. The effect of volume fraction and the effect of ply angle were discussed in detail in the first and second part, respectively.

3.1 *Effect of Fibre Volume Fraction*

Physical Properties

Physical properties like fibre volume fraction, resin content, density, hardness of prepared laminates were evaluated as per the ASTM standards, and data is given in Table 1.

Mechanical Properties

Mechanical properties of cross-ply composites are given in Table 2. Tensile strength of C2 and C3 composites has reduced by 25 and 40% respectively as compared to C1 composite. The flexural strength (3-point as well as 4-point bending) also reduced with a decrease in fibre volume fraction. Moreover, 3-point strength of all the laminates was found to be higher than 4-point bend strength. However, no significant change in ILSS was observed.

Table 1 Physical properties of cross-ply composites

Property	Composite 1	Composite 2	Composite 3
Code of the composite	C1	C2	C3
No. of plys	21	20	18
Resin content (%)	22	28	35
Volume fraction (V_f)	0.60	0.53	0.45
Thickness (mm)	5.35	5.45	5.5
Density (g/cc)	1.95	1.84	1.75
Hardness (Barcol)	30–35	35–40	30–35
Areal density-AD (Kg/m^2)	10.4	10.0	9.7

Table 2 Mechanical properties of cross-ply composites

Property	Fibre volume fraction (V_f)		
	0.6	0.53	0.45
Tensile strength (MPa)	283	212	168
Young's modulus (GPa)	18.4	16.6	15.4
Elongation at break (%)	1.95	1.87	1.75
Flexural test_3-point bending			
Strength (MPa)	173	152	133
Modulus ((GPa)	22.3	21.5	18.3
Flexural test_4-point bending			
Strength (MPa)	135	127	108
Modulus ((GPa)	28.2	26.5	25.4
ILSS (MPa)	11.5	11.7	11.6

Low Velocity Impact Properties

Force–displacement curves for the composites C1, C2 and C3 are shown in Fig. 3. Closed loop and displacement in the negative direction indicates partial penetration and rebound of impactor and open loop indicates complete penetration of impactor into the test specimen. Oscillations or kinks in the curves represent the progressive damage of the laminates either by matrix crack or fibre breakage. From the figure it is clear that, impactor has rebounded at all the three impact energies of C1. The displacement is found to be increased with an increase in impact energy while the max. force or peak force is remain constant. In case of C2 laminate, partial penetration at 100 and 150 J and complete perforation at 200 J impact energy was observed, while C3 laminate perforated even at lower impact energy, i.e. 150 J. This suggests that penetration resistance of glass composites decreases with a decrease in fibre volume fraction. The slope of the F-D curve indicates the modulus of the laminates and found to be similar for all the three laminates.

Energy-time histories of C1–C3 laminates are shown in Fig. 4a–c. C1 composite laminate has absorbed total impact energy. Part of the impact energy utilized for

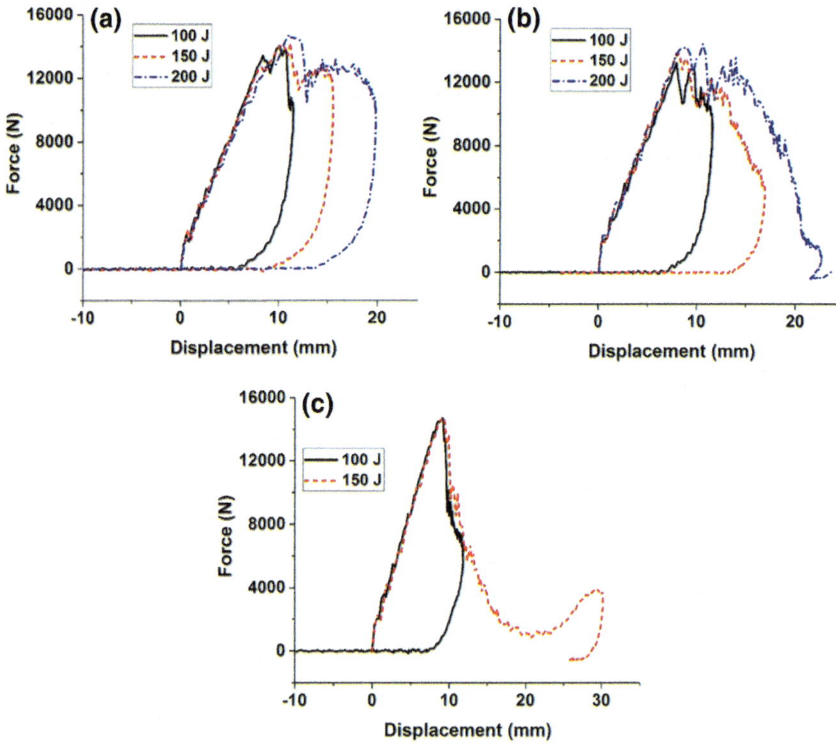

Fig. 3 Force–displacement curves for **a** composite 1, **b** composite 2 and **c** composite 3

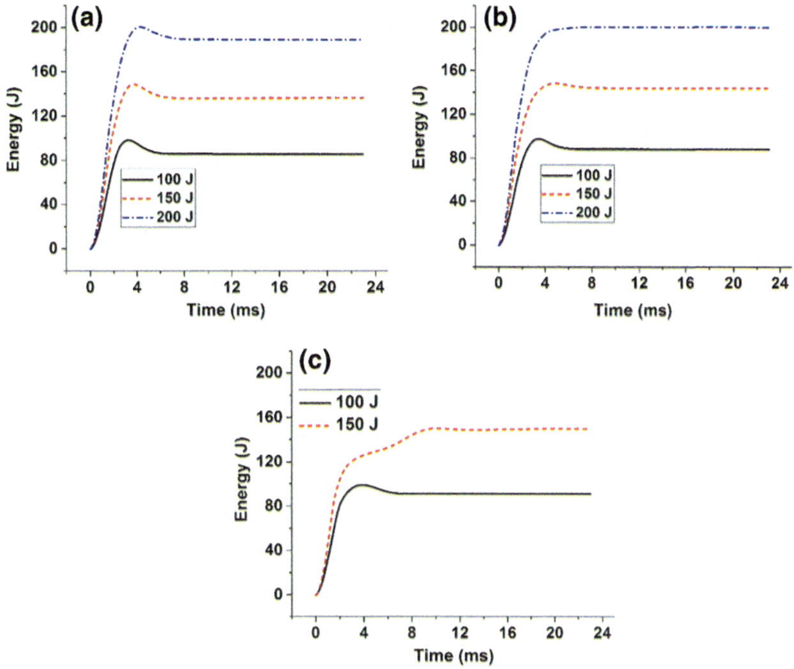

Fig. 4 Energy-time histories of **a** composite 1, **b** composite 2 and **c** composite 3

elastic deformation (>90%) and remaining is used for rebound of the impactor. In case of C2 laminate, as explained above, rebound of impactor took place at 100 and 150 J by undergoing matrix crack, fibre breakage and elastic deformation. C3 laminate perforated at 150 J by absorbing only 125 J and remaining energy is dissipated for plastic deformation and further down ward travelling of impactor after perforation. Impact at 200 J was not performed for C3 laminate.

Figure 5 shows comparison of force–displacement and energy-time curves of C1, C2 and C3 laminates impacted at 150 J. This clearly shows that C1, C2 laminates have resisted the penetration of impactor and resulted in rebound and C3 laminate is completely perforated. Energy-time curves comparison shows that C1 laminate has better penetration resistance with higher rebound energy and C2 laminate is at threshold of penetration and total energy is utilized for elastic and plastic deformation, while C3 laminate has shown least penetration resistance and absorbed only 125 J, i.e. only 83% of impact energy.

Comparison of max. force and max. displacement of three composites at all impact energies is shown in Figs. 6 and 7, respectively. Max. force is found to be within the range of 13–15 kN. It increased marginally with an increase in impact energy for individual composites. For instance, max. force of C1 laminate at 100 J is 14 kN, and it increased by 2 and 9% at 150 and 200 J, respectively.

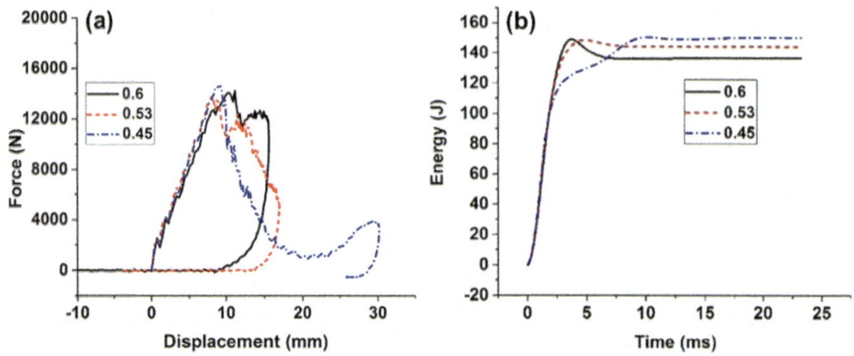

Fig. 5 Comparison of three composites at 150 J impact energy

Fig. 6 Comparison of max. force of three composites at all impact energies

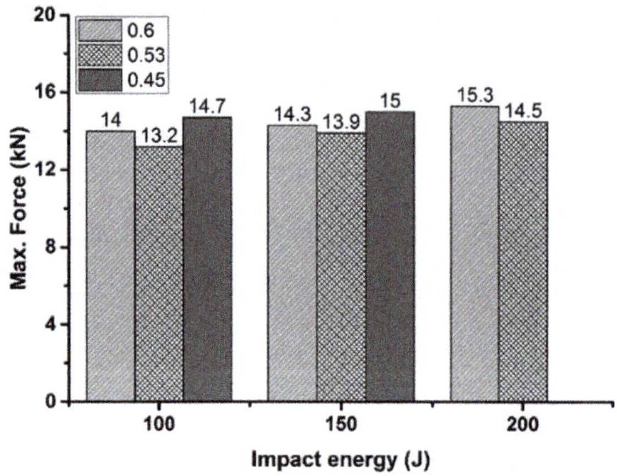

Fig. 7 Comparison of max. displacement of three composites at all impact energies

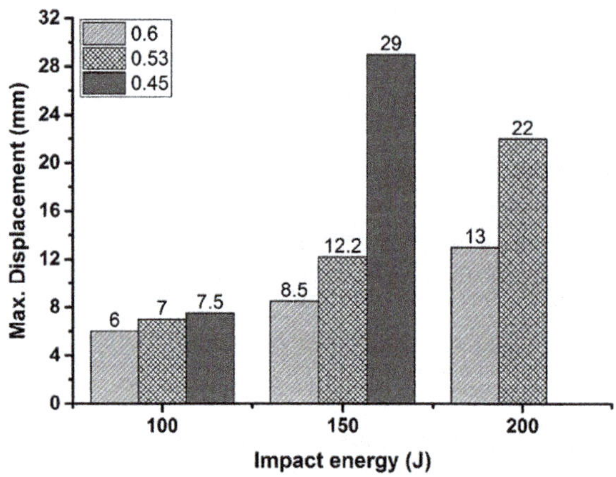

Max. displacement has increased with a decrease in fibre volume fraction. At lower impact energy (100 J), the variation is found to be linear and at higher impact energy it is nonlinear. At 100 J impact energy, the increase in max. displacement as compared to C1 laminate is 17 and 25% for C2 and C3 laminates, respectively, whereas at 150 J, the increase is 43 and 240%, respectively. Max. displacement has also increased with an increase in impact energy. For example, max. displacement of C2 laminate is 7 mm at 100 J and it rose to 12.2 mm (1.7 times) and 22 mm (3.1 times) at 150 and 200 J, respectively. This is due to reduction of fibre volume fraction which is prime load bearing ingredient of composite and thereby reduction in penetration resistance of the composite.

Impacted laminates were visually observed to find out the various types of damage modes. Photograph of front and rear side of laminates impacted at 150 J energy is presented in Fig. 8. Indentation mark can be seen at the point of impact for all the laminates. In case of C1 and C2 laminates, matrix cracking and delamination are the major damage modes. In addition to these two damage modes, fibre breakage was also observed in C3 composites.

3.2 *Effect of Ply Angle*

Physical Properties

Physical properties of angle ply laminates are given in Table 3. As mentioned earlier, these laminates were prepared with same number of woven fabric layers as that of C1 laminate and maintained the same fibre volume fraction. As expected all these laminates have same thickness and density. But the hardness was found to be slightly higher.

Mechanical Properties

Mechanical properties of angle ply composites are given in Table 4. From the table it is clear that tensile strength of angle ply laminates is lower than that of control (C1) laminate and has further reduced with an increase of ply angle. This can be attributed to reduction in fibre content of composite in loading direction (pulling direction) as we increase the ply angle. However, 3-point bend strength and 4-point bend strengths are found to be similar for all angle ply laminates.

Low Velocity Impact Properties

Force–displacement and energy-time curves of HC_15, HC_30, HC_45 along with C1 laminate are shown in Figs. 9 and 10. As explained earlier, C1 laminate has not perforated up to 200 J energy, whereas angle ply laminates have shown deviation from the C1 laminate. HC_15 where fabric orientation is changed for every 15o, resisted upto150 J and perforated at 200 J. Unlike C1 laminate where the fibre is in 0 and 90o orientation, HC_15 have wide distribution of fabric in all directions and will act as quasi isotropic material. This will result in an increase in bending stiffness

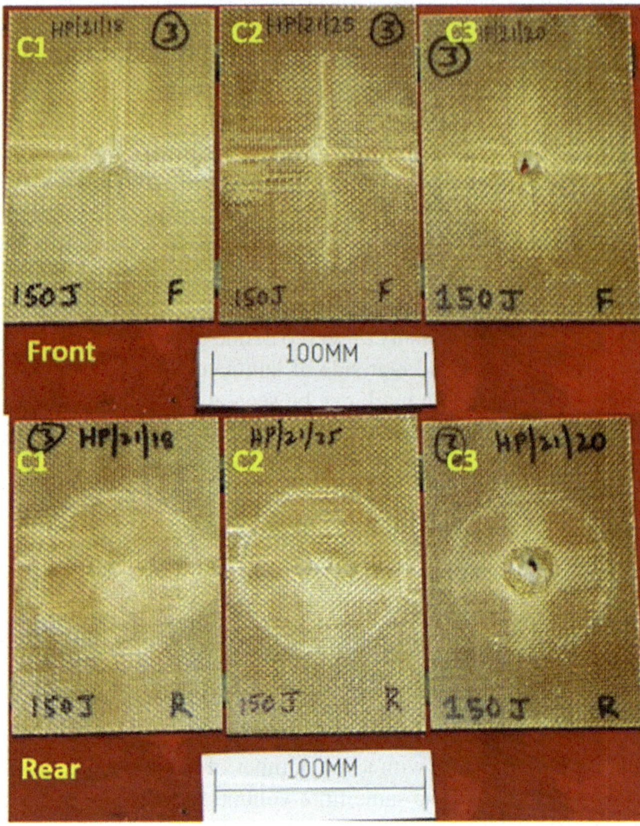

Fig. 8 Photograph showing front and rear side of impacted laminates (C1, C2 andC3)

Table 3 Physical properties of angle ply laminates

Property	Composite 1	Composite 4	Composite 5	Composite 6
Code of the composite	C1 (control)	HC_15	HC_30	HC_45
No. of fabric layers	21	21	21	21
Resin content (%)	22	22	22	22
V_f	0.60	0.60	0.60	0.60
Thickness (mm)	5.35	5.34	5.32	5.34
Density (g/cc)	1.95	1.96	1.95	1.96
Hardness	30–35	35–40	35–40	35–40

Table 4 Mechanical properties of angle ply composites

Property	Control (C1)	HC_15	HC_30	HC_45
Tensile strength (MPa)	283	214	183	156
Young's modulus (GPa)	18.4	17.6	16.5	15.3
Elongation at break (%)	1.95	1.83	1.75	1.7
Flexural test_3-point bending				
Strength (MPa)	173	130	128	127
Modulus (GPa)	22.3	15.4	15.8	14.5
Flexural test_4-point bending				
Strength (MPa)	135	117	123	110
Modulus (GPa)	28.2	19.5	20	19.5
ILSS (MPa)	11.5	9.6	9.7	10.4

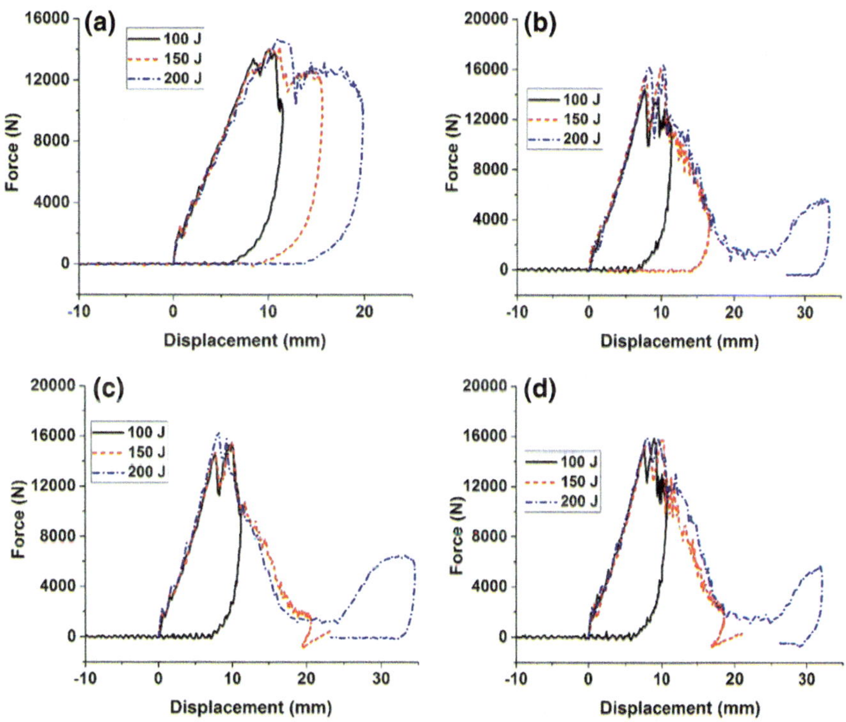

Fig. 9 Force–displacement curves for **a** C1, **b** HC_15, **c** HC_30 and **d** HC_45

of the material and thereby increase in brittleness of overall laminate. Hence, the penetration resistance of the HC_15 laminate decreased. The absorbed energy at 200 J impact is found to be only 160 J.

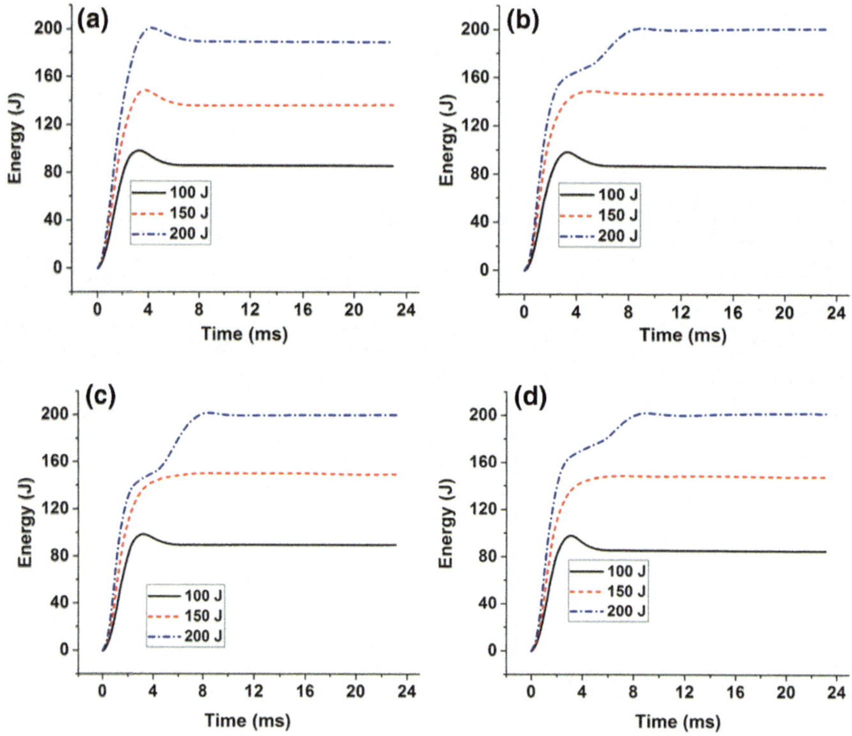

Fig. 10 Energy-time histories of **a** control, **b** HC_15, **c** HC_30 and **d** HC_45

Other two angle ply laminates, i.e. HC_30 and HC_45, behaved similarly. At 150 J impact, they are at the verge of perforation and completely perforated at 200 J impact by absorbing only 140 J and 170 J energy, respectively.

Figure 11 shows comparison of force–displacement and energy-time curves of C1 and angle ply composites at 150 J impact energy. Max. displacement is found to be 12, 16, 18 and 22 mm for C1, HC_15, HC_30 and HC_45 composites, respectively. From this graph, it is clear that slope which indicates the stiffness of laminate has increased for angle ply laminates as compared to C1 laminate. Max. force also found to be marginally higher for angle ply laminates.

Photograph of front and rear side of C1, HC_45, HC_30 and HC_15 composite laminates impacted at 150 J energy is presented in Fig. 12. Damage modes like matrix cracking, delamination and fibre breakage were observed in angle ply laminates also. However, it is worth to note that unlike cross-ply laminates, angle ply laminates have shown matrix cracking in multiple directions. This is due to alignment of fibres in different angles.

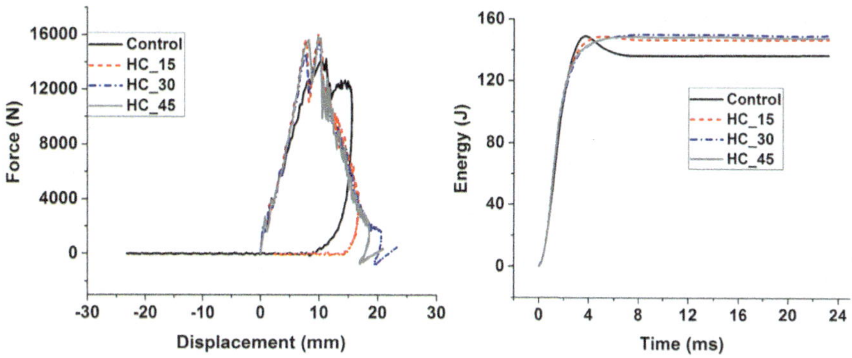

Fig. 11 Comparison of C1 and angle ply composites at 150 J impact energy

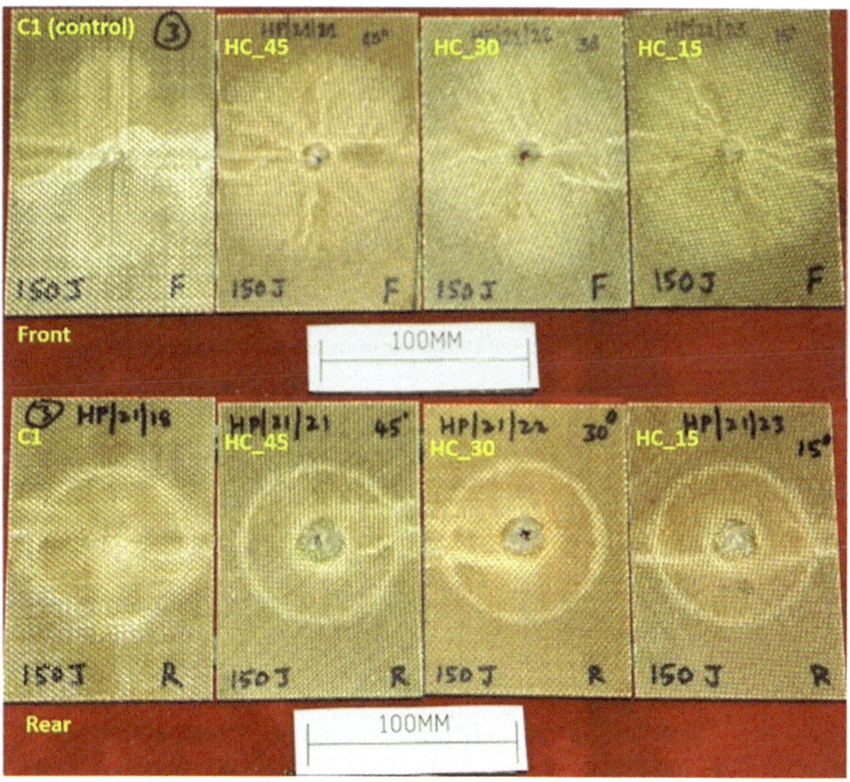

Fig. 12 Photograph of front and rear side of impacted laminates (C1, HC_45, HC_30 and HC_15)

4 Conclusions

E-glass/epoxy composite laminates with varying fibre volume fractions (V_f: 0.6, 0.53 and 0.45) and with different ply angle, viz. 15, 30 and 45^0 were designed and manufactured. All these composites were extensively studied for their physical, mechanical and low velocity impact properties and following conclusions were drawn.

- Mechanical properties of glass composites have strong dependence on fibre volume fraction and lay-up sequence. Cross-ply laminates showed superior tensile and flexural properties than their angle ply counterparts having the same fibre volume fraction.
- Low velocity impact revealed that the displacement increases with an increase in impact energy for a particular laminate and penetration resistance increases with an increase in fibre volume fraction.
- Change of ply angle resulted in increase in stiffness of the material and thereby increase in brittleness of overall laminate. Hence, the penetration resistance of the angle ply laminates of present study has decreased as compare to cross-ply laminates of equal fibre volume fraction.
- Visual observation of impacted laminates showed that matrix cracking and delamination are the major damage modes at lower impact energy and in addition to these, fibre breakage is also taking place at higher impact energy.

Acknowledgements Authors are grateful to the Director, Defence Metallurgical Research Laboratory (DMRL), Hyderabad for his encouragement to publish this work. Authors also acknowledge the support provided by the colleagues of Directorate of Armour Technologies (DoAT).

References.

1. Quaresimin M, Ricotta M, Martello L, Mian S (2013) Energy absorption in composite laminates under impact loading. Comp Part B: Engg 44 (1):133–140
2. Padaki NV, Alagirusamy R, Deopura BL (2008) Low velocity impact behaviour of textile reinforced composites. Ind J Fibre Text Res 33:189–202
3. Belingardi G, Vadori R (2002) Low velocity impact tests of laminate glass-fibre-epoxy matrix composite material plates. Int J Impact Eng 27:213–229
4. Kersys A, Kersiene N, Ziliukas A (2010) Experimental research of the impact response of E-Glass/epoxy and carbon /epoxy composite systems. Mater Sci 16(4):324–329
5. Yang FJ, Cantwell WJ (2010) Impact damage initiation in composite materials. Compos Sci Tech 70:336–342
6. Amaro AM, Reis PNB, deMoura MFSF, Neto MA (2013) Influence of multi-impacts on GFRP composites laminates. Comp Part B: Engg 52:93–99
7. Riccio A, Di Felice G, Saputo S, Scaramuzzino F (2014) Stacking sequence effects on damage onset in composite laminate subjected to low velocity impact. Procedia Eng 88:222–229
8. Sreekantha Reddy T, Rama Subba Reddy P, Madhu V (2019) Low velocity impact studies of E-glass/epoxy composite laminates at different thicknesses and temperatures. Def Technol 15:897–904

9. Kallagunta H, Tate JS (2020) Low-velocity impact behavior of glass fibre epoxy composites modified with nanoceramic particles. J Compos Mater 54(16):2217–2228
10. Liao B, Zhou J, Shigang A, Lin Y, Li X, Cao Y, Xiao D (2021) Comparison of laminate thickness on the low velocity impact behaviours for Z-pinned composite laminates. Int J Mech Sci 204
11. Durdu O, Ozdemir O, Kemiklioglu U (2021) Low-velocity impact behaviours of thermoplastic composites reinforced polypropylene glass fabric with different embedded delaminations. Plast Rubb Compos 51(1):35–46
12. ASTM (2007) Standard test method for measuring the damage resistance of a fiber-reinforced polymer matrix composite to a drop-weight impact event. D7136 / D7136M, West Conshohocken, PA.1–16

Estimation of Uncertainty of Measurement for Experimental Determination of Longitudinal Modulus for Carbon/Carbon Composites

Venkata Naga Mohan Manchiraju, Atul Ramesh Bhagat, and Vijay Kumar Dwivedi

1 Introduction

Uncertainty of measurement is a non-negative parameter characterizing the dispersion of the quantity values being attributed to a measurand, based on the information used [1]. Defense, space, and aviation sectors need materials that withstand very high temperatures. Carbon/carbon composites are known materials that can withstand very high temperatures of more than 3000 °C. Researchers work on the numerical determination of properties for these materials, as the design of components is based on these properties. The numerical determination of these properties is normally validated using experimental determination of the properties. The experimental results have an uncertainty of measurement, which will have an impact on the validation of numerical methods. In the present study, uncertainty of measurement is evaluated for experimental determination of longitudinal modulus for unidirectional carbon/carbon composites.

Several guides are published by national and international bodies for the estimation of uncertainty of measurement [2–4]. Some authors have [5, 6] worked on rules for the uncertainty of measurement. Several researchers have worked on the uncertainty of measurement for various fields of science and engineering [7–9]. Pendrill et al. [10] have studied the use of uncertainty of measurement for decision-making in testing and inspection. Statistical techniques such as ANOVA and DOE [11] and

V. N. M. Manchiraju (✉) · A. R. Bhagat
Advanced Systems Laboratory, Dr. APJ Abdul Kalam Missile Complex, DRDO, Ministry of Defence, Government of India, Hyderabad, TS, India
e-mail: mvnmohan33@gmail.com

V. N. M. Manchiraju · V. K. Dwivedi
Department of Mechanical Engineering, IET, GLA University, Mathura, UP, India
e-mail: bvijay.dwivedi@gla.ac.in

M. R. Manohara Babu et al. (eds.), *Recent Advances in Composites Engineering and Research*, Springer Proceedings in Materials 39,
https://doi.org/10.1007/978-981-99-8807-5_21

the Monte Carlo method [12] are used by some researchers. Hall et al. [13] have worked on a Python package useful for the uncertainty of measurement. Salicone et al. [14] have discussed a mathematical approach based on possibility theory for handling uncertainty.

In the current study, uncertainty of measurement is evaluated for experimental determination of longitudinal modulus for unidirectional carbon/carbon composites. The regression analysis is used for the estimation of uncertainty of measurement.

2 Unidirectional Carbon/Carbon Composite Material

Composite materials are made of two or more materials. These materials are chosen in such a way that when they are combined, the resulting material will have desired properties. Carbon/carbon composites are classified under fiber-reinforced composites. Fiber-reinforced composites are made of two materials, namely fiber and matrix. In carbon/carbon matrix, both the fiber and matrix are made of carbon. The combination of these materials called preforms is subjected to heat treatment to form carbon/carbon blocks, which are further used to fabricate components. The fabrication process of carbon/carbon composite is shown in Fig. 1.

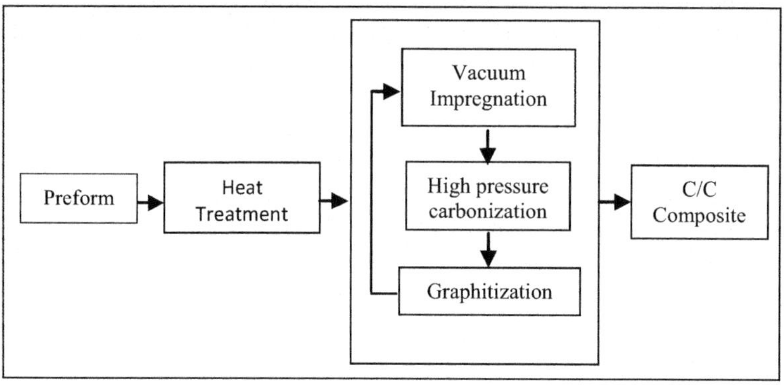

Fig. 1 Fabrication process of carbon/carbon composite

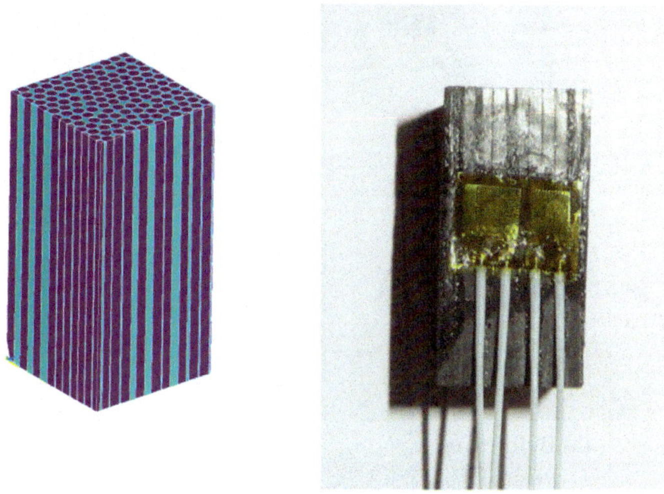

(a) Configuration (b) Test sample

Fig. 2 Configuration of and real and test specimen

3 Experimental Evaluation of Longitudinal Modulus

3.1 Test Method

Experimental determination of longitudinal modulus was performed on unidirectional carbon/carbon composite. The compression test samples were fabricated for experimenting. The compression test was performed on the uni-axial testing machine. The load was applied to the test specimen until its failure. The data of load is acquired during the experiment. Strain data is acquired through a strain gauge bonded on the test specimen. Using the acquired data, the stress and strain are plotted. The slope of the plot is evaluated using a linear regression model. The method of estimating longitudinal modulus using the experimental results is given in Table 2.The configuration of the test sample and the real test sample are shown in Fig. 2. The test setup is shown in Fig. 3.

3.2 Equipment for Testing

A universal testing machine was used to measure the load applied to the sample. The machine has a resolution of 1 N and was calibrated with an uncertainty of measurement of ±0.3%. The machine's accuracy conforms to class 1 as per Indian standards. The strain was measured using a bi-axial strain gauge with an uncertainty of measurement of ±3%. Digital calipers were used to measure the dimensions of

(a) Before Test (b) After Test

Fig. 3 Test setup

Table 1 Details of the equipment used for testing

Measurement	Test equipment
Load applied on the sample	Universal testing machine
	Uncertainty of measurement: ± 0.3%
	Accuracy: Class 1
	Resolution: 1 N
Strain on the sample	Bi-axial strain gauge
	Uncertainty of measurement ± 3%
Dimensions of the sample	Digital calipers
	Uncertainty of measurement: ± 0.016 mm
	Accuracy: 0.01mm
	Resolution: 0.01 mm

the test sample. The details of the equipment used for the testing are tabulated in Table 1.

3.3 Test Data

The data of the test includes stress and strain until the failure of the test specimen. Stress and strain data acquired during the test are plotted. The plot of stress and strain data is shown in Fig. 4. The slope of the stress–strain plot gives the longitudinal modulus of the test specimen.

Table 2 Formulae for regression model and Type A evaluation

Parameter	Formula	Value
Slope	Equation (2)	116.5 GPa
Empirical covariance	Equation (3)	−6.25
Standard deviation of strain (y)	Equation (4)	35.15
Standard deviation of stress (x)	Equation (5)	299.11
Correlation coefficient	Equation (6)	0.99
Standard deviation of slope	Equation (7)	0.0017
Type A uncertainty	Equation (8)	0.000184

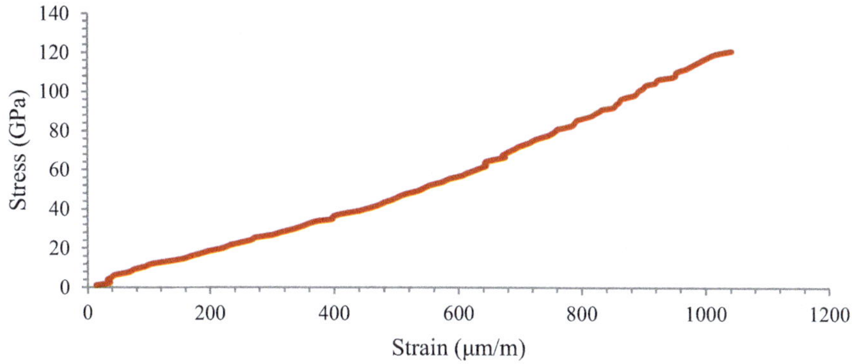

Fig. 4 Stress and strain plot of the test specimen

4 Uncertainty of Measurement

Uncertainty of measurement is estimated using NABL guidelines [4]. The uncertainty of measurement is estimated based on Type A and Type B methods of evaluation. Type A evaluation is the method of estimating uncertainty using statistical analysis of a series of observations. Type B uncertainty is evaluated through means other than statistical evaluation of a series of observations.

4.1 *Methodology for Estimation of Uncertainty of Measurement*

The first step to evaluate the uncertainty of measurement is to express the relation between output quantities in terms of input quantities using a mathematical model. In the current study, the output quantity is longitudinal modulus and the input quantities are force, strain, and dimensional measurements. Then, the sources of all uncertainties are identified and quantified. The quantified components of the uncertainty of measurement are combined and further expanded to report at a given confidence level.

4.2 Type A Uncertainty

The stress and strain data acquired during the experiment were plotted and shown in Fig. 4. The Type A uncertainty is evaluated using the regression model [15]. The linear regression model follows Eq. (1). The x in the equation relates to stress, and y is the strain. The data of stress and strain is denoted by x_i and y_i with n number of data. The slope of the curve is the longitudinal modulus and is estimated using Eq. (2) considering all the points from experimental data. The uncertainty of the slope is also estimated using the formula given below and the values are given in Table 2. The uncertainty of the slope is considered while estimating combined uncertainty.

$$y = mx + c, \tag{1}$$

$$m = \frac{n\sum_{i=1}^{n} x_i y_i - \sum_{i=1}^{n} x_i \sum_{i=1}^{n} y_i}{n\sum_{i=1}^{n} x_i^2 - \left(\sum_{i=1}^{n} x_i\right)^2}, \tag{2}$$

$$S_{xy} = \frac{1}{n-1}\left[\sum_{i=1}^{n} x_i y_i - \frac{\sum_{i=1}^{n} x_i \sum_{i=1}^{n} y_i}{n}\right], \tag{3}$$

$$S_x = \sqrt{\frac{1}{n-1}\left[\sum_{i=1}^{n} x_i^2 - \frac{\left(\sum_{i=1}^{n} x_i\right)^2}{n}\right]}, \tag{4}$$

$$S_y = \sqrt{\frac{1}{n-1}\left[\sum_{i=1}^{n} y_i^2 - \frac{\left(\sum_{i=1}^{n} y_i\right)^2}{n}\right]}, \tag{5}$$

$$r = \frac{S_{xy}}{S_x S_y}, \tag{6}$$

$$S_m = \sqrt{\frac{\left(1 - r^2\right) S_y^2}{(n-2) S_x^2}}, \tag{7}$$

$$U_A = \frac{S_m}{\sqrt{n}}. \tag{8}$$

The formulae used for the regression model and the values are given in Table 2.

4.3 Type B Uncertainty

Type B uncertainty is evaluated using data available from calibration certificates, manufacturers' specifications, etc. In the current study, the sources of uncertainty are load measurement, strain measurement, and dimensional measurement. Sensitivity coefficients are estimated to obtain a differential change in longitudinal modulus due to differential changes in load, strain, and dimension measurement. Sensitivity coefficients are partial derivatives used to describe how the output estimate varies with changes in the values of the input estimates. Essentially, sensitivity coefficients show you how the variables in an equation or function are related to the calculated result. Sensitivity coefficients are just a multiplier used to convert your uncertainty components to the right units and magnitude for your uncertainty analysis. The sensitivity coefficients are estimated and tabulated in Table 3.

$$\frac{\partial E}{\partial F} = \frac{\partial}{\partial F}((F/A)/\varepsilon) = \frac{1}{A\varepsilon}, \tag{9}$$

$$\frac{\partial E}{\partial \varepsilon} = \frac{-F}{A\varepsilon^2}, \tag{10}$$

$$\frac{\partial E}{\partial a} = \frac{-F}{A\varepsilon^2}, \tag{11}$$

$$\frac{\partial E}{\partial b} = \frac{-F}{ab^2\varepsilon}. \tag{12}$$

Type B uncertainties from various sources are calculated using the following formula, and the values are given in Table 4.

$$U_{\text{B1}} = C_{\text{L}} \times U_{\text{L}}, \tag{13}$$

$$U_{\text{B2}} = C_{\varepsilon} \times U_{\varepsilon} \tag{14}$$

$$U_{\text{B3}} = C_{\text{a1}} \times U_{\text{a1}}, \tag{15}$$

Table 3 Sensitivity coefficients

Sensitivity coefficient for measurements	Formula	Value
Sensitivity coefficient for load measurement C_{L}	Equation (9)	6.67
Sensitivity coefficient for strain measurement C_{ε}	Equation (10)	-1.12×10^8
Sensitivity coefficient for length measurement C_{a}	Equation (11)	−9718.14
Sensitivity coefficient for width measurement C_{b}	Equation (12)	−9718.14

Table 4 Type B uncertainties from various sources

Uncertainty of measurement due to	Formula	Value (MPa)
Load measurement uncertainty	Equation (13)	349.85
Strain measurement	Equation (14)	−3498.53
Length measurement uncertainty	Equation (15)	−77.75
Width 1 measurement uncertainty	Equation (16)	−77.75
Length measurement instrument resolution	Equation (17)	−28.05
Width 2 measurement uncertainty	Equation (18)	−28.05

$$U_{\mathrm{B4}} = C_{\mathrm{b1}} \times U_{\mathrm{b1}}, \tag{16}$$

$$U_{\mathrm{B5}} = C_{\mathrm{a2}} \times U_{\mathrm{a2}}, \tag{17}$$

$$U_{\mathrm{B6}} = C_{\mathrm{b2}} \times U_{\mathrm{b2}} \tag{18}$$

4.4 Combined Uncertainty

The combined uncertainty is calculated by using Eq. 19, and value is 2028 Mpa.

$$U_c = \sqrt{U_A^2 + U_{B1}^2 + U_{B2}^2 + U_{B3}^2 + U_{B4}^2 + U_{B5}^2 + U_{B6}^2}. \tag{19}$$

4.5 Expanded Uncertainty

The expanded uncertainty is calculated by using Eq. 20, and value is 4056 Mpa

$$U_E = k \times U_c, \tag{20}$$

where k is the coverage factor at 95% confidence level.

4.6 Uncertainty Budget

The uncertainty budget is given in Table 5.

Table 5 Uncertainty budget for experimental determination of longitudinal modulus

Source of uncertainty	Value	Type	Probability distribution	Sensitivity coefficient	Uncertainty (MPa)	Degrees of freedom
U_{B1}	±0.3%	B	Normal	6.67	349.85	∞
U_{B2}	±3%	B	Normal	−1.12 × 108	3498.53	∞
U_{B3}	±0.016 mm	B	Normal	−9718.14	77.75	∞
U_{B4}	±0.016 mm	B	Normal	−9718.14	77.75	∞
U_{B5}	0.01 mm	B	Rectangular	−9718.14	28.05	∞
U_{B6}	0.01 mm	B	Rectangular	−9718.14	28.05	∞
U_A	±0.0015 Mpa	A	Normal	1	0.000184	85
Expanded uncertainty at 95% confidence level					4056	

4.7 *Reporting Results*

The longitudinal modulus of unidirectional carbon/carbon composite is 116.5 ± 4.06 GPa with a confidence level of 95.45% with coverage factor k = 2.65.

5 Conclusions

The longitudinal modulus of unidirectional carbon/carbon composites has been experimentally determined. The current study depicts the method of evaluation of uncertainty of measurement for the evaluation of longitudinal modulus for composite materials. The relationship between the stress and strain for the experiment has been depicted using regression analysis. Using this model, the longitudinal modulus is determined, and the uncertainty of measurement has also been evaluated. Researchers working on composite materials and their properties can use this model for the evaluation of uncertainty. This study is further useful when numerical methods are used for the prediction of elastic properties and further validated with experimental methods, the uncertainty of measurement shall be considered.

6 Future Work

Multiple experiments can be performed and evaluate longitudinal modulus for each experiment. The reproducibility of evaluation of longitudinal modulus can also be estimated. This future work will further corroborate the uncertainty analysis.

References

1. Joint Committee for Guides in Metrology (2012) International vocabulary of metrology—basic and general concepts and associated terms (VIM). JCGM 200. Retrieved from http://www.bipm.org/utils/common/documents/jcgm/JCGM_200_2012.pdf
2. Joint Committee for Guides in Metrology (2012) International vocabulary of metrology—basic and general concepts and associated terms (VIM). JCGM 200. Retrieved from http://www.bipm.org/utils/common/documents/jcgm/ JCGM_200_2012.pdf
3. Ellison SLR, Williams A (2012) Eurachem/CITAC guide: quantifying uncertainty in analytical measurement, Third edition, ISBN 978–0–948926–30–3. Retrieved from www.eurachem.org.
4. National Accreditation Board for testing and calibration laboratories (2016) Guide for estimation and expression of uncertainty if measurement. NABL-141
5. Lisiecki J, Kłysz S (2007) Estimation of measurement uncertainty. Res Works Air Force Inst Technol 22:81–114. https://doi.org/10.2478/v10041-008-0004-4
6. Farrance I, Frenkel R (2012) Uncertainty of measurement: a review of the rules for calculating uncertainty components through functional relationships. Clin Biochem Rev 33(2):49–75. PMID: 22896744; PMCID: PMC3387884
7. Mondal P (2010) Understanding uncertainty of measurement
8. (2018) Evaluation of measurement uncertainty for the determination of the mechanical resistance of the brick samples by using uni-axial compressive strength test Klodian Dhoska, Saimir Tola, Agus Pramono, Indrit Vozga. Int J Metrol Qual Eng 9:12. https://doi.org/10.1051/ijmqe/2018012
9. Kornilov NV, Pronyaev VG, Grimes SM (2022) Systematic distortion factor and unrecognized source of uncertainties in nuclear data measurements and evaluations. Metrology 2:1–18. https://doi.org/10.3390/metrology2010001
10. Pendrill LR (2014) Metrologia 51:S206
11. Maddulety K A research paper on measurement-uncertainty analyses through DoE and ANOVA. Int eJ Math Eng 2(2):1000–1011
12. Smith I, Luo Y, Hutzschenreuter D (2022) The storage with in digital calibration certificates of uncertainty information obtained using a monte carlo method. Metrology 2:33–45.https://doi.org/10.3390/metrology2010003
13. Hall BD (2022) The GUM tree calculator: a python package for measurement modelling and data processing with automatic evaluation of uncertainty. Metrology 2:128–149. https://doi.org/10.3390/metrology2010009
14. Salicone S, Jetti HV (2021) A general mathematical approach based on the possibility theory for handling measurement results and all uncertainties. Metrology 1:76–92. https://doi.org/10.3390/metrology1020006
15. Gabauer W (2000) The determination of uncertainties in tensile testing. Voest-Alpine Stahl Gmbh, Linz, Austria

Experimental Characterisation and Failure Behaviour of Inserts Embedded in Stiffened Honeycomb Sandwich Panel

Krushna Chandra Dakua, G. Srinivasa Rao, Raghunatha Behara, M. N. M. Patnaik, and D. Poomani

1 Introduction

Generally, the sandwich structure consists of a face sheet and core. The face sheets are carried in-plane loads and are of high stiffness materials and the core that carries transverse shear loads are light materials like honeycomb which are mainly used to sustain the strong facings. In honeycomb sandwich panels, the function of the face sheet is similar to the flange in an I beam, and the function of the core is similar to the web of an I beam [1]. Sandwich structures have very high specific strength and specific stiffness, but they are weak in joining. Sandwich plates are unsuited for carrying localised loads, for the simple reason that the core does not have the stiffness necessary to distribute the forces effectively [1]. Therefore, a sandwich structure should choose an appropriate type of insert. The major role of inserts is to transfer the loads into sandwich structures.

The load transfer mechanism is very complex in sandwich panels so an experimental evaluation is more viable for the evaluation of joint strength [2]. In sandwich constructions, the panels have to be connected to other panels, and other decks which are potential locations of failure. There are different methods exist for introducing localised loads into sandwich structures. Different types of inserts such as partially potted, fully potted and through-thickness inserts. The partially potted inserts have little mass saving also lower load-carrying capability compared to fully potted inserts which transfer loads to both the face skins. The through-thickness inserts can be used for achieving higher in-plane loads [3].

K. C. Dakua (✉) · G. Srinivasa Rao · R. Behara · M. N. M. Patnaik · D. Poomani
Structures Group, U R Rao Satellite Centre, Old Airport Road, Vimanapura Post, Bangalore 560017, India
e-mail: krushna@ursc.gov.in

M. R. Manohara Babu et al. (eds.), *Recent Advances in Composites Engineering and Research*, Springer Proceedings in Materials 39,
https://doi.org/10.1007/978-981-99-8807-5_22

The inserts are more prone to tensile loads normal to the sandwich surface and shear loads parallel to the surface, localised compression and bending loads are generally avoided [4]. Torsional loads are experienced during assembly and which are not very significant during the external loading. Another method for introducing localised stiffness, and strength is by the addition of doublers in the insert bonding area to strengthen the panel. In the present paper, the experimental evaluation is carried out using three types of coupons which are namely inserts mounted on a sandwich coupon (Type-1), Insert mounted on a sandwich coupon with a doubler (Type-2) and insert mounted on a sandwich coupon with doubler and bolt (Type-3) which are explained in detail in the following Figs. 1, 2 and 3. The knowledge of load-carrying capability of a particular insert is very important for the designer of a sandwich panel. In this work, the insert strength capability increase was studied experimentally by providing externally localised doublers and further enhancing the load-carrying capability by providing a bolt.

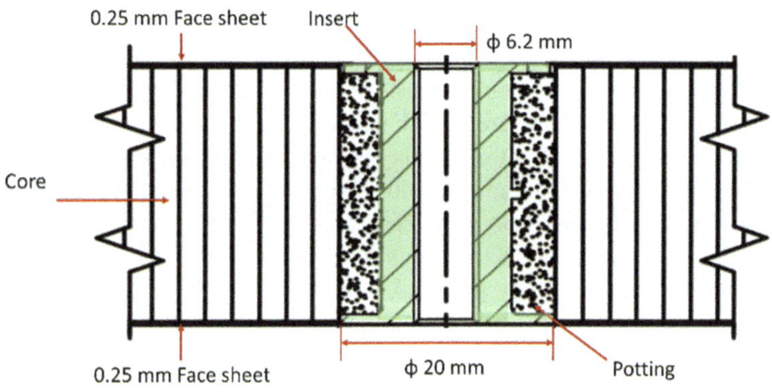

Fig. 1 Type-1 coupon

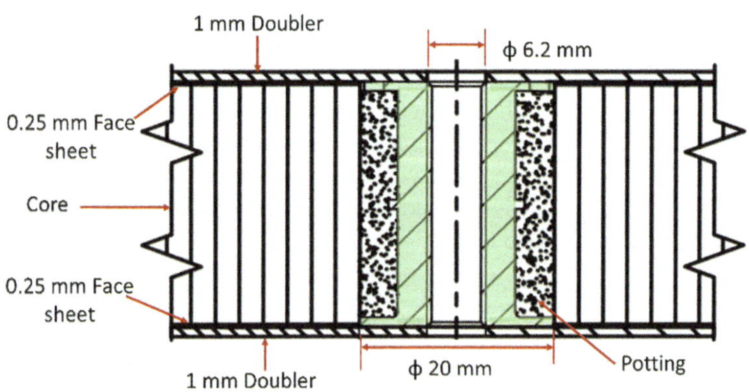

Fig. 2 Type-2 coupon

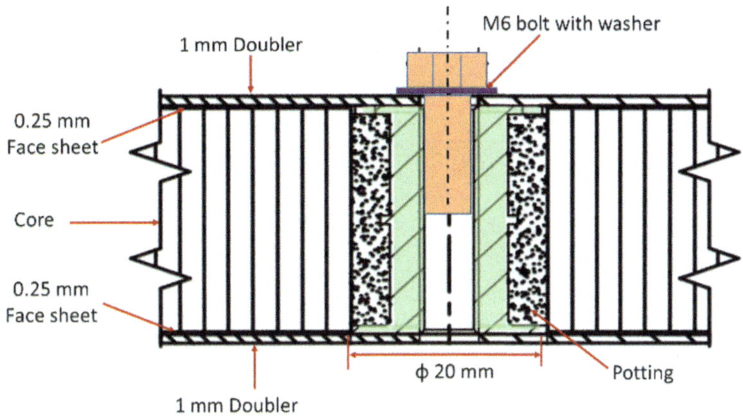

Fig. 3 Type-3 coupon

2 Experiments

2.1 Materials and Structure Description

The following experimental study was focused on honeycomb sandwich panel coupons of 3 different type of configurations with through-thickness inserts as shown in Figs. 1, 2 and 3.

The sandwich panel is composed of aluminium face skins and an aluminium honeycomb core joined together by an adhesive. These coupons were fabricated through the co-curing process in an autoclave. The through-thickness insert was made of aluminium alloy and complies with ISRO standards. The material details, geometrical details and mechanical properties are described in Tables 1, 2 and 3. The insert bonded to the honeycomb panel after fabrication. The position of the insert relative to the adjacent honeycomb geometry is random. In Type-2 and Type-3 configuration, an additional 1 mm doubler bonded on the sandwich coupon with the help of HYSOL 9392 adhesive. These sandwich coupons were carried out in-house in the structures group of the U R Rao satellite centre.

Table 1 Sandwich structure component description

Component	Description
Face sheet/doubler	AA 2024 T81
Core	CR III-3/16–4953.76 -0.0007P-2.0 Density (32 kg/m^3)
Adhesive	Redux 312 (Film), Hysol 9392 (RTV)
Potting	Stycast (Potting Resin)
Insert	Aluminium alloy (AA2024 T3)

Table 2 Geometrical configuration of three types of coupon

Sl. No.	Detailed dimensions	Type-1	Type-2	Type-3
1	Height of the core (mm)	25	25	25
2	Core cell size (mm)	3.2	3.2	3.2
3	Face sheet thickness (mm)	0.25	0.25	0.25
4	Aluminium doubler	–	1 mm	1 mm
5	M6 bolt used	–	–	SS bolt used
6	Size of coupon (mm^2)	100 × 100	100 × 100	100 × 100

Table 3 Mechanical properties of sandwich coupon and adhesives

Face skin properties		Epoxy	
E_{skin} (N/mm^2)	72,000	$\nu_{potting}$	0.3
G_{skin} (N/mm^2)	27,000	$r_{potting}$ (mm)	10
t_{skin} (mm)	0.25		
Honeycomb core properties		*Potting properties*	
E_L (N/mm^2)	0.01	$E_{potting}$ (N/mm^2)	2000
E_w (N/mm^2)	0.01	$\sigma_{potting}$ (N/mm^2)	65
E_T (N/mm^2)	310		
τ_{wt} (N/mm^2)	0.34		
h_{core} (mm) (insert test)	25		
G_{LW} (N/mm^2)	0.01		
G_{LT} (N/mm^2)	186		
G_{WT} (N/mm^2)	90		
τ_{LT} (N/mm^2)	0.72		

2.2 Experimental Set-Up Procedure

The pull-out behaviour of the inserts with sandwich honeycomb panels was studied; however, there is no standard method exists for the pull-out test. There are some recommendations in ESA insert design handbook, and some aerospace industries have their own test methods [5]. In this experimental evaluation, the test fixture made up of stainless-steel material with a circular hole 80 mm diameter was provided to reduce the edge effects. The specimen is placed above the fixture as shown in Fig. 4. The specimen was loaded in out-of-plane loading, i.e. in vertical direction using M6 stud/pulling rod. The test was carried out in INSTRON 5969B-50KN universal testing machine with a machine speed of 1 mm/min [6]. The load and displacements were recorded during the test.

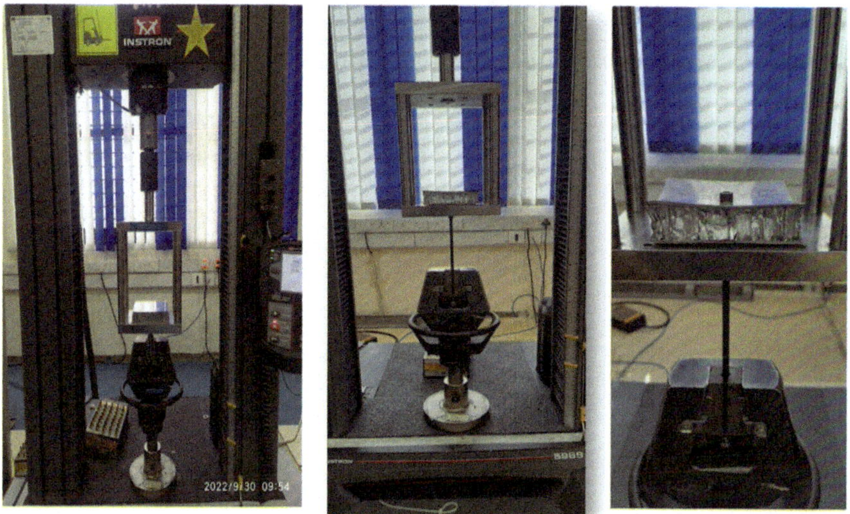

Type-1
(Sandwich Coupon)

Type-2
(Sandwich coupon with doubler)

Type-3
(Sandwich coupon with doubler and M6 bolt & washer)

Fig. 4 Test set-up of all the three types of coupons

2.3 *Experimental Results and Failure Mechanisms*

2.3.1 Type-1 Configuration

These three numbers of coupons were tested for characterising the failure behaviour. Table 4 shows experimentally tested values up to the linear range and maximum failure load of the sandwich coupons. It is observed from tests that the average load is up to linear range 1565 N where whereas individual coupon loads are approximately 1315,1758 and 1621 N (Figs. 5, 6 and 7). The coefficient of variation of pull-out strength is well within 10% of the data. The scatter is largely due to varying effective potting radius in the specimen which has a significant impact on the pull-out strength of the sandwich insert. The varying potting radius is a result of a random number of honeycomb cells being opened during the machining operation on the honeycomb panel observed in general [3]. The average maximum failure load was around 2000 N where as individual coupon loads are approximately 1771, 2148 and 2080 N, respectively. The stiffness values also given in Table 4 are well-matched within each coupon.

The linear elastic shear deformation as shown in Fig. 8 followed by curve flattening and rapid stiffness degradation shortly after the peak force as shown in Fig. 8. The shear buckling of the cell walls at the interface between potting and core is a general phenomenon in a typical honeycomb sandwich panel with insert. In honeycomb sandwich coupon due to the buckling of the core material the shear deformation starts which is very evident by a 45-degree angle to the surrounding core material

Table 4 Loads and stiffness values for Type-1 coupons

Type-1	Maximum load at linear range	Maximum failure load	Stiffness (within linear range) K1
Units	(Newton)	(Newton)	(Newton/mm)
Coupon-1	1315	1771	4301.8
Coupon-2	1758	2148	4238.2
Coupon-3	1621	2080	4798.9

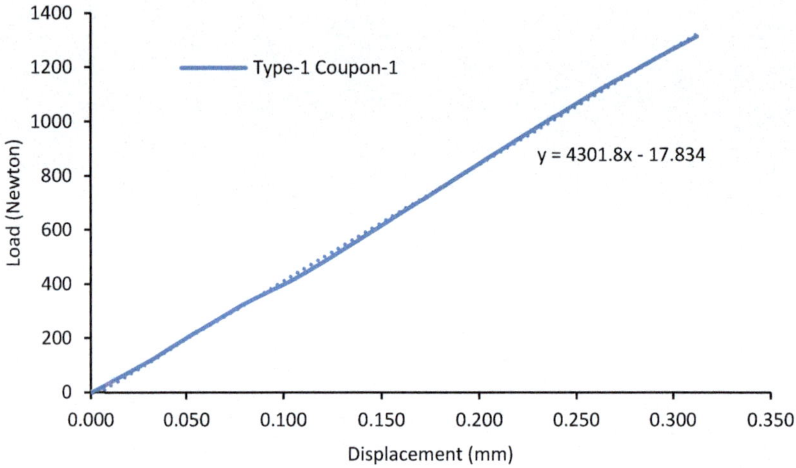

Fig. 5 Load versus displacement (Coupon-1)

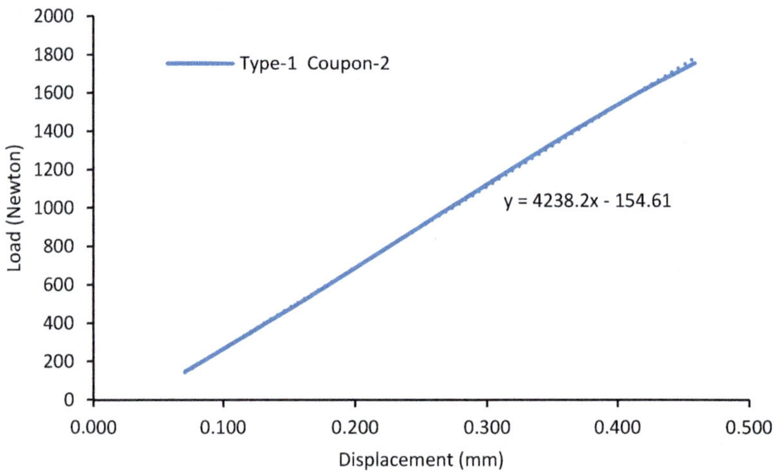

Fig. 6 Load versus displacement (Coupon-2)

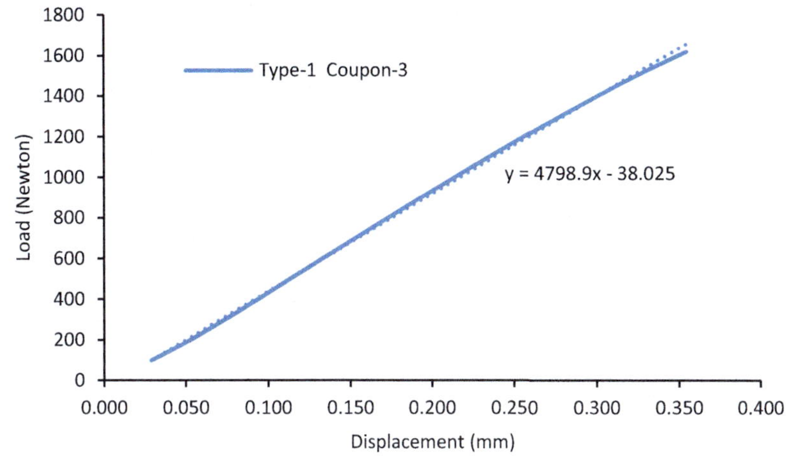

Fig. 7 Load versus displacement (Coupon-3)

this indicates the good bond between the core material and the adhesive material [7]. The next mechanism involved is rapid stiffness degradation, i.e. there is a sudden failure initiated by tensile failure of the core in out of out-of-plane direction below the potting. The face sheet failure and debonding of core to the face sheet occurs after the peak force.

The first slope change of the load–displacement curve always denotes the start of the failure in the honeycomb core at the boundary of the potted region, after this the structure will still continue to take additional load, as the composite face over

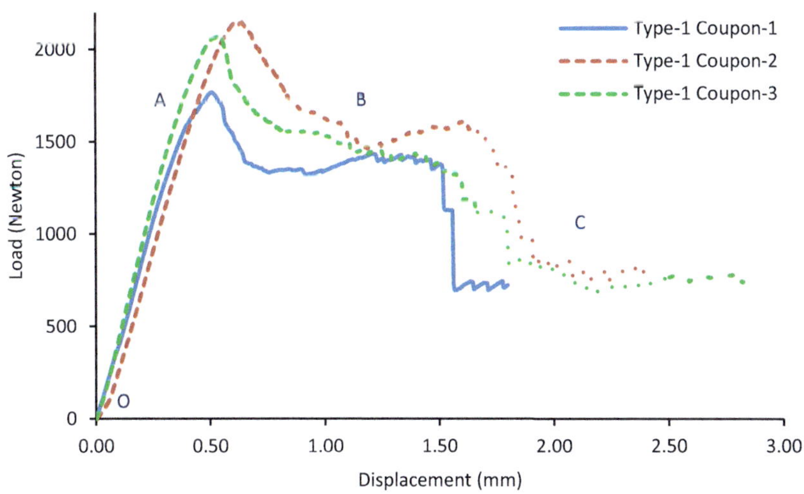

Fig. 8 Failure regimes of sandwich coupon without doubler

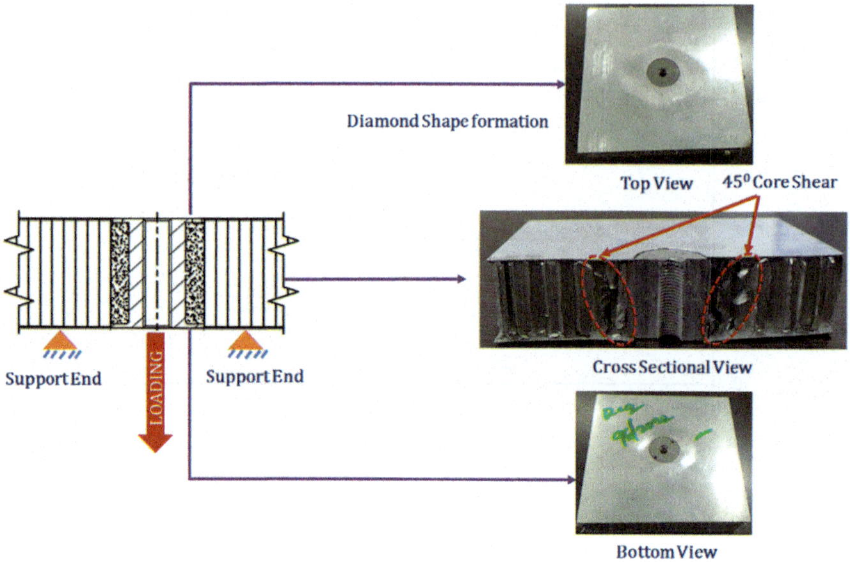

Fig. 9 Failure modes of sandwich coupon without doubler

the potting material does not experience significant damage at the first slope change, but further it will continue to carry the load until there is sudden load drop and face sheet cracking.

After the experimental evaluation, the specimen was cut into two pieces for detailed failure analysis as shown in Fig. 9. It was observed that 45 deg core shear happened in the honeycomb core normal to the applied pull-out load. The diamond shape was observed on the top face sheet which is opposite to the loading direction as shown in Fig. 9.

The sequence of failure mechanism is shown in load–displacement graphs in Fig. 8. The load–displacement curve which has three distinct regions (OA, AB, BC) is explained below (Fig. 10).

Three Distinct Failure Regimes

- **OA**—Elastic shear deformation of the core material near to the insert element.
- **AB**—Failure of core-face sheet bonding around the insert element, membrane (Diamond) formation of the face sheet.
- **BC**—Disintegration of insert element and face sheet.

2.3.2 Type-2 Configuration

In this, only one coupon was tested for characterising the failure behaviour. Table 5 shows the experimentally tested values up to the linear range and maximum failure load of the sandwich coupon. The failure behaviour as shown in Fig. 11, i.e. shows the

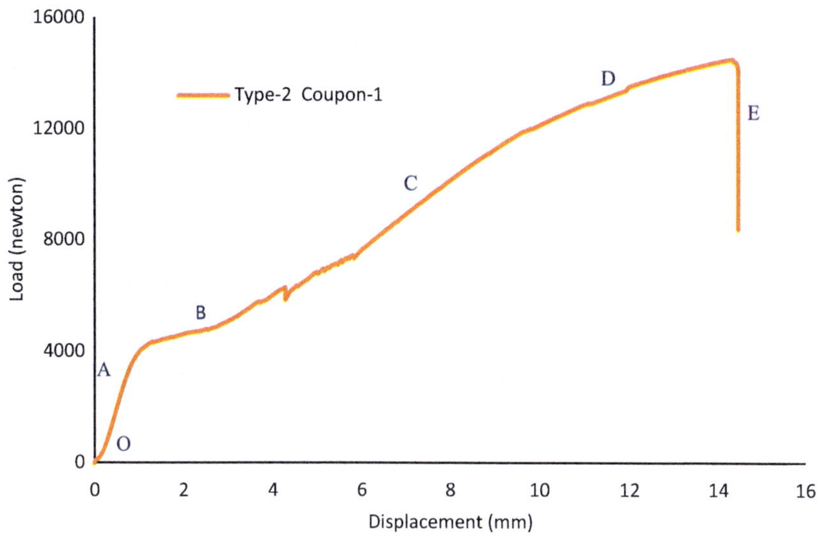

Fig. 10 Load versus displacement graph (Type-2 coupon)

Table 5 Load and stiffness value for Type-2 coupon

Type-2	Max. load at linear range	Maximum failure load	Stiffness (linear range-OA region) K1	Stiffness (membrane formation-CD region) K2
Units	(Newton)	(Newton)	(Newton/mm)	(Newton/mm)
Coupon-1	3777	14,541	4953.7	1137.7

insert disintegration from the top face sheet which shows that cup formation on the top face sheet and cone formation on the bottom face sheet of the sandwich coupon. The sequence of failure mechanism shown in Fig. 14 in the load–displacement curve has five distinct regions [8].

2.3.3 Type-3 Configuration

These two coupons tested for characterising the failure behaviour. Table 6 shows the experimentally tested values, i.e. up to the linear range and maximum failure load of the sandwich coupons. It is observed from the tests that the average load is up to the linear range of 3752 N. The stiffness values also given in the table are well-matched with each other. The maximum failure load is around 15,000 N which is almost 4 times the load-carrying capability compared to linear elastic region. The failure behaviour is shown in Fig. 13, i.e. shows the insert intact with the top face sheet which shows cup formation on the top side of the sandwich coupon. The bottom

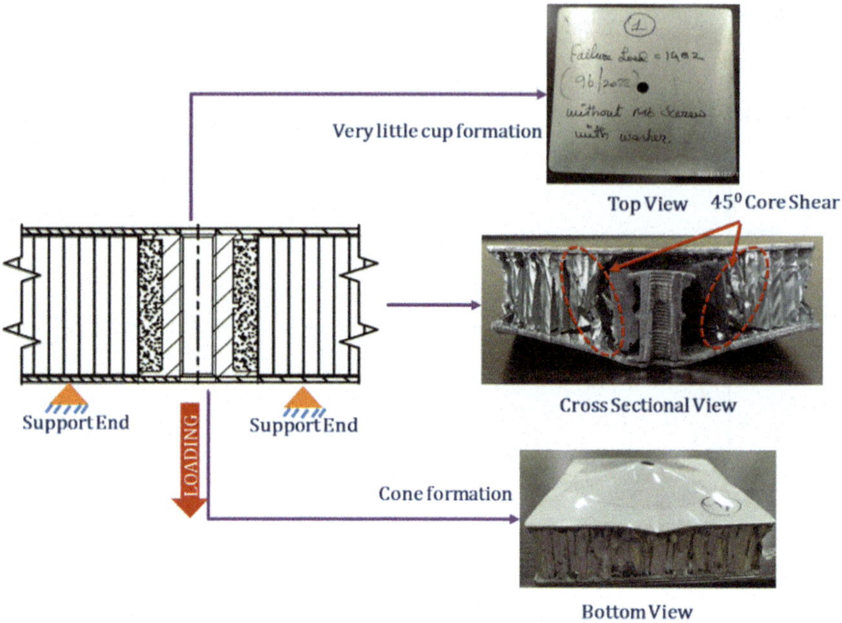

Fig. 11 Failure modes of sandwich coupon with doubler

side shows cone formation on the bottom face sheet of the sandwich coupon with a face sheet crack. The stiffness parameters after the first slope change up to failure load are well-matched with each other as shown in Fig. 12. The sequence of failure mechanism shown in Fig. 14 load–displacement curve has 5 distinct regions (OA, AB, BC, CD and DE) as explained below [8].

The failure regimes of Special Through-the-thickness (ttt)-Inserts along with doubler and bolt as shown in Fig. 14.

- **OA**—Elastic shear deformation of the core material near to the insert element.
- **AB**—Irreversible response - Plastic shear deformation of core material (instability of the core cell walls started).
- **BC**—Failure of core-face sheet bonding around the insert element, membrane
- (Diamond) formation of the face sheet and doubler.

Table 6 Loads and stiffness values for Type-3 coupons

Type-3	Max. load at linear range	Maximum failure load	Stiffness (within linear range) K1	Stiffness (membrane formation zone) K2
Units	(Newton)	(Newton)	(Newton/mm)	(Newton/mm)
Coupon-1	3680	16,261	4874.8	1472.7
Coupon-2	3824	14,625	4681.3	1467.7

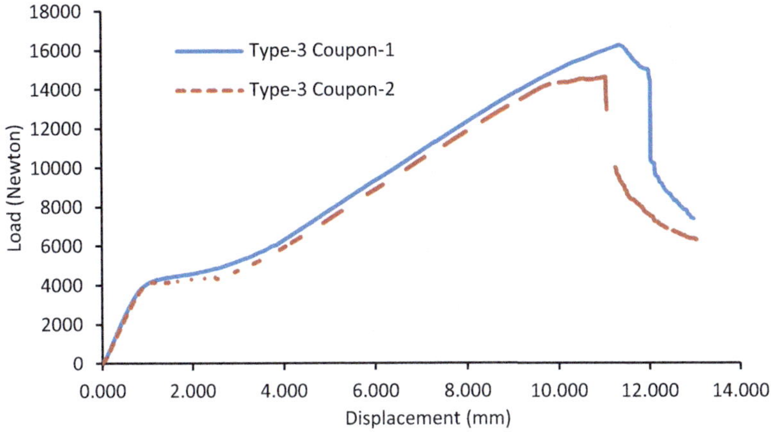

Fig. 12 Load versus displacement graph (Type-3 coupon)

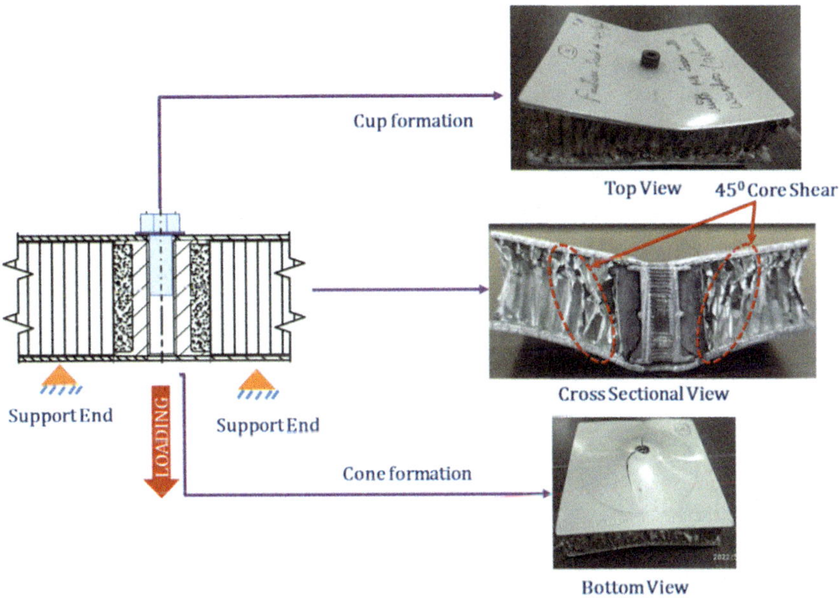

Fig. 13 Failure modes of sandwich coupon with doubler M6 bolt and washer for Type-3 Coupon-1 and Coupon-2

- **CD**—Membrane formation of the face sheet and doubler continue to cup and cone shape. The region-(CD) and region-(DE) show the face sheets and doublers contributing the load carrying capacity.
- **DE**—Disintegration of insert element, face sheet and doubler tearing.

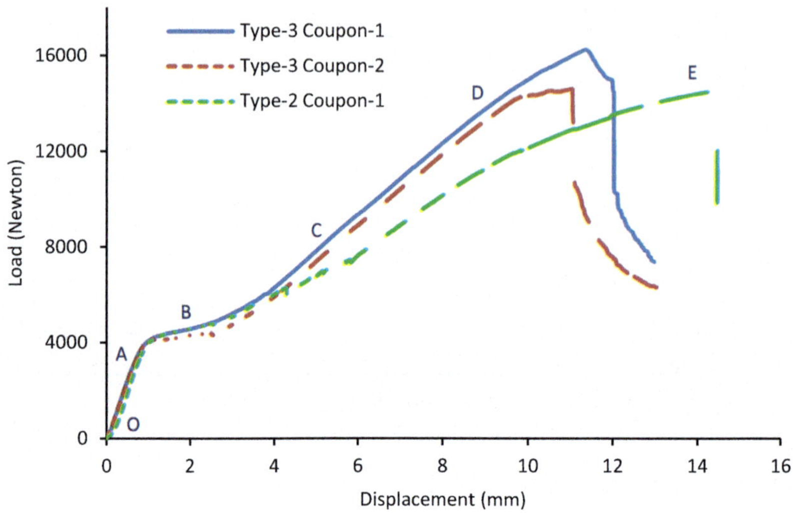

Fig. 14 Load versus displacement graph (Type-2 and Type-3 coupons)

3 Observations

It is observed from tested data initially all types of coupons follow the same slope up to elastic region which is approximately 1766 N. The Type-2 and Type-3 coupons together are linear up to 3777 N as shown in Fig. 15. By introducing a doubler the out-of-plane tensile load carrying capability is increased significantly by 2 times, whereas stiffness (K1) up to linear elastic range marginally increased around 10% which was given in Tables 4 and 5. Also, the stiffness (K2) in the (membrane formation, cup and cone region) further improved by 5% in the Type-3 specimen compared to the Type-2 specimen but strength increased by 10% up to maximum failure load.

4 Conclusions

The failure behaviour of the honeycomb sandwich structure with potted inserts embedded in different honeycomb sandwich structures was investigated experimentally under pull-out loading conditions. This study was conducted to investigate the effect of face sheet thickness and boundary conditions on the failure loads of a sandwich insert joint. Based on the experimental evaluation, it was observed that by introducing a doubler locally near the insert region the strength carrying capability is almost increased two times up to the linear region, and 7 times up to the maximum failure load region where as stiffness marginally improved by 10% only. The failure modes in different regimes are shown and discussed. Based on the load–deflection

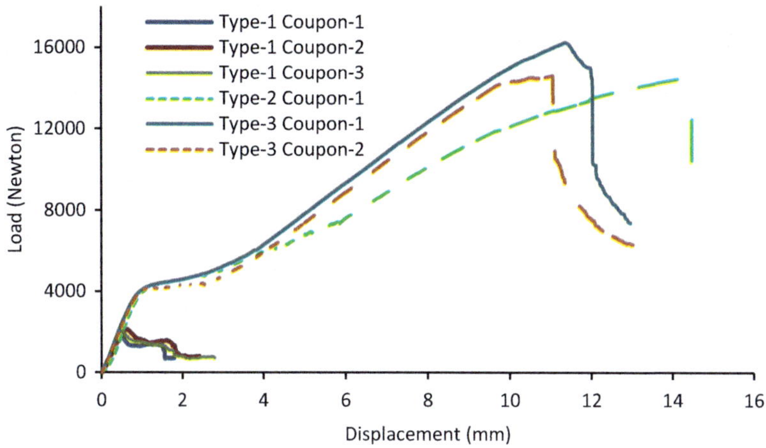

Fig. 15 Combined load versus displacement graph (Type-1, Type-2 and Type-3)

curves presented, practically design engineers can arrive at the strength and stiffness for optimal safe design structure.

Acknowledgements The authors wish to acknowledge the Structural Fabrication Division of Structure group, URSC for fabrication of test specimens to carry out the activities. We gratefully acknowledge the encouragement provided by Dr. Alok Kumar Shrivastava, Deputy Director and Director URSC for publishing this work.

References

1. Zenkert D (1997) The Handbook of Sandwich Construction, free download copy from online, First edition:1997© Dan Zenkert
2. Song K-H, Choi J-Y, Kweon J-H, Choi J-H, Kim K-S (2008) An experimental study of the insert joint strength of composite sandwich structures. Compos Struct 86:107–113
3. Seemann R, Krause D (2018) Numerical modelling of partially potted inserts in honeycomb sandwich panels under pull out loading. Compos Struct 203:101–109
4. Heimbs S (2009) Marc pein: failure behavior of honeycomb sandwich corner joints and inserts. Compos Struct 89:575–588
5. Insert Design Handbook (1987) ESA PSS-03-1202. European Space Agency, Paris
6. Instron 5960 Series operator's guide M10–16247-EN. Retrieved from www.instron.com
7. Thomas LAPID[1], cees BIL[1], Greg HANLON[2] Honeycomb Panel Insert Bolt Pull-Out Testing of a New fast curing Adhesive,1) School of Engineering, RMIT university, Melbourne, Australia, 2) Five rings Aerospace Pty Ltd, East Keilor, Australia
8. Wolff J (2017) German aerospace center DLR: analytical formulation for strength prediction of orthogonal loaded, cold bonded insert load introductions in sandwich elements, ICCS20, Paris, CNAM, 07.09.2017, Session "Sandwich Structures"

Characterization of Al-PUR Sandwich Panel by Mechanical Testing and Rocket Launch-Induced Dynamic Environment Testing

G. Vaibhav, M. Lokesh Babu, V. V. Ramakrishna, R. Ram Kumar, C. Anil Kumar, S. Sankaran, and A. Rajarajan

1 Introduction

As a part of Human Spaceflight programme (Gaganyaan mission), a clean room, namely 'White room', is envisaged at launch pad for final checks of crew just before the entry into Crew Module and also to maintain the clean environment inside the Crew Module when its hatch is opened. White room should have the capability to resist/withstand dynamic launch environment of acoustics, vibration, heat flux, over pressure, elevated temperatures, etc., and also needs to be lighter in weight to minimize the load on the existing assembly platform at launch pad (as shown in Fig. 1).

Hence, composite materials are considered as the material of construction due to their superior specific strength. The present paper proposes a novel sandwich panel made up of polyurethane PUR foam material (30 mm thick) as core with Aluminium Al-6061 (1.2 mm thick) sandwiched on either face of the foam core, i.e. sandwich panel total thickness is 32.4 mm.

In order to minimize the weight of the white room being established on the edge of platform at launch pad, composite panel with Carbon Fibre Reinforced Plastic (CFRP) facings of 1.25 mm thick each and Aluminium honeycomb core of 10 mm thick was studied as its material of construction. Though CFRP composite material has high tensile/shear properties and very light weight, it was not recommended because.

G. Vaibhav (✉) · M. Lokesh Babu · V. V. Ramakrishna · R. Ram Kumar · C. Anil Kumar · S. Sankaran · A. Rajarajan
Indian Space Research Organization, SDSC-SHAR, Sriharikota 524124, India
e-mail: vai.gog.25@gmail.com

M. R. Manohara Babu et al. (eds.), *Recent Advances in Composites Engineering and Research*, Springer Proceedings in Materials 39,
https://doi.org/10.1007/978-981-99-8807-5_23

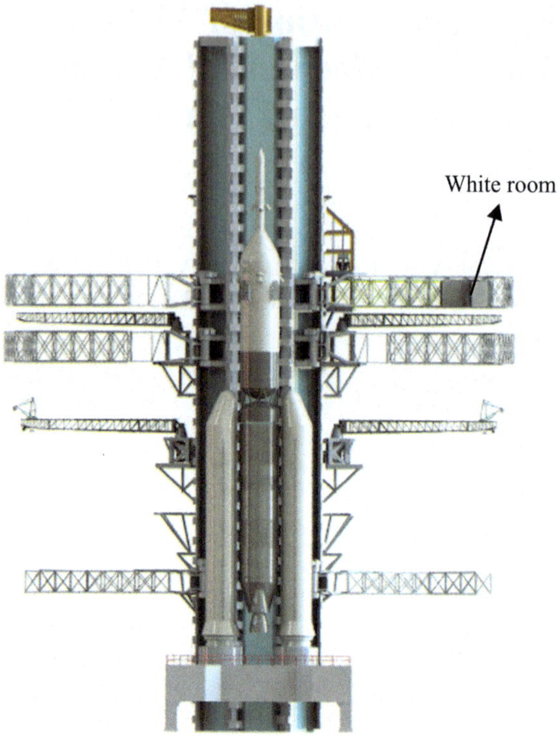

Fig. 1 White room location at second launch pad (SLP)

- Difficulty in interface generation during in situ drilling of CFRP composite face sheet; the carbon layers and honeycomb core construction are shredded and de-bonded resulting in loss of sandwich properties locally, whereas sandwich panel drilling is easy and no de-bonding is observed.
- Dust accumulation on carbon fibre surface due to rough surface finish compared to aluminium surface; smooth surfaces are only acceptable as white room is planned as 100,000 class categories.
- Cost of white room is three times higher if the walls are made of CFRP rather than Aluminium sandwich panel as processing of CFRP face sheets is much complex than Aluminium face sheets.

In comparison, the proposed material is better in terms of density, generating interfaces, and cost per unit surface. Hence, this study is taken up to determine the properties of sandwich panel and to verify its integrity in rocket launch-induced environment.

2 Mechanical Testing

Mechanical testing plays a significant role in both the design and quality control of sandwich panels. This includes tests to determine parameters such as the interaction of bending moment and support reaction, the behaviour of connections as well as basic material properties such as the compressive and shear strength of the core. It is important that the components of sandwich panel form an integral structure such that the bond between them is reliable and can transfer the required forces. The material tests that are specific to sandwich panels are tests on core materials and tests whose outcome depends on the bond between the core and the faces. The following tests are carried out:

- **Core shear test**, as per ASTM C-273 [1], to find shear strength and shear modulus of core.
- **Flatwise tensile strength test**, as per ASTM C-297 [2], to find flatwise tensile strength of the core to facing bond.
- **Four-point bend test** with quarter point loading, according to ASTM C-393 [3] to find flexural stiffness, core shear strength and shear modulus, facing compressive and tensile strengths.
- **Flatwise compression test**, according to ASTM C-365 [4] to find compressive strength and compressive modulus of the sandwich core.

The construction of white room is such that the sandwich panels are held together within the Aluminium frames (as shown in Fig. 2). The loads expected by the panels are shear loads between the face sheets and the core material, flexural stress on the panel due to wind loads and self-weight of the structure. The flatwise tensile strength test evaluates the bond strength between the face sheets and the core material which gives an opportunity to enhance the properties required at failure region. The flatwise compression test allows to estimate the maximum compression load that can be applied on the panel while applying torque at the interfacing fasteners, local loads. Hence studying the sandwich panel by these respective tests only to obtain the required basic structural properties for computational load analysis on white room.

The relevant failure modes of metal-faced sandwich panels expected during the panel testing are as follows:

- Tensile failure of the faces.
- Wrinkling failure of the faces (due to compression).
- Shear failure of the core or the adhesion between the core and face of panel.
- Crushing failure of the face and core at the supports.

Universal tensile testing machine (UTM) is used to carry out the mechanical tests which requires loading rate of 0.5–2 mm/min and has a maximum capacity of 10 kN. Also, the test equipment gives the force–displacement (F-S) curve (or) the stress–strain curve (σ-ε) and the parametric values pertaining to the test by choosing the appropriate standard of testing as per ASTM [1–4].

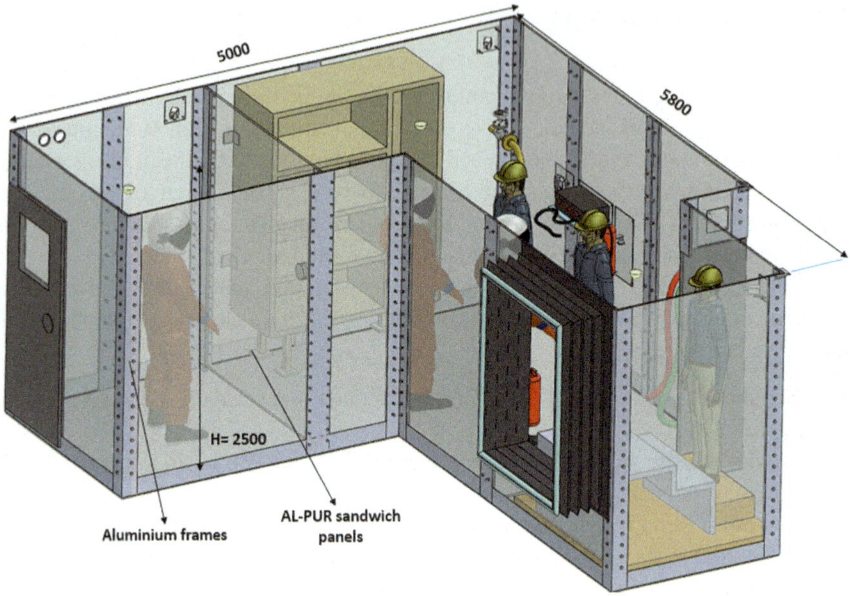

Fig. 2 White room configuration

2.1 Specimen Preparation

Sample specimens for the mechanical tests are prepared from two different lots of master panels received from the supplier which differ in terms of porosity, bond quality, etc.; the sandwich panel in both lots is made up of polyurethane (PUR) foam material (30 mm thick) as core with Aluminium Al-6061 (1.2 mm thick) sandwiched on either face of the foam core so that the sandwich panel total thickness becomes 32.4 mm. The specimen as required (given in Table 1) was prepared by suitably holding and cutting the master panel using the jig saw cutting machine. The cutting speed was high and low feed rate was given to avoid panel delamination while preparing the specimens. The specimen lengths are customized to suit specific requirements to suit the objectives of the material development and hence are different from the ASTM standards [1–4] as shown in Table 1. However, it is known that the length of the specimen insignificantly changes (0.01%) the strength of the material with respect to the tests we are proposing. Suitable adaptors are realized for each test to hold the specimen while testing.

The dimensions finalized are based on the space constraint in UTM machine (maximum width available is 400 mm) and complying to standards. The difference in densities of specimens [6] is due to the difference in material processing techniques and processing conditions maintained by the supplier (Table 2).

Table 1 Specimen dimensions for mechanical tests

Ref. No.	Test (standard no.)	Dimensions as per standard	Dimensions finalized for the test as per standards and testing possibility
[1]	Core shear test (ASTM C-273)	L—360 mm W—60 mm; T—32.4 mm	L—360 mm; W—60 mm; T—32.4 mm
[2]	Flatwise tensile strength test (ASTM C-297)	Minimum specimen facing areas should be 625 mm^2 and maximum of 10,000 mm^2	L—65 mm; W—65 mm; T—32.4 mm
[3]	Four-point bend test (ASTM C-393)	L—650 mm W—90 mm; T—32.4 mm	Span Length—360 mm (L—400 mm); W—45 mm; T—32.4 mm
[4]	Flatwise compression test (ASTM C-365)	Minimum specimen facing areas should be 625 mm^2 and maximum of 10,000 mm^2	L—27 mm; W—27 mm; T—26 mm

Table 2 Specimen Identities for mechanical tests

Sl. No.	Test	Specimen from lot-1 (core density—36 kg/m^3)	Specimen from lot-2 (core density—30 kg/m^3)
1	Core shear test	S1	S2, S3
2	Flatwise tensile strength test	T1	T2, T3
3	Four-point bend test	B1	B2, B3, B4
4	Flatwise compression test	C1	C2, C3, C4, C5

2.2 Core Shear Test

The standard ASTM C-273 [1] is used to determine the core shear strength and shear modulus of the core material. A sandwich panel specimen of length—360 mm, width—60 mm, and thickness—32.4 mm is prepared and the specimen is bonded [7], Ch. 13.2, pp. 318–319] to adapters with Araldite AW106 and Hardener HV953U compounds mixed in the 80:20 ratios which were cured for more than 8 h by holding and applying the clamping force.

The specimen is assembled to the UTM with the adapters which hold the sandwich panel at its opposite faces [5, 6] such that the line of action of force passes through the diagonally opposite corners of the core as shown in Fig. 3. The test is carried out with pre-load of 5N and cross-head travel speed of 2 mm/min. Two different density specimens were tested, which are prepared from two different lots as mentioned in Table 2 [6].

The results are shown in Table 3 and Fig. 4 (σ-ε graph). After pre-loading (Pt.1), the shear stress increases till first peak of 0.06 MPa (Pt.2 as shown in Fig. 4) and it decreases with increase in shear strain, because of the presence of air pockets/ porosity which results in shearing of layers even with minimal load which shows

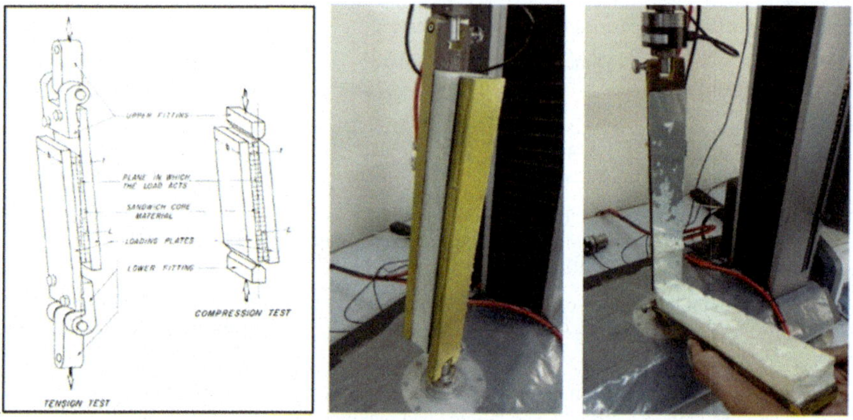

Fig. 3 Core shear test loading, setup, and failure

Table 3 Core shear test results

Sample No.	Shear strength (MPa)	Maximum load (N)	Shear modulus (MPa) (range 100–1500 N)
S1	0.1250	2700	1.57
S2	0.0987	2130	1.58
S3	0.0710	1340	1.34

the viscoelastic behaviour of the core material. The load needed to deform the specimen further, increases as the layers within core are now closely packed and the bond between core to face shear reaches it threshold to de-bond (Pt.3 to Pt.4). Hence, the shear stress increases and fails at ultimate strength (Pt.4) where the sample breaks. The test aborted at the instant when rapid load reduction was seen, which is a failure condition that occurred at the interface bond between core and face sheet as shown in Fig. 3. The breaking strength noticed for two specimens is different as they are taken from two different panels which vary in foam density, face to core bonding quality, etc.

2.3 *Flatwise Tensile Strength Test*

This test method determines the flatwise tensile strength of the core, the core-to-facing bond, or the facing of an assembled sandwich panel. This test method consists of subjecting a sandwich construction to a uniaxial tensile force normal to the plane of the sandwich. The force is transmitted to the sandwich through thick loading blocks, which are bonded to the sandwich facings [7], [Ch. 13.2, pp. 319–321]. The only

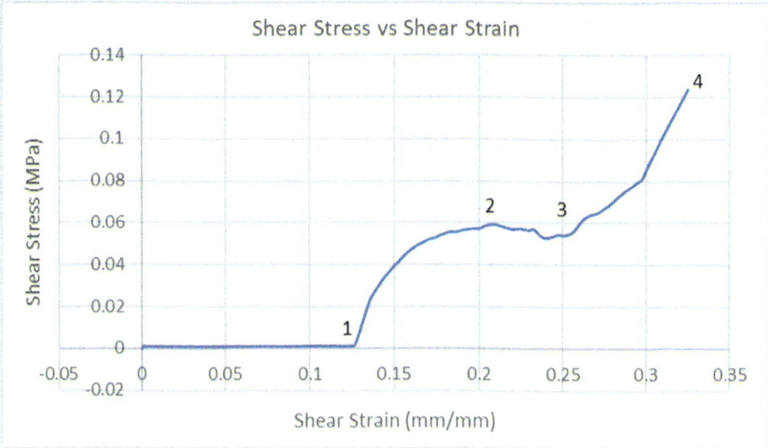

Fig. 4 Typical shear stress (MPa) versus shear strain (mm/mm) for sample-S1

acceptable failure modes are those which are internal to the sandwich construction. Failure of the loading block-to-sandwich bond is not an acceptable failure mode.

Three specimens of length—65 mm, width—65 mm, and thickness—32.4 mm are prepared from different master panels to test for flatwise tensile strength as per ASTM C-297 [2]. The adapters are bonded with specimen using Araldite AW106 and Hardener HV953U compounds mixed in the 80:20 ratios which were cured for more than 8 h by holding and applying the clamping force. The bonded specimens are assembled in UTM as shown in Fig. 5. Test is initiated with a pre-load condition of 5N and cross-head travel speed of 0.5 mm/min.

The force required for initial strain is high indicating the capability offered by the PUR foam to resist tensile loads (up to Pt.2). Then with minimal load, more strain is achieved due to porosity (from Pt.2 to Pt.3) and tensile stress increases as the effective area for resisting decreases as seen in Fig. 6 and ultimately breaks at peak point (Pt.4), i.e. flatwise tensile strength of the composite. The test aborted when rapid load reduction is detected, which is a failure condition that occurred within the core layer. The results are shown in Table 4 and Fig. 6.

2.4 Four-Point Bend Test

Out of several test methods to measure the shear strength and shear modulus, the recommended method for rigid plastic foam core panels is to load a short strip of panel, with thin metal faces, which is four-point bending [7], [Ch. 13.2, pp. 316–318]. The dimensions of specimen according to ASTM C-393 standard [3] are span length—360 mm (L—400 mm), width—45 mm, thickness—32.4 mm.

Fig. 5 Flatwise tensile test setup and failure obtained

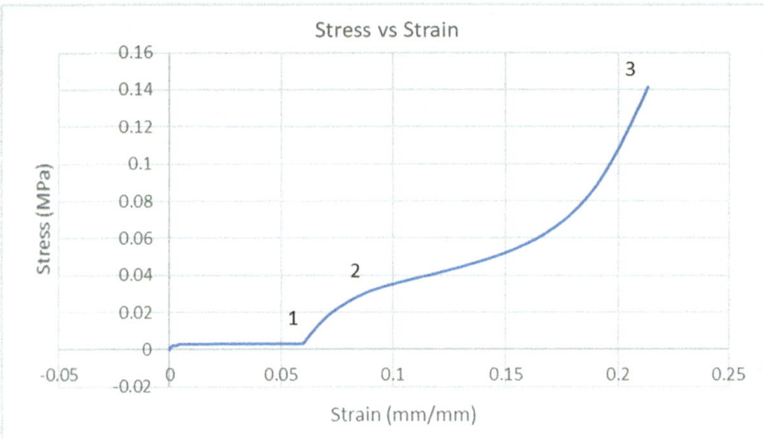

Fig. 6 Tensile stress (MPa) versus strain (mm/mm) for sample-T2

Table 4 Flatwise tensile test results

Sample No.	Flatwise tensile strength (MPa)	Maximum load (N)
T1	0.20	846
T2	0.141	597
T3	0.16	677

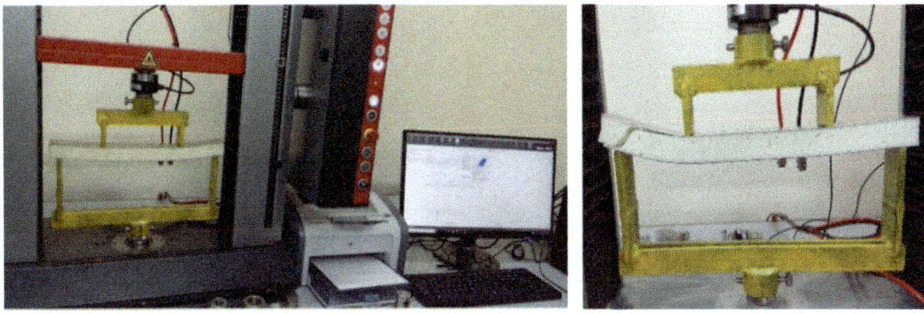

Fig. 7 Four-point bend test and failure obtained

Table 5 Four-point bend test results

Sample No.	Maximum load (N)	Core shear strength (MPa)	Face flexural stress (MPa)
B1	558	0.2	14.89
B2	401	0.14	10.70
B3	419	0.15	11.18
B4	427	0.15	11.41

Adapters are configured in such a way that the support should be given at 20–25 mm from both ends and the loading should be given at L/4 from support points, where 'L' is span length of the specimen. Loading pads should not have knife edges, as it can induce local crushing on the material while loading and also should avoid line contact while bending. The adapters are hinged to the top and bottom interfaces of UTM and the test setup is shown in Fig. 7.

The test is performed with 0.5 mm/min speed of cross-head travel while testing the panel specimen in the UTM. Mode of failure observed is between interface bond of core and cover layer, which propagated to core as shown in Fig. 7.

The panel resists bending force initially, and once the crack develops in the foam, its resistive capability decreases with larger strains until failure (Table 5, Fig. 8).

2.5 *Flatwise Compression Test*

This test is carried out to determine the compressive strength and modulus of sandwich cores as per ASTM C-365 [4]. These properties are determined in a direction normal to the plane of facings as the core would be placed in a structural sandwich construction. Four specimens (only core) of square cross-Sect. 27 mm x 27 mm are prepared with core thickness of 26 mm from different master panels. The specimen is placed between the two parallel stiff loading plates [6, 7], [Ch. 13.2, pp. 320–322] of a suitable compression testing machine and loaded in increments as shown in Fig. 9.

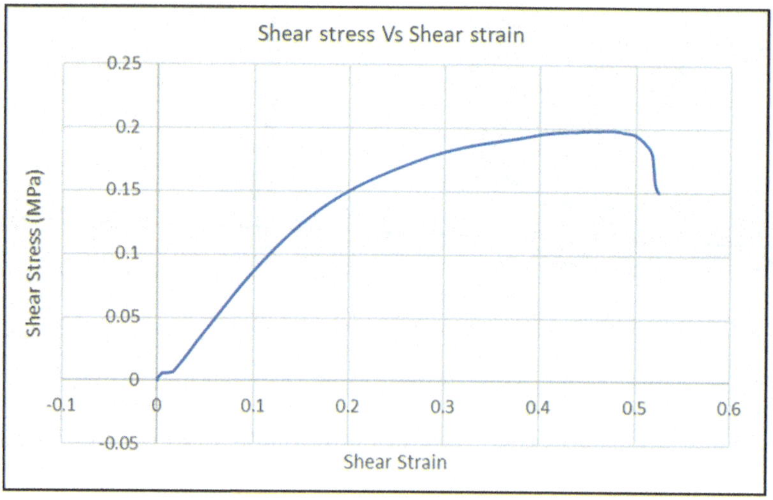

Fig. 8 Shear stress (MPa) versus shear strain for sample-B1

Fig. 9 Flatwise compression test and failure obtained

The speed of cross-head travel maintained while testing the panel specimen in the UTM is 0.5 mm/min. Test results are shown in Table 6 and Fig. 10.

Mechanical testing results of sandwich panel are used to determine properties/strengths of white room panel. As per the loads (cyclonic wind loads, equipment hanging inside/on the white room) to be induced on sandwich panel, the corresponding stresses are evaluated and it is found safe, i.e. lesser than the value of strength evaluated by tests.

Table 6 Flatwise compression test results

Sample No.	Force @ 2% deformation (N)	Flatwise compressive strength @ 2% deformation (MPa)	Flatwise compressive modulus (MPa) (range 40–110 N)
C1	255, Ex	0.126	5.8
C2	41.1, Ex	0.0563	2.90
C3	29.5, Ex	0.0405	2.18
C4	43.9, Ey	0.0562	2.91
C5	23.2, Ez	0.0318	1.48 [30–70 N]

where 'Ex' is the specimen tested in the direction parallel to foam rise and 'Ey' and 'Ez' are the specimens tested perpendicular to foam rise. Forces in cases 1 and 2 are different as case-1 is test with face sheets and case-2 is just the foam alone

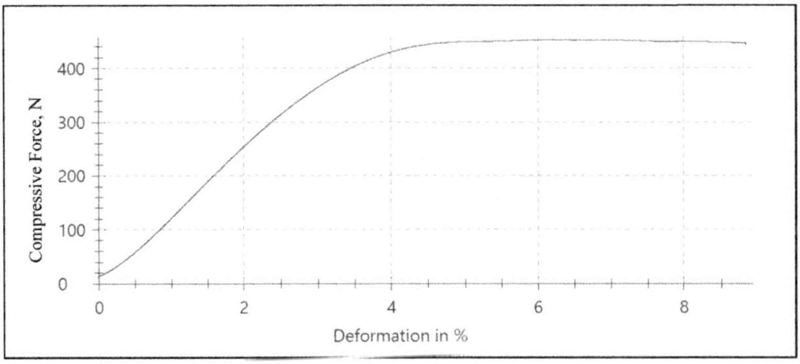

Fig. 10 Compressive load (N) versus deformation (%) for sample-C1

3 Thermal Test

Composite panel curing temperature for adhesive bonding of Al facing with PUR core is 80 °C and the maximum temperature observed at the location where white room is envisaged, during LVM3, M1 mission launch is 61 °C on front face and 47.3 °C on back face (on CFRP panel). On this basis and to verify the above parameters on Al-PUR sandwich panel, it is exposed to 70 and 80 °C in thermal humidity testing chamber (shown in Fig. 11) for 2 min. To monitor the temperature attained by composite panel, two nos. of K-type thermocouples are bonded on either side of composite panel of dimension 350 × 110 × 32.4 mm as shown in Fig. 11.

After hot soaking the composite panel for elevated temperatures for 2 min, it is taken out and cooled at ambient temperature as shown in Fig. 12. Visually no abnormalities are found and panel is tested in ultrasonic testing (to check de-bonding) and it is found normal.

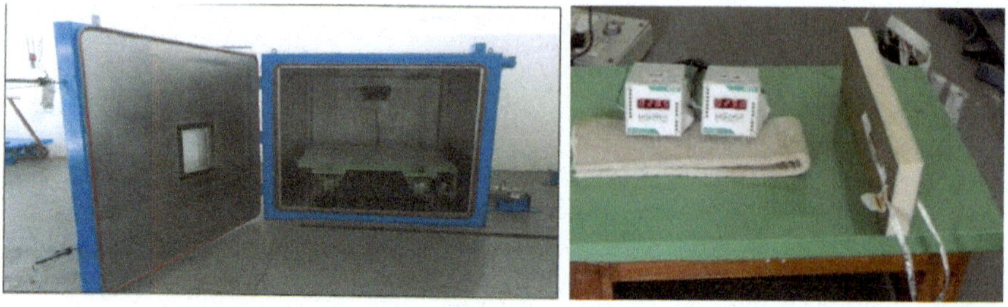

Fig. 11 Thermal humidity testing chamber and bonding of thermocouple to specimen along with data monitoring

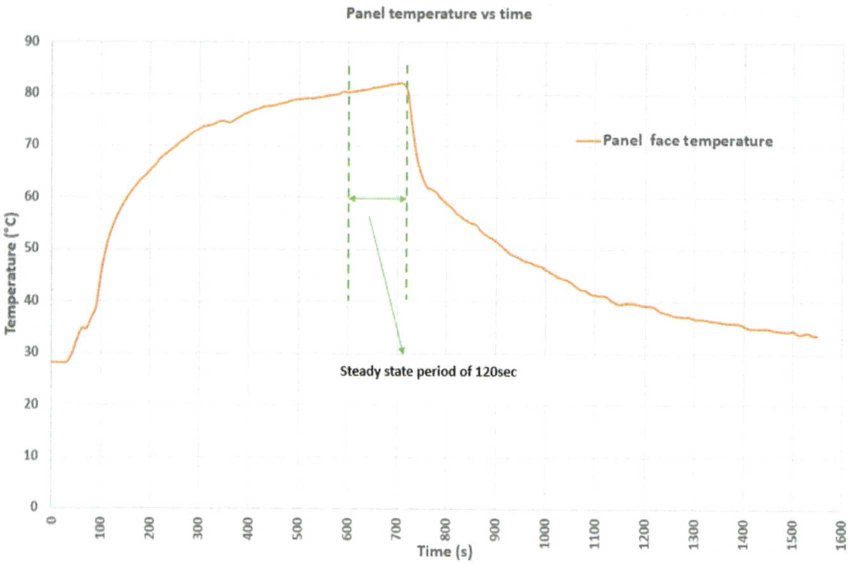

Fig. 12 Panel face temperature (°C) versus time(s)

4 Dynamic Testing

The proposed sandwich panel for Gaganyaan white room has to endure intense dynamic environments, i.e. over pressure, acoustics, vibrations, and unsteady heat flux during lift-off of launch vehicle. In order to ensure its structural integrity under such environment, measurements were taken on the panel by positioning it in appropriate locations during static firing of solid rocket motor and the actual lift-off of a launch vehicle.

4.1 Dynamic Environment Exposure During Static Test

S139 is the solid booster stage made of hydroxyl-terminated polybutadiene (HTPB) propellant of 139 tonnes which gives high acceleration during lift off. The stage burns for 110 s and gives a maximum thrust of 4800 kN. It is used as first stage in Polar Satellite Launch Vehicle (PSLV) and Geo-Synchronous Satellite Launch Vehicle (GSLV) so exposing sandwich panel to static test of this stage is similar to rocket launch environment. The static test is performed horizontally on the test bed to evaluate the ballistic performance of stage like thrust, chamber pressure and demonstrate the segment joints, igniter interfaces, overall integrity of stage.

Composite panel is tested under S139 static test exhaust to study the after-effects on panel. Al-PUR composite panel of size 700 × 425 × 32.4 mm is prepared and mounted in the downstream of nozzle exit at $13D_e$ (30.9 m) and 23° with respect to jet axis at a height of 3.2 m equal to the centre height of the jet. Plan view of measurement location is shown in Figs. 13, 14.

The measurements planned during static test firing are:

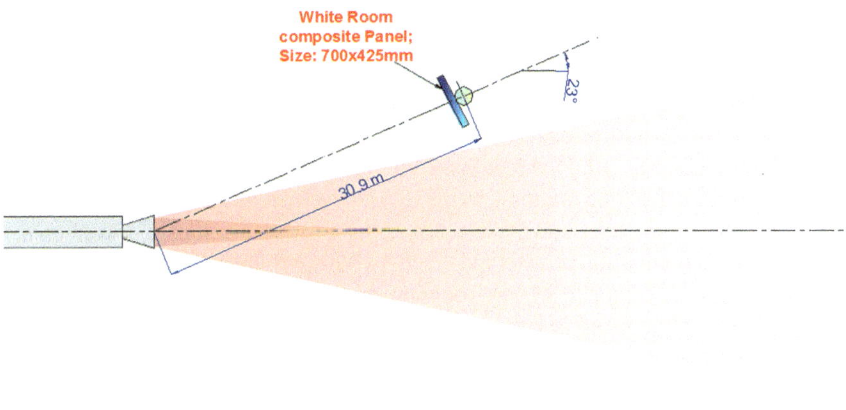

Fig. 13 Location of panel with respect to nozzle exit

Fig. 14 Test setup with panel location

- Heat flux (1 no.) Slug type gauge with 150 W/cm^2 measurement range.
- Unsteady Pressure (1 no.) Kulite make, XCS-062 sensor with 15psiG range.
- Temperature (3 nos.) K-type thermocouples are used for panel and air temperatures.
- Vibration (1no) Uni-axial accelerometer, Endevco model, 25 mV/g sensitivity.
- Acoustics (1 no.) Endevco microphone, model 2510.

After static test firing and data acquisition, the results obtained are shown in Table 7 and Fig. 16.

The heat flux is measured during the entire duration of firing as shown in Fig. 17. Maximum heat flux of 3.6 W/cm^2 was experienced by the panel at T + 0 s (just after firing) and has gradually reduced to 1 W/cm^2 by the end of firing. The reduction in heat flux may be attributed to the gradual increase of panel/ambient air temperature during the motor firing.

Unsteady pressure measurement on the panel was carried out using Kulite sensor, (model: XCS-062). Sensor was mounted on top of the panel and is fixed using an adhesive tape. Figure 15 shows that an 'overall sound pressure levels (OASPL)' of 165–170 dB is experienced by the panel during the course of motor firing. Also, an initial ignition transient load known as 'ignition over pressure (IOP)' of around 220 mbar is acted on the panel at T + 0.3 s as shown in Fig. 18.

Front wall temperature (GT1) has reached a maximum value of 210 °C at end of firing, back wall temperature (GT3) has read a maximum value of 90 °C, and air temperature (GT2) has raised from around 30 °C from T + 0 s to ~120 °C at end of firing as shown in Fig. 19. Post-test inspection of the panel shown high temperature exposure of foam for long duration on the jet side face of the panel and delamination between aluminium sheet and PUR foam is seen at bottom location as shown in Fig. 20. This delamination is observed due to exposure of the panel to the worst-case scenario, i.e. solid motor firing environment for a long duration of 110 s, whereas in actual case, white room is exposed to vehicle dwell time of not more than 8 s at launch pad. Even the lateral distance from the exhaust jet to the panel location is more in actual case.

Table 7 Sensors data from static test firing

Sl. No.	Properties	Maximum value
1	Heat flux	3.6 W/cm^2
2	Overall sound pressure level (OASPL)	165–170 dB
3	Ignition over pressure	220 mbar
4	Front wall temperature	210 °C
5	Back wall temperature	90 °C
6	Air temperature	120 °C
7	Vibration	400 g (saturated)
8	Acoustics	Saturated

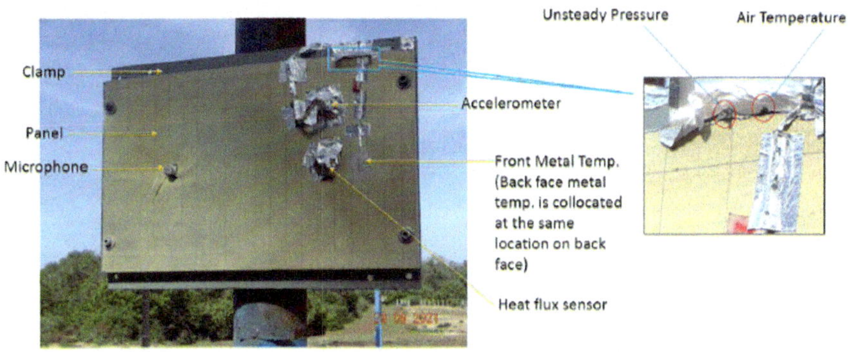

Fig. 15 Sensors' mounting scheme

Fig. 16 S139 ST-04 static firing

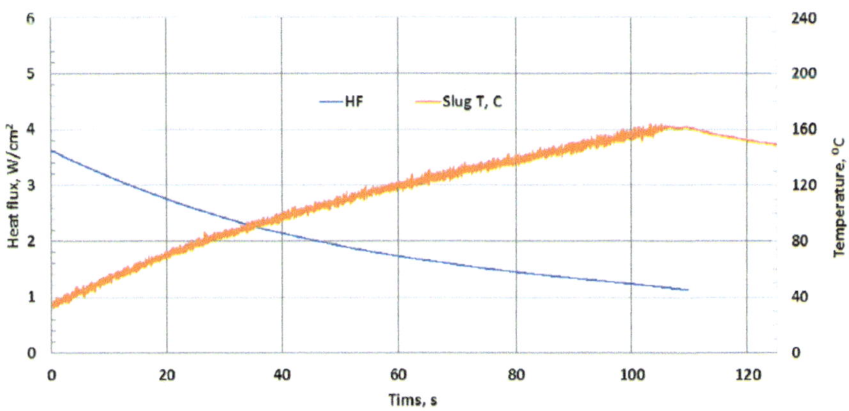

Fig. 17 Heat flux versus time

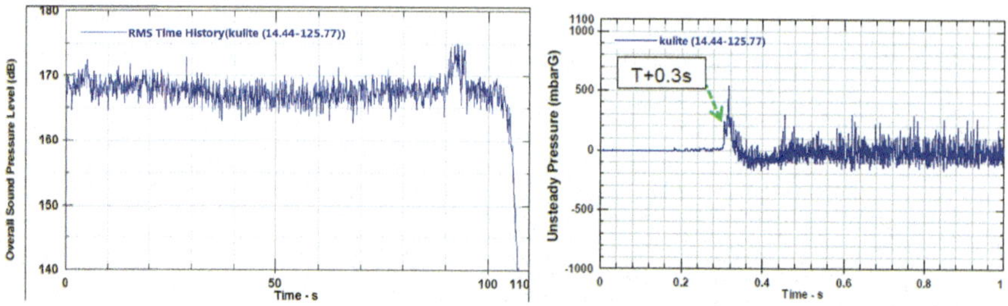

Fig. 18 Overall sound pressure level and unsteady pressure versus time

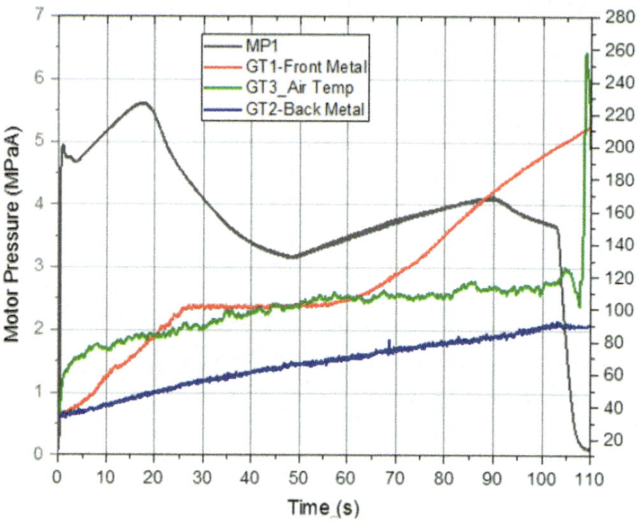

Fig. 19 Motor pressure, panel, and air temperatures versus time

Fig. 20 Before and after static firing effects on panel

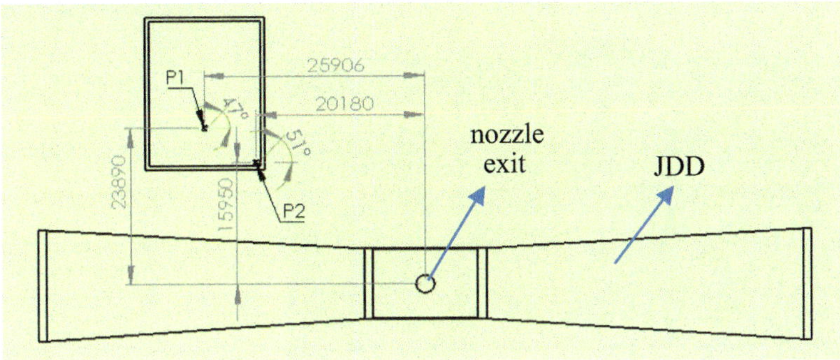

Fig. 21 Panel location with respect to nozzle centre

4.2 *Rocket Launch-Induced Environment*

The sandwich panel specimen was exposed to environment of S139 motor static test where the burn duration is 110 s which differs from actual dwell time of 6 s during its lift-off of Gaganyaan launch vehicle from second launch pad (SLP). Hence, it is recommended to subject the panel to the dynamic environment during actual launch and evaluate the integrity of panel for acoustics, temperature, and vibration. Panel mounting location is selected to simulate L/D_e of the white room location on UT platform, which is at a distance of 25.9 m along MST track and 23.9 m along the Jet Deflector Duct (JDD) from the nozzle exit, at which earlier measured data during the launch of PSLV C26 & C28 missions are available as shown in Fig. 21.

The acoustic levels predicted at that location are about 150 dB, overpressure of 10 mbar, and temperature in excess of 150 °C for about 10 s duration. Except acoustic level, other two environments (over pressure and thermal) will be simulating close to that of earlier measurements. Acoustic spectrum at panel location has a similar frequency distribution; however, the amplitude is on lesser by ~15 dB. Hence, the measured vibration levels on composite panel (during C52 launch) are extrapolated for the earlier measurements' condition and the composite panel integrity will be assessed for a given panel size. The measurements planned and the sensors mounting plan are shown in Fig. 22.

The composite panel of size 800 × 500 × 32.4 mm was prepared from a panel of 2.5 × 1.2 × 32.4 mm size by jig saw cutting machine. All the sensors with connectors were mounted on the panel and mounted on the frame. The frame was assembled with the pole at an angle of 47° facing the vehicle as shown in Fig. 23.

After exposing to dynamic launch environment of PSLV C-52 launch, the data acquired are analyzed and results obtained are shown in Table 8 and Fig. 24.

It is observed that from T + 10 s, the air temperature and temperature on the front face of the panel measured using K-type thermocouple and heat flux start to increase. Panel face temperature reaches a maximum value of 55.4 °C and is constant (as shown in Fig. 25) for a period of time till the surrounding air temperature (245 °C)

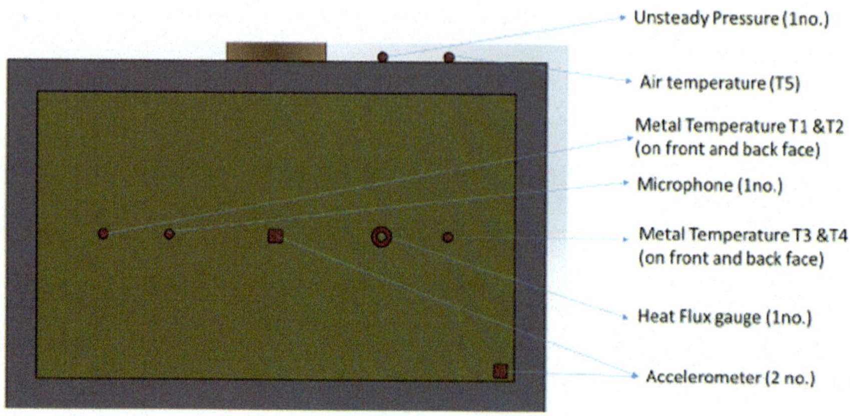

Fig. 22 Sensor mounting scheme

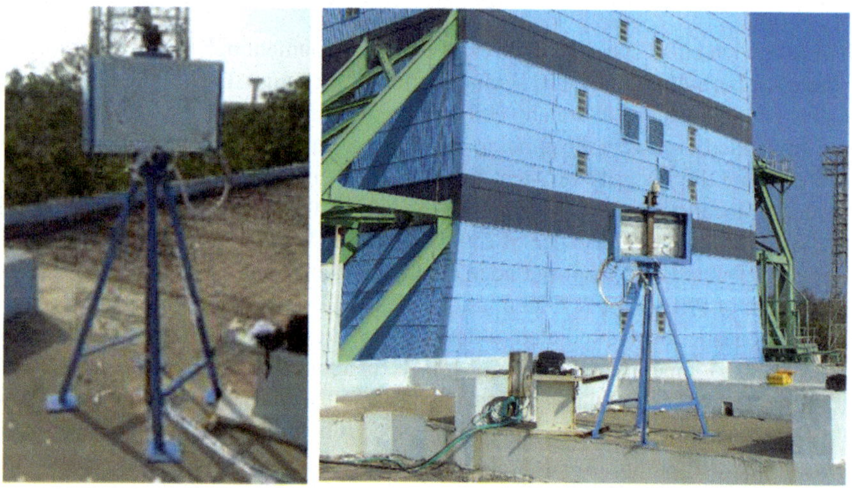

Fig. 23 Panel fixing arrangement

Table 8 Sensors data from PSLV launch

S No.	Properties	Maximum value
1	Heat flux	4.4 W/cm^2
2	Unsteady pressure	10 mbar
3	Front wall temperature	55.4 °C
4	Air temperature	245 °C
5	Accelerometer	67.5 g
6	Acoustics	157.7 dB

Fig. 24 Location of panel with respect to vehicle and launch

is higher than the panel front face temperature. Around T + 10 s, the maximum value of heat flux of 4.4 W/cm^2 is observed on the panel as shown in Fig. 25.

The maximum ignition over pressure/unsteady pressure was observed around 0.35 s which is nearly 10 mbar. The panel has experienced a maximum acoustic environment of 157.7 dB (OASPL) as shown in Fig. 26. Post-test inspection of the panel has shown exposure to medium-level temperatures, and no delamination is observed post-launch.

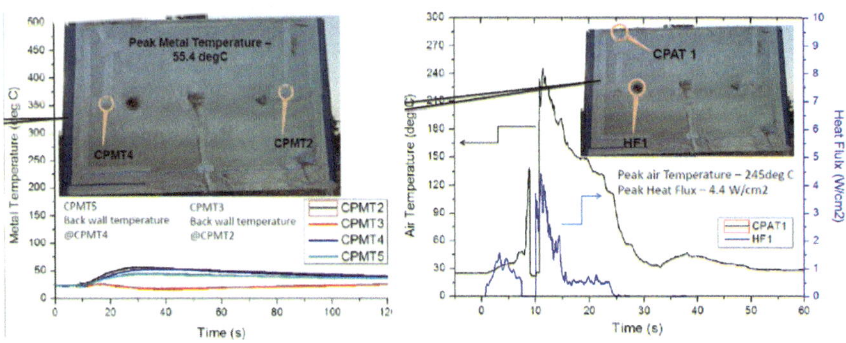

Fig. 25 Panel, air temperatures (°C), and heat flux (W/cm^2) versus time (s)

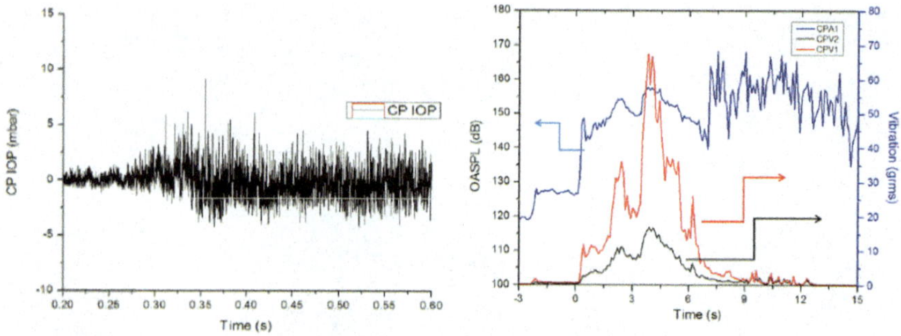

Fig. 26 Unsteady pressure (mbar) and OASPL (dB) versus time (s)

5 Conclusions

- Specimens of sandwich panel prepared from two master panels made up of polyurethane (PUR) foam material (30 mm thick) as core with Aluminium Al-6061 (1.2 mm thick) sandwiched on either face with total thickness of 32.4 mm is tested as per standards [1–4] for mechanical properties, viz., core shear strength, shear modulus, flatwise tensile, and compressive strengths. The various properties evaluated, shows that with increase in density of the core material and effective bonding between core and facings increases the strength by ~15% as can be inferred from the specimens tested from two different lots of different densities (Table 9).

- Sandwich panel is exposed in thermal humidity testing chamber at elevated temperatures of 70 and 80 °C. The thermal soaking at reference severe conditions yielded no delamination; integrity is verified using ultrasonic testing.
- Sandwich panel is subjected to rocket launch-induced environment during S139 static test firing as well as PSLV rocket launch. The dynamic parameters measured, viz., acoustics (157.7 dB), vibration (67.5 g), heat flux (4.4 W/cm^2), over pressure (10 mbar), air temperature (245 °C), and facing temperatures (55.4 °C), are much benign for the panel required for white room at launch pad.

Table 9 Properties of composite panel

S No.	Properties	Maximum value (Mpa)
1	Shear strength	0.125
2	Ultimate shear strength	0.2
3	Shear modulus	1.58
4	Flatwise tensile strength	0.2
5	Facing bending stress	14.89
6	Core flatwise compressive modulus and stress	5.8 and 0.126

References

1. ASTM C273/ C273M-11 (2011) Standard test method for shear properties of sandwich core materials, ASTM International. Retrieved from www.astm.org.
2. ASTM C297/C297M–15 (2015) Standard test method for flatwise tensile strength of sandwich constructions, ASTM international. Retrieved from www.astm.org.
3. ASTM C393/C393M-11 (2011) Standard test method for core shear properties of sandwich constructions by beam flexure, ASTM international. Retrieved from www.astm.org.
4. ASTM C365/C365M-11 (2011) Standard test method for flatwise compressive properties of sandwich cores, ASTM International. Retrieved from www.astm.org.
5. Escusa G, Cruz F, Sena Cruz J, Pereira E, Valente I, Barros J (2017) Shear behaviour of polyurethane foam presented at the APFIS2017—6th Asia-Pacific Conference on FRP in structures, Singapore, 19–21st July 2017
6. Witkiewicz W, Zielinski A (2006) Properties of the polyurethane (PU) light foam. In: Advances in materials science, vol 6, no 2(10)
7. Davies JM (2001) Mechanical testing. In: Lightweight sandwich construction, published on behalf of CIB Working Commission W056 Sandwich Panels, Manchester, UK, 2001, Ch. 13.2, pp 316–322

Measurement of Coefficient of Thermal Expansion of Composite Structures Using Fiber Bragg Grating Sensor

Vijay Kumar, C. M. Revathi, Srividya Ganti, Nitesh Gupta, and M. J. Augustin

1 Introduction

The coefficient of linear thermal expansion is a material property indicative of the extent to which a material expands upon heating. Structural design necessitates the determination of the Coefficient of Thermal Expansion (CTE) to account for the issues that may arise due to thermal expansion after structural assembly during service life. Coefficient of Thermal Expansion is defined as

$$\alpha = \left(\frac{\mathrm{d}l}{l\mathrm{d}T}\right), \tag{1}$$

where $\mathrm{d}l$ is the change in length for the temperature change $\mathrm{d}T$ and l is the original length.

There are various conventional measurement techniques for the determination of the CTE, namely dilatometry [1], interferometry [2] and thermomechanical analysis [3]. The dilatometry and thermomechanical analysis rely on measuring the displacement occurring to the rods or probes attached to the specimen under test. These methods require the specimen to be placed inside a cell, which is limited in size in general. Also, to measure the CTE in different directions, which is a requirement in composite specimens, additional specimens are to be prepared in that direction. Another issue with mechanical dilatometers is that CTE of the cell—in which

V. Kumar (✉) · C. M. Revathi · S. Ganti · N. Gupta · M. J. Augustin
Advanced Composites Division, National Aerospace Laboratories, HAL Airport Road, Kodihalli, Bangalore, Karnataka 560017, India
e-mail: vijaykumar@nal.res.in

M. J. Augustin
e-mail: mjaugustin@nal.res.in

M. R. Manohara Babu et al. (eds.), *Recent Advances in Composites Engineering and Research*, Springer Proceedings in Materials 39,
https://doi.org/10.1007/978-981-99-8807-5_24

samples are placed, changes over time which can affect the accuracy of the measurement. Interferometry requires optical measurements; this depends a lot on the surface preparation of the specimen [4]. Thermomechanical analysis is similar to dilatometry, but controlled loads are applied to the specimen, and displacement of the push road in contact with the sample is measured as a function of temperature. These devices require two calibrations as per ASTME2113, ASTME1363 standards: one for calibrating the length change and second for temperature calibration of the thermomechanical analyzers [5]. A review of different CTE measurement methods is provided in the review paper [6, 7].

As $\frac{dl}{l}$ is the linear strain, accurate measurement of thermally generated strain with temperature can be used for the estimation of the CTE. Bonding strain gauge and Fiber Bragg Grating Sensors for metals like Aluminum and Invar are reported in literature for measurement of CTE [8, 9]. Use of FBG sensor for monitoring the CTE for the coating of FBG at cryogenic is reported in [10]. The advantages of using Fiber Bragg Grating Sensors are, their immunity to the Electromagnetic Interference (EMI), small in size and weight [10] and can be surface bonded on any specimen in any orientation. These features make it as suitable choice for the measurement of CTE. Also, the sensor technology has matured in terms of availability of sensors and data acquisition system at moderate costs. Any calibrated temperature chambers can be used for the heating the specimens instrumented with the FBG sensors. Also, multiple specimens can be tested at same time due to the multiplexing capability of the Fiber Bragg Grating sensors. This make the testing faster when many samples need to be characterized.

This paper presents a method for the CTE measurement of composite specimens using Fiber Bragg Grating (FBG) sensors. FBG sensors consist of periodic refractive index variation made on the core of optical fiber. When a broadband source is given to the FBG, one wavelength will get reflected back. So FBG sensors are typically a wavelength-based sensor, wherein a change in wavelength gives information about the external perturbation, such as strain, temperature as per the equation [11]

$$\Delta\lambda_B = \lambda_B(1 - \rho_\alpha)\Delta\varepsilon + \lambda_B(1 + \xi)\Delta T \tag{2}$$

ξ is the thermo-optic coefficient and ρ_α is the effective strain-optic coefficient, which is a function of the Poisson ratio ν of the fiber and the Pockel constant ρ_{ij} as

$$\rho_\alpha = n_{\text{eff}}^2/2[\rho_{12} - \vartheta(\rho_{11} - \rho_{12})] \tag{3}$$

In the proposed CTE measurement technique, the FBG sensor is surface bonded on the test specimen. The specimen is then subjected to increasing temperature in a temperature-controlled chamber (calibrated). The reflected wavelength from the FBG is measured using a suitable data acquisition system. This reflected wavelength measured during the experimental test is the sum of wavelength shifts due to the thermally induced stresses/strains on the structure in addition to the temperature-induced shift. In order to determine strains due to the thermal expansion of the structure, the effect of wavelength shifts due to temperature induced on the structure has to be compensated. This is achieved by using a bare FBG sensor (not bonded to the specimen) during the experiments.

Experimental studies are carried out to characterize the CTE values for BHOR and AS914 prepreg systems using FBG sensors. Next sessions will discuss methodology for the preparation of test specimen, sensor installation, and experimental results for various studies.

2 Methodology

The following methodology is proposed for the CTE measurement:

- The test specimens are prepared in the required dimension (based on the studies, it is found that a size of 55 × 30 mm is best suited for the measurement)
- FBG sensors are bonded to the specimen using cyanoacrylate on the test specimens.
- Specimens with FBG's surface bonded is placed in the chamber
- FBGs are connected to sm130 interrogator through the feedthrough facility in chamber.
- FBG baseline is taken at room temperature (T).
- Temperature is increased to T + ΔT.
- The FBG readings are monitored till they stabilize. This ensures that specimen get socked in the set temperature. The values are recorded at this temperature for few seconds.
- Increase temperature by ΔT and record the data (This is continued till 70 °C as the cyanoacrylates changes its behavior after 70 °C).
- Take average of the reading of FBG for each temperature.
- Calculate the wavelength shifts for FBGs. Subtract the bare FBG wavelength shift from specimen FBG wavelength shift for each temperature.
- Convert the temperature-compensated wavelength shift to strain (conversion factor 1.23 pm/με) and plot the temperature versus strain. The slope is estimated as the CTE.

The schematics for the experimental setup are presented in Fig. 1.

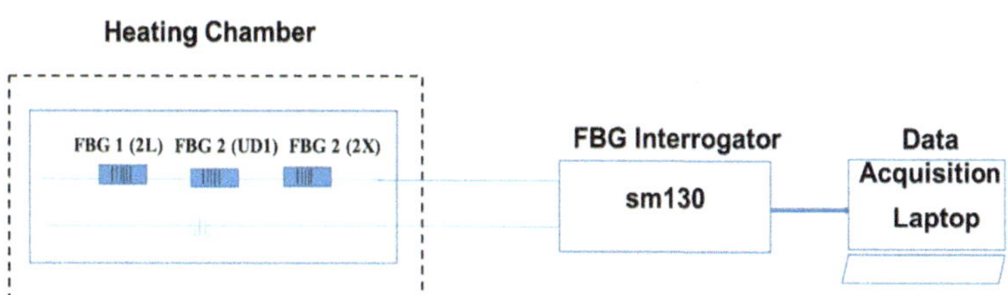

Fig. 1 Schematic of the test setup

As shown in figure, the test specimen with the FBG sensors bonded on the surface is kept in the heating chamber, where controlled heating can be carried out.

The sensors are connected to the instruments through the feedthrough access cover in the chamber to the data acquisition device (Micron Optics sm130) which is controlled through a laptop. The data acquisition is carried out through the OEM software in the data acquisition laptop.

2.1 Sample Preparation and Sensor Installation

Different samples are used in the study. The list of specimens used in the study along with the details of composition, layup, and dimension is provided in Table 1. As the method is new to find the best suitable specimen size, different sizes are being used.

Initially, to verify the methodology, an aluminum specimen was used in the study. Further unidirectional specimens and quasi-isotropic specimens were used in the study using IM7-8552, IM7M56, Bhor prepreg system UC-162-42S-M15N-38. Studies were extended to the BD prepreg, where the CTE along the warp and weft direction are measured.

During fabrication, an extra fabric layer called a peel ply is laid on the surface of the specimens to give a better surface finish. This layer will be peeled off when being used. This provides a fresh, clean, roughened surface of matrix resin to which the sensors can be bonded. To install the sensors, this peel ply is removed from the specimen, and sensors are installed on the lines which are marked. The advantage of this method is that in the single CTE test, both directions' sensors can be installed

Table 1 Specimen details

Si. No.	Material	Layup	Dimension
1	Aluminum	–	183 × 34 × 3 mm
2	Hexply 8552/33%/134/IM7-12K	[+45/45/0/90]3s QI	50 × 50 × 3.12 mm
3	Hexply 8552/33%/134/IM7-12K	$[0]_{24}$ UD	50 × 10 × 3.12 mm
4	Hexply M65/35%/134/IM7-12K	[+45/−45/0/90]3s. QI	55 × 30 × 3.12 mm
5	Hexply M65/35%/134/IM7-12K	[+45/−45/0/90]3s	45 × 10 × 3.12 mm
6	Hexply M65/35%/134/IM7-12K	$[0]_9$	250 × 100 × 1.2 mm
7	Bhor UC-162-42S-M15N-38	$[0]_{12}$	55 × 30 × 2 mm
8	Hexply 914/40%/285H5/AS4C-3K/ 1170 mm	$[0/90]_7$	55 × 30 × 2.1 mm

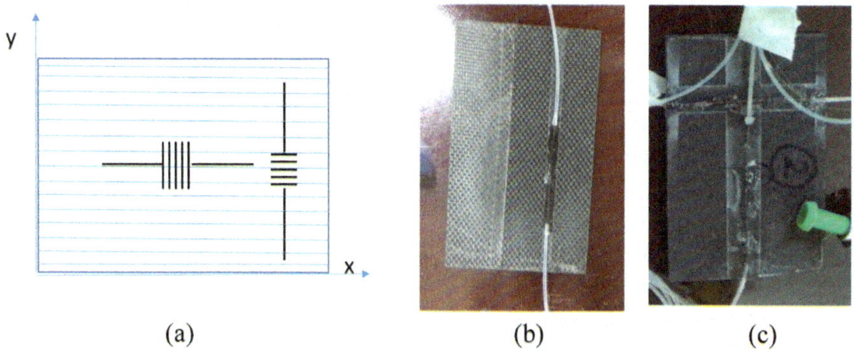

Fig. 2 Sensor installation schematics and the specimens with the sensors installed

and monitored simultaneously in one test. One sensor is installed along the length following the center axis and another sensor along the breadth is slightly offset from the center so that the sensing regions are not crossing on the specimen. Flash breaker tapes are pasted on both sides of the line marking by giving a small gap on the specimen where the sensor fiber can be placed and bonded. This gap is gently rubbed with emery paper. Then, fiber is placed on the prepared surface and cyanoacrylate is applied over the fiber and applied pressure gently so that the adhesive is cured. A schematic of the sensor installation scheme and few specimens with the sensors installed is presented in Fig. 2. Figure 2a shows the schematic for the location of the FBG sensors. The fiber direction in the CFRP is indicated as lines along x-direction. Two FBG sensors are installed on the specimen one along the x–direction, i.e., the fiber direction on the specimen and other perpendicular to the fiber direction. Figure 2b shows the sensor installed on the x-direction and Fig. 2c on x- and y-direction.

Teflon tubes (L28) are kept on both sides of the sensing region to safeguard the fiber from breaking and making the handling easy. A small part of the tube is also bonded to the specimen as shown in Fig. 2. This will prevent the fiber from breaking at the edges. Considering all these lengths to be bonded, avoiding the crossing of the fiber, it is found that specimens of size 55 × 30 mm are optimum for tests. The length is kept high to avoid the crossing of the perpendicular sensing region of the fiber over the sensing region along the fiber direction. While bonding the sensor, this region as bare fiber will get bonded to the specimen. Then around 5–7 mm is inserted into the Teflon tube which is bonded to the specimen. This makes 30 mm specimen size good enough for the sensor bonding in one side. The flash breaker tapes are removed after the sensors are installed. The use of flash breaker tapes ensures that the sensors are installed on the required installation line and restricts the spreading of the adhesive on the specimen surface.

2.2 Experimental Studies

Once sensors are installed on the specimens, they are kept on a clean smooth surface inside heating chamber. This eliminates the chances of friction, or due to some stickiness, the material is prevented from expanding freely. In the experimental study, Hylam sheet is used to place the specimens in the chamber. While placing, care must be taken that the specimens are not under strain or touched by due to the nearby specimens. To prevent specimen from getting pulled, the other part of the fiber is secured by sticking to the Hylam sheet. The bare FBG for the temperature compensation is placed on the sheet. Figure 3 presents the specimens in heating chambers and the complete experimental setup. The specimens with sensors bonded are kept on Hylam sheet whose surface is smooth. The temperature sensor in the chamber monitors the chamber temperature. Before starting the chamber, the FBG wavelengths of the sensors installed on the specimens and the free FBG (for finding the wavelength shift due to temperature alone) are recorded. This is saved as the baseline. The temperature is raised by 5° or 10° and the wavelengths are recorded. This is repeated for different temperatures. The baseline value of each sensor is subtracted from each of the sensor values. The wavelength shift in specimen sensors will be due to the temperature and temperature-induced strain. The bare FBG will be seeing the wavelength shift due to temperature alone. The wavelength shift due to temperature alone is subtracted from all the sensors on the specimen, which is then wavelength shift due to temperature-induced strain. This wavelength shift is converted to strain (strain conversion factor 1.23 pm/$\mu\varepsilon$) and plot the temperature versus strain. The slope is estimated as the CTE.

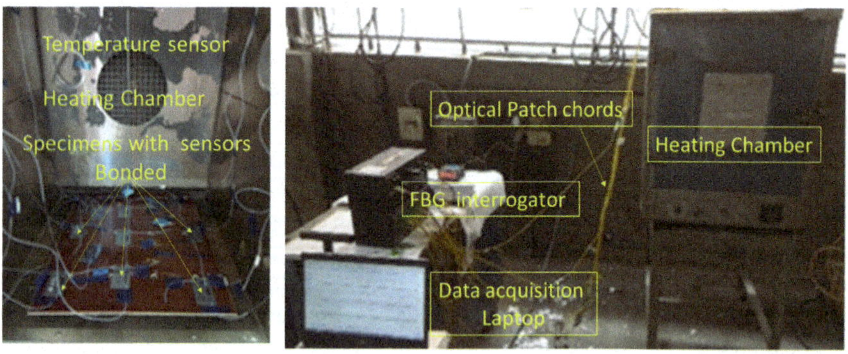

Fig. 3 Specimens with the sensors installed in the heating chamber and experimental setup

3 Results and Discussion

The data are recorded during the test and are analyzed as discussed above. Following figures represent the result for the specimen provided in Table 1. Three specimens are used in the study, and experiment is repeated at least three times (after the chamber get cooled). The average of the CTE measured across the trials across the specimen is taken to be the CTE of the material. Figure 4 represents the results for aluminum 2024.

It can be seen that the CTE values along the length and breadth direction gives out same values, confirming that there is no constraint over the specimen size. The average CTE measured for Aluminum 2024 material is 22.27 × 10^{-6}/ °C, with standard deviation of 0.24.

Figure 5 summarizes the results for Hexply 8552/33%/134/IM7-12K for the specimens with serial number 2–4 in Table 1. In agreement with the studies carried out on aluminum though different dimensions are there for the Quasi Isotropic (QI), they gave the same results for the x- and y-direction. For unidirectional (UD) specimen, x is the fiber direction and y is the matrix direction.

The average values estimated for all the specimens across the trials are presented in Table 2.

Similar studies are carried out for Hexply M65/35%/134/IM7-12K specimens with Quasi Isotropic and unidirectional layup sequence. The average values CTE obtained are presented in Table 3.

These values agree with the manufactures data sheet.

The studies were extend to characterize the BHOR prepreg (UC-162-42S-M15N-38). Three specimens from three batches of fabrication (Batch I, II, and III) are being used in the study. In all specimens, FBG sensors are installed in both the directions and all specimens are UD. The summary of the results is presented in Table 4.

It can be seen that the values are in range and across each batch.

In the studies so far unidirectional prepreg tapes were used. The studies were carried out using the bidirectional prepreg materials. The FBG sensors are bonded along the warp and weft direction. Using the methodology described above, the CTE values are estimated. Figure 5 presents a typical graph showing the variation

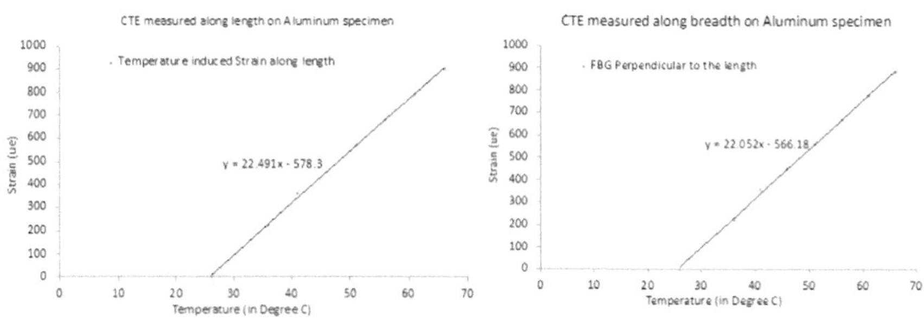

Fig. 4 CTE measurement in aluminum

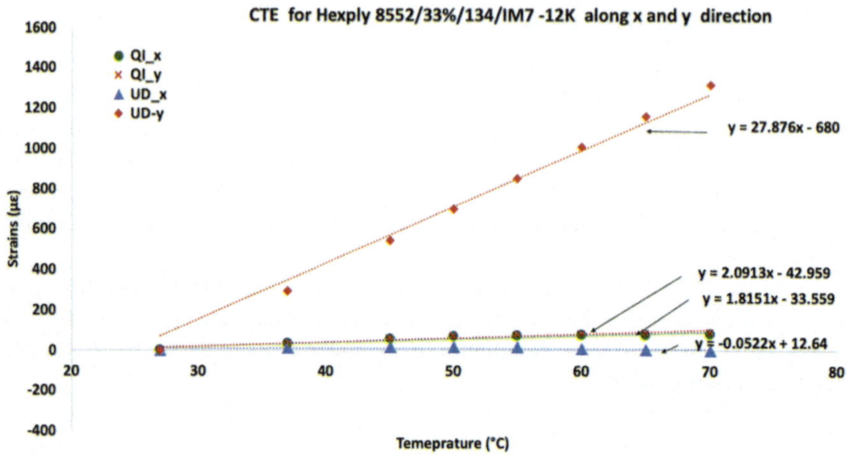

Fig. 5 CTE measurement for the Hexply 8552/33%/134/IM7-12 K specimens

Table 2 CTE values measured for Hexply 8552/33%/134/IM7-12 K in (/°C)

Configuration → Direction↓	QI	UD
X-direction	1.8×10^{-6}	-0.0522×10^{-6}
Y-direction	2.1×10^{-6}	27.876×10^{-6}

Table 3 CTE values measured for Hexply M65/35%/134/IM7-12 K in (/°C)

Configuration → Direction↓	QI	UD
X-direction	1.58×10^{-6}	-0.54×10^{-6}
Y-direction	1.36×10^{-6}	25.77×10^{-6}

Table 4 CTE values measured for Bhor UC-162-42S-M15N-38 in (/°C)

Batch	X-direction (0°)	Y-direction (90°)
Batch I	-1.6273×10^{-6}	29.518×10^{-6}
Batch II	-1.6575×10^{-6}	30.282×10^{-6}
Batch III	-1.3071×10^{-6}	30.889×10^{-6}

of thermal-induced strain versus temperature, and Table 5 presents the average CTE value obtained using four specimens over four trials. The SD across the trials is ~0.8.

The studies show that FBG-based thermally induced strain is capable of measuring the CTE of different specimens reliably.

Table 5 CTE values measured for Hexply 914/40%/285H5/AS4C-3K/1170mm (/°C)

Specimen	Warp direction	Weft direction
Hexply 914/40%/285H5/AS4C-3K/1170 mm	3.6 × 10–6	3.8 × 10–6

4 Conclusions

In the present work, Fiber Bragg Grating sensors are used to measure the thermally induced strain and hence used to calculate the Coefficient of Thermal Expansion for different composite specimens. Experimentally, it is found that the measurement across different samples and across the trials is in agreement with manufacture's data. In the present study, the temperature is increased up to 70 °C as the adhesive property limits to go beyond. It is also found that the size of the specimen is not having any constraint over the estimation compared to the conventional CTE measurement and measurement can be done on many samples and any direction as any calibrated heating chambers can be used for the purpose. As Fiber Bragg Grating Sensors and related data acquisition systems are being commercially available now a days, this method of CTE can be used for fast and easy measurement of CTE.

References

1. Hunkel M, Surm H, Steinbacher M (2018) Chapter 3—dilatometry. In: Vyazovkin S, Koga N, Schick C (eds) Handbook of thermal analysis and calorimetry Elsevier science, vol 6, pp 103–129, ISSN 1573–4374, ISBN 9780444640628. https://doi.org/10.1016/B978-0-444-64062-8.00019-X.
2. Jorge C, Heinrich T, Thilo S, Martin G, Stefano L, Weise D, Ulrich J, Claus B (2009) Interferometry based high-precision dilatometry for dimensional characterization of highly stable materials. Meas Sci Technol 20.https://doi.org/10.1088/0957-0233/20/9/095301
3. Catherine ML (2020) Literature review of thermomechanical analysis, its instrument components and applications. United States. https://doi.org/10.2172/1633565
4. Luo W, Xue S, Zhao C, Zhang M, Li G (2020) Robust interferometry for testing thermal expansion of dual-material lattices. Materials 13(2):313. https://doi.org/10.3390/ma13020313
5. Richard EC (2019) Thermomechanical analysis. In: Materials characterization ASM handbook, ASM international, vol 10, p 324–334. https://doi.org/10.31399/asm.hb.v10.a0006674
6. James J, Spittle JA, Brown SGR, Evans R (2001) A review of measurement techniques for the thermal expansion coefficient of metals and alloys at elevated temperatures. Meas Sci Technol 12:R1. https://doi.org/10.1088/0957-0233/12/3/201
7. Roose S, Heltzel S (2011) High-precision measurements of thermal expansion at cryogenic temperature on stable materials MacroScale 2011. In: Proceedings recent developments in traceable dimensional measurements
8. Kang Sang-Guk, Kang Donghoon, Kim Chun-Gon, Hong Chang-Seon (2003) Measurement of CTE change in a composite laminate with aging under space environment using fiber optic sensors. J Korean Soc Aeronautic Space Sci 31:21–26. https://doi.org/10.5139/JKSAS.2003.31.10.021
9. Measurement of thermal expansion coefficient using strain gages, Technical note TN-513–1, Micro measurements. Retrieved from http://www.vishaypg.com/docs/11063/tn5131tn.pdf

10. Esposito M, Buontempo S, Petriccione A, Zarrelli M, Breglio G, Saccomanno A, Szillasi Z, Makovec A, Cusano A, Chiuchiolo A, Bajko M, Giordano M (2013) Fiber Bragg Grating sensors to measure the coefficient of thermal expansion of polymers at cryogenic temperatures. Sens Actuat A Physic 189:195–203. ISSN 0924–4247. https://doi.org/10.1016/j.sna.2012.09.016
11. Her SC, Lin WN (2020) Simultaneous measurement of temperature and mechanical strain using a fiber Bragg grating sensor. Sensors (Basel). 20(15):4223. https://doi.org/10.3390/s20154223

Low-Frequency NMR Relaxometry Studies of Composite Cylinders with EPDM Insulation

D. Ramesh Babu, Srinivas Kuchipudi, Sanjaya K. Sahoo, and R. N. Rao

1 Introduction

Fiber-reinforced composite materials are used for different industrial and aerospace structural applications because of their strength to weight advantages. The composite cylinders are lined with insulation rubber materials like ROCASIN/EPDM to protect the structure from thermal or corrosive environment in certain applications [1]. The insulation liner is adhesively bonded to the composite casing using suitable bonding agents. The interface bonding shall be achieved without any debonds to ensure proper functioning of system as per the design requirement [2]. Hence, it is very much essential to ensure the bond quality between different interfaces to meet the functional requirements under the specified environmental conditions.

Various Nondestructive Testing (NDT) methods can be used to study the defects in the composite casing insulation interfaces [3]. But there are limitations depending on applications like accessibility, defect detection and quantification etc. NMR is found to be an excellent noninvasive tool to scan and detect the interface defects in a quantitative manner using single side access [4]. In the present study, the application of low-frequency single-sided NMR technique has been explored as NDT tool to study the insulation thickness, interface defects and the bond quality in adhesively bonded EPDM to composite structures. In general, the interface debonds can be due to the following reasons.

1. Poor surface preparation

D. Ramesh Babu (✉)
Defence Research and Development Laboratory, Kanchanbagh P.O, Hyderabad 500058, India
e-mail: duppalaramesh@gmail.com

S. Kuchipudi · S. K. Sahoo
Advanced Systems Laboratory, Kanchanbagh P.O, Hyderabad 500058, India

R. N. Rao
National Institute of Technology, Warangal 506004, India

M. R. Manohara Babu et al. (eds.), *Recent Advances in Composites Engineering and Research*, Springer Proceedings in Materials 39,
https://doi.org/10.1007/978-981-99-8807-5_25

2. Absence of adhesive in a particular zone or
3. Weak chemical bonding or
4. Inadequate adhesive layer thickness or
5. Lack of mechanical intactness due to improper curing

The extent of debonds and air gaps shall be fully evaluated and studied for the acceptance of the components for the intended usage. During the service or storage, there will be change in elastic properties/degradation of elastomers with the passage of time depending on the storage environment/conditions.

Hence, the characteristics of elastomers, interface bond quality and acceptable debonds, if any, shall be periodically assessed to ascertain its serviceability or early detection of defects to ensure the life extension. The minimum interface bond strength is specified based on the service loads and debond may result if the interface bond strength is inadequate or the service load exceeds the achieved bond strength. Also, inadequate insulation thickness or mechanical failure may lead to failures in hot gas flow applications [1]. Hence, the bond quality is to be ensured correctly for the proper functioning of such systems under the prevailed environmental conditions. It is found that NMR technique is a very useful and accurate method to study the interface properties and quantify the rubber thickness and also to compare the elastic properties of different types of insulation material system [5]. Other NDT methods like Acousto-Ultrasonic method (AUS) and Digital Radiography (DR) results for the same samples are also presented in this chapter to validate with the NMR results in the detection of debonds. This chapter reports and extends application of low-frequency NMR for relaxometry studies of EPDM bonded adhesively with Glass composite structures.

2 Experimental Details

In the present case, GFRP *flat plate* and *cylindrical* samples with EPDM insulation bonded with adhesive and in the cured condition are considered for the study. The sample details of the present study are presented below.

a. GFRP flat plate bonded with EPDM insulation (Schematic is shown in Fig. 1)
   (EPDM Rubber thickness measured as 2.2 mm on 4 mm E-glass laminate) of 280 mm length and 125 mm width.
b. GFRP cylinder bonded with EPDM (Schematic is shown in Fig. 2a and b)
   (EPDM Rubber thickness measured as 4.5 mm inside E-glass cylindrical shell of thickness 10 mm) of 280 mm length and 240 mm diameter.

2.1 Flat Sample Preparation

Flat glass fiber epoxy resin laminate was made through filament winding on a flat mandrel, where the flat mandrel rotates and glass fiber passes through epoxy resin

Fig. 1 GFRP flat plate bonded with EPDM insulation, the B-Zone indicates debond

(a) (b)

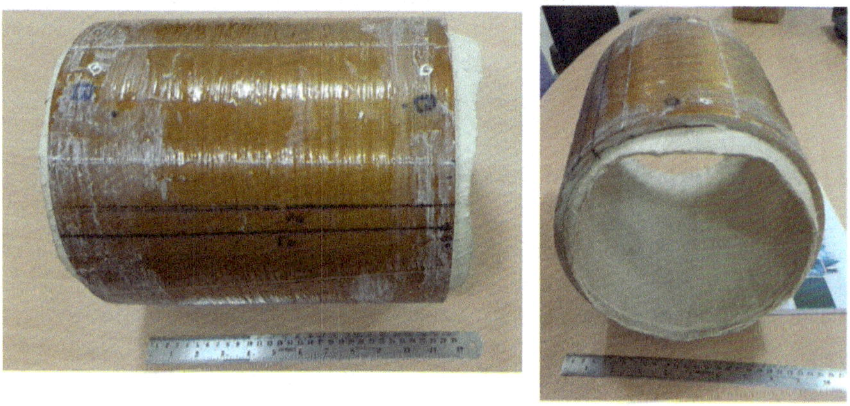

Fig. 2 **a** GFRP cylinder bonded with EPDM. **b** GFRP cylinder debond at B-zone

bath and winding was carried out till the desired thickness of laminate achieved [6]. Subsequently, the composite along with the flat mandrel cured in the oven and the required laminate of approximately 4 mm thickness, 280 mm length and 125 mm width extracted through cutting process. EPDM rubber liner of 2 mm thickness bonded on one side of the flat laminate leaving intentional debond in the zone of interest (Zone B) and subsequently cured. The flat sample photograph is shown in Fig. 1.

2.2 Cylindrical Sample Preparation

GFRP cylinder is made through filament winding process where the glass fiber impregnated in the epoxy resin bath is wound on to the cylindrical metallic mandrel of 240 mm diameter. Approximately, 10 mm thickness was built up through the filament winding layers over a length of 280 mm and subsequently cured in the oven and composite cylindrical shell is extracted out from the mandrel. Later, the EPDM rubber sheet of two layers of each approximately 2 mm thickness is lined inside the composite cylinder with deliberate debond created in zone B. subsequent to the layup, the rubber was vulcanized as per the standard curing procedure [1].

Hence, the samples are prepared with known defects and both good and defective zones have been studied using NMR technique and other alternative NDT techniques of Acousto-Ultrasonic and Digital Radiography for comparison and validation of NMR method.

The photograph of the cylindrical sample is shown in Fig. 2.

Bond quality studies and approximate EPDM rubber thickness estimation were carried out on above two types of samples comparing results obtained in bonded and debonded zones. This study was also extended to characterization of EPDM using the relaxation times.

3 Low Frequency NMR Technique and Test Setup

Single sided low-field NMR uses permanent magnets for generating direct current (DC) magnetic field (B_0) for magnetizing the polymer materials. It generates a magnetic field outside the magnets, unlike high-field NMR systems, wherein the magnetic field is generated within the core. In the case of low-field single-sided NMR; B_0 is superimposed by an AC magnetic field (B_1) for creating a slice of sensitive volume, which moves inside the materials using a precise stepper motor (high precision lift) system [4, 7, 8].

Proton based Portable NMR equipment PM25 with operating frequency of 13.35 MHz is based on inside out principle, where the sample is kept outside the magnet. The test setup is shown in Fig. 3, where the flat and cylindrical samples are kept on the holder plate above the magnet. The sensitive volume is generated over lateral size of 40 $\times$ 40 mm cut out where the test sample is placed. The control volume gradually moved up in predefined step sizes of 100 μm to scan the total thickness of sample under study. In this equipment, the maximum thickness that can be scanned is up to 25 mm, the height adjustment can be done by selecting the suitable spacers [9]. The numbers of scans employed for the present experiment are 1024.

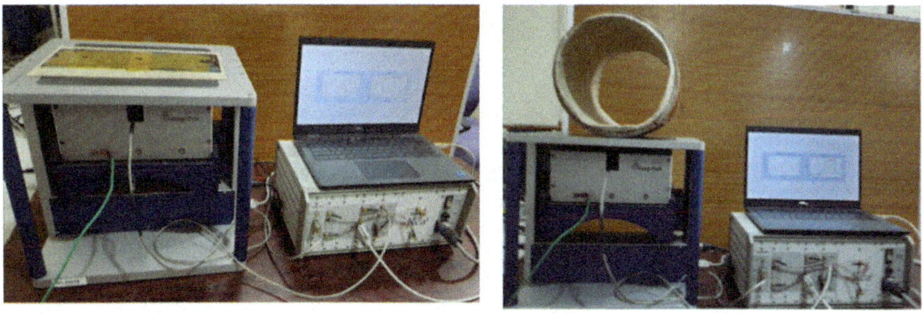

Fig. 3 Profile NMR setup with flat and cylindrical sample on the top

4 Alternate NDT Methods

The same samples were studied with other alternative methods like **AUS** and **DR** method and the test results are compared with NMR data to validate NMR results.

4.1 Experimental Details of Acousto-Ultrasonic System (AUS)

AUS uses low-frequency stress waves using piezo-electric transducer and the energy is transmitted through transmitter to the structure for evaluating the adhesive bond quality. It uses contact type transmitter and receiver on the same side of the inspected surface. The receiver is kept at a fixed distance from the transmitter and it catches the signal through pitch-catch method [5, 10–12]. It requires a reference signal for the comparison of defect zone. The low-frequency Acousto-Ultrasonic testing includes the tone burst excitation pulser and broad band receiver system (10 kHz–20 MHz) as shown in the Fig. 4 (AUS experimental setup).

The testing of GFRP *flat plate* and *cylindrical* samples with EPDM insulation is carried out using single-sided access pitch-catch method [13]. For GFRP *flat plate,* the parameters employed are 200 kHz frequency of probes with probe to probe distance of 100 mm and gain of 20 dB, scanned on good zone and de bonded zone.

For GFRP *cylindrical* samples with EPDM insulation, the parameters employed are 50 kHz frequency of probe and with probe to probe distance of 200 mm and gain of 32 dB, scanned on good zone and de bonded zone.

4.2 Experimental Details of Digital Radiography

Radiography Testing is well known and widely used NDT method to study the interface defects, where the interface defects are to be detected in tangential shots, that

Fig. 4 Portable AUS setup

is, the debonds/air gaps aligned parallel to the beam direction. RT gives qualitative information about the presence of debonds; however, there is a possibility to miss out very small gaps or kissing debonds kind of defects based on the sensitivity and direction of the taken shot. The cylindrical sample has been studied with digital X-ray facility in the tangential direction of the defect region. The following are the specifications of DR facility used [14].

Specifications:

X-ray Energy: 225 kV.

X-ray source: One Micro focus tube (3-4 μm) CT.

:One Mini focus tube (200 μm) DR.

Inspection volume: Maximum: 600 mm dia. and 1250 mm height.

: Minimum: 2 × 2 mm.

X-ray power source: 1000W for mini focus.

Detector: Flat panel detector (static 15FPS and dynamic 30FPS).

In the present case of DR (digital radiography) scanning, source radiation of 180 kV and 1.5 mA is employed with 512 ms integration time (Fig. 5).

Fig. 5 DR and CT experimental setup of cylindrical sample

5 Results and Discussion

5.1 NMR Results

Figure 6a shows the representative flat plate sample without any defect. Figure 6b shows the corresponding NMR data in terms of amplitude versus depth. From the figure, it can be seen that the amplitude raise observed on the depth scale from 4200 to 6500 μm indicates the thickness of ~2.3 mm which has been closely matching with physical dimension of EPDM rubber used in the sample. The flatness in the amplitude data indicates the homogeneity of rubber sample under consideration [4]. The sharp rise and fall indicate intactness of the achieved adhesive interface bonding.

Similarly, Fig. 7a shows the representative flat plate sample with intentional debond and Fig. 7b shows the corresponding NMR data in terms of amplitude versus depth. In this case, the EPDM insulation has been lifted off from the GFRP face as shown in Fig. 7a. The curvature effect is seen in the corresponding NMR plot of Fig. 7b. The amplitude rise is observed from 5200 to 7500 μm because of the debond, whereas it is supposed to start from 4200 μm. The shift in the amplitude rise from 4200 to 5200 μm, that is, about 1 mm indicates the air gap. The curvature effect of debonded rubber is seen in the rise fall characteristics.

It is evident from comparison of both plots, that the NMR clearly distinguishes the bond and debond zones. The EPDM thickness has been captured accurately and detection of debonds/airgaps present and their extent can be correlated.

In the second experiment, GFRP cylindrical samples are taken with EPDM rubber bonded inside the cylinder. Here also intentional debond is introduced in one orientation for study and characterization.

The cylindrical sample in good zone is shown in Fig. 8a and the corresponding NMR scan result in Fig. 8b. The amplitude pick up is observed from 10,800 to

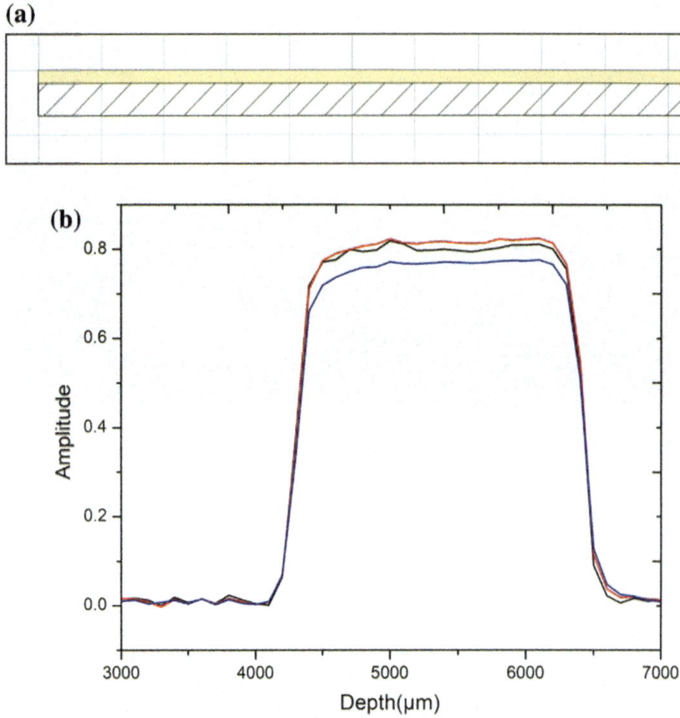

Fig. 6 **a** Flat sample schematic view of (no debond case) (EPDM). **b** NMR data of flat sample (no debond case)

15,300 μm on the depth scale corresponding to 4.5 mm, which is matching well with the physical quantity.

When there is no debond in the cylindrical sample, i.e., the rubber is intact with the cylindrical wall and the curvature effect is seen in the above figure corresponds to the cylinder geometry. The amplitude rise of depth profile at 10,800–15,300 indicates the rubber thickness of 4.5 mm.

The debond representative sketch and corresponding NMR data is plotted in Fig. 9a and b, respectively.

In the debond region, the rubber is not intact with the cylinder and lifted off from the wall surface, which will appear like more or less flat rubber specimen, the corresponding behavior is seen in the plot. The amplitude pickup observed from 13,000 to 17,600 μm, which has observed from 10,800 μm in the intact sample above. It indicates a debond/air gap from 10,800 to 13,000 μm is about 2.2 mm.

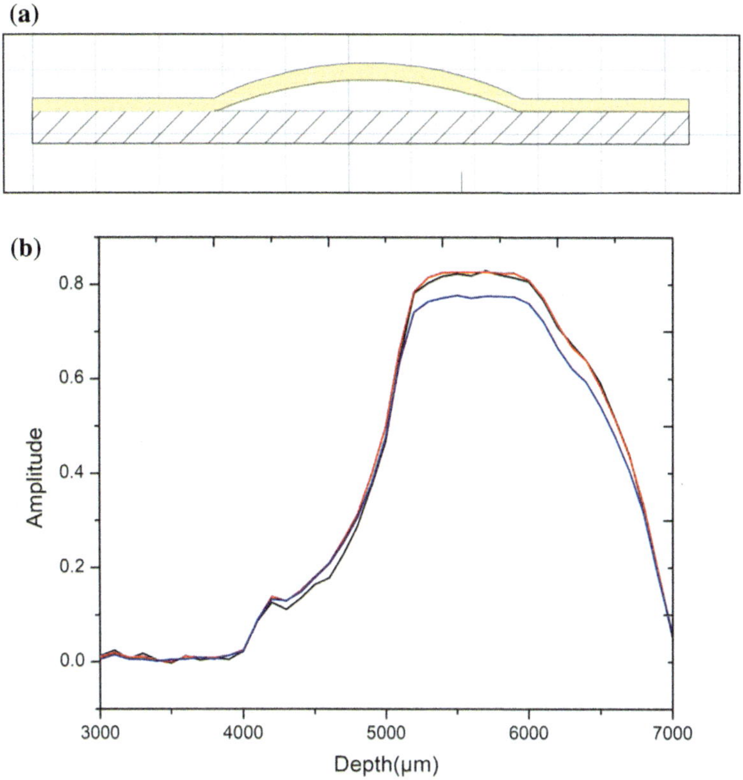

Fig. 7 **a** Flat sample schematic view of (debond case) (EPDM). **b** NMR data of flat sample (debond case)

5.2 NMR Results of Rocasin Rubber

NMR study was extended to 2 mm thick Rocasin rubber for comparison with EPDM behavior and is shown in Fig. 10. The thickness of 2 mm has been captured similar to EPDM, whereas the amplitude is different than EPDM signature. The EPDM amplitude recorded approximately about 0.8 (arbitrary units), where as in case of Rocasin, it is about 0.4 which is proportional to the number of protons (proton density) present in the material.

Hence, the NMR technique clearly distinguishes between debond and intact samples and captures the rubber thickness and also curvature effects based on geometry of the specimen. The same samples were studied using AUS and DR methods in support of NMR observations qualitatively [13].

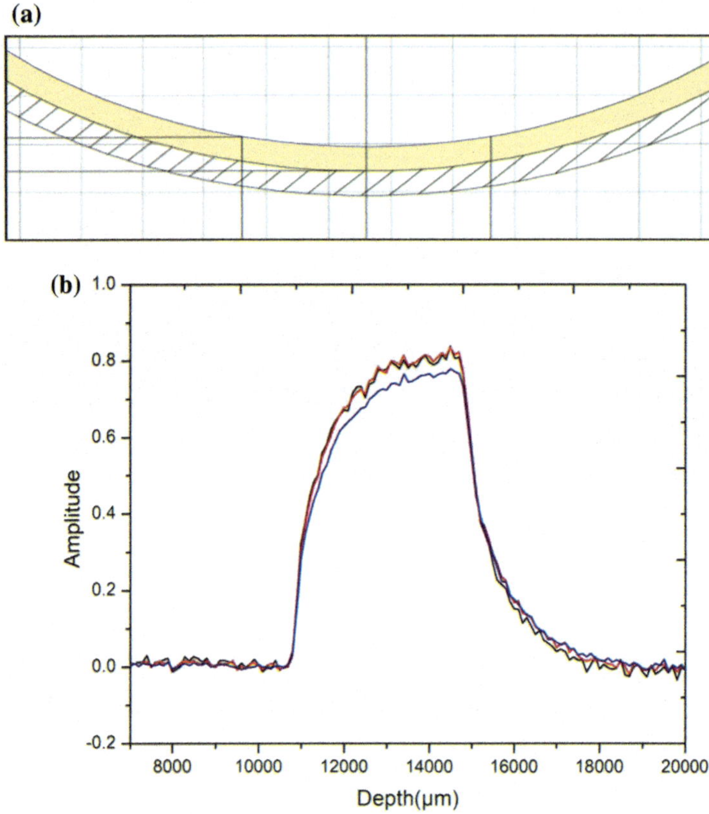

Fig. 8 **a** Cylindrical sample schematic view (no debond case). **b** NMR data of cylindrical sample (no debond case) (EPDM)

5.3 Acousto-Ultrasonic Results

AUS signature captured for flat and cylindrical samples with and without debonds are shown in Figs. 11 and 12.

From the above data, Fig. 11a represents good zone and the Fig. 11b is the debond indication in comparison with the reference peak. It clearly indicates the increase in amplitude than the reference peak which indicates the presence of debond at the zone.

From the above data, Fig. 12a represents good zone and is treated as reference peak for the comparison of debonded signal. In Fig. 12b, it clearly indicates the higher amplitude than the reference peak which indicates the presence of debond at the zone. The debonds captured by the AUS method with the reference peak, where it needs to be complimented with other NDT methods in quantification and further analysis.

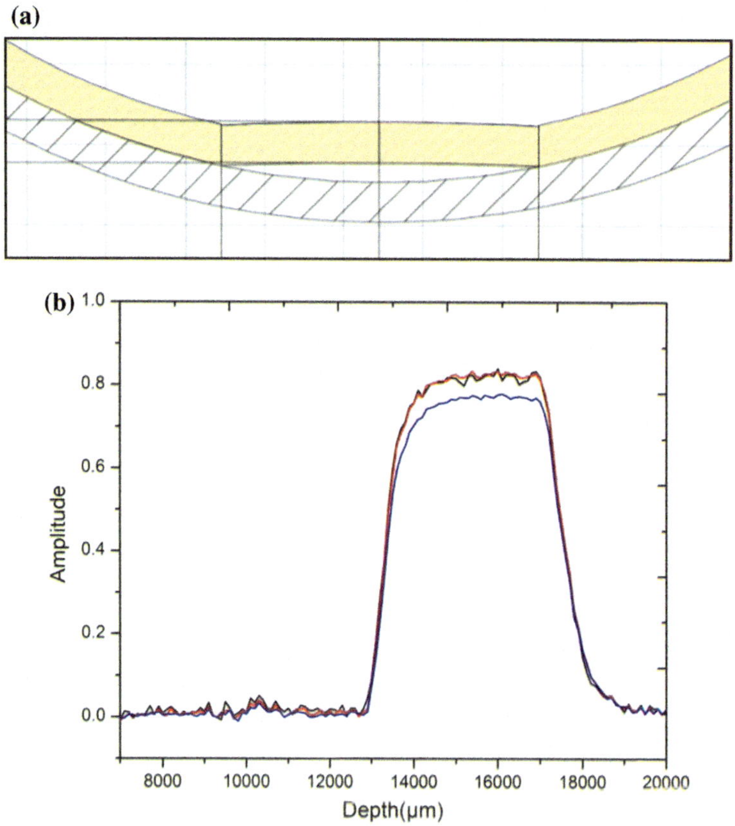

Fig. 9 **a** Cylindrical sample schematic view (debond case). **b** NMR data of cylindrical sample (debond case) (EPDM)

5.4 *Digital Radiography Results*

DR scanning is carried out for cylindrical sample using DR set up and debond is prominently seen in Fig. 13 which reconfirms NMR results. The DR technique indicated the debond very clearly but, it is only qualitative information. The debond in the flat plate case cannot be captured clearly as the defect is planar indication in the RT, hence not considered.

5.5 *Relaxation Studies*

The relaxation studies are useful for the material characterization in terms of chemical nature, elastic properties and material degradation over time and degree of cross

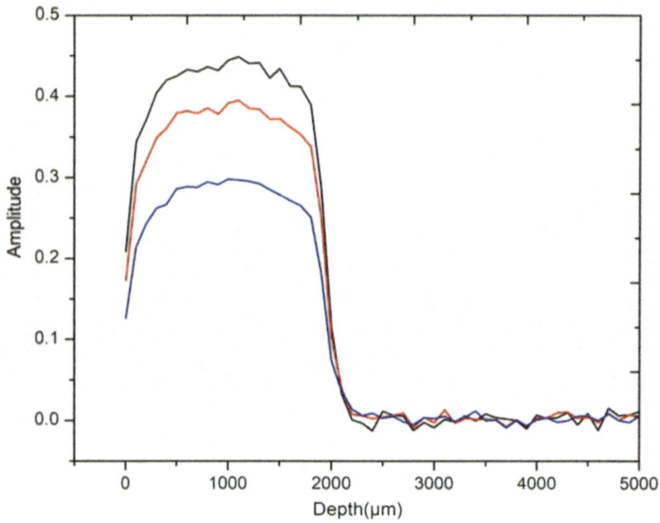

Fig. 10 NMR data of Rocasin rubber sheet

linking [15]. The bond quality assessment is also possible to evaluate using NMR technique. T_{2short} is related to polymerized rubber with cross linked side chains, whereas T_{2long} relates to the free side chains [5]. The transverse relaxation time (T_2) has been proved to be informative in characterizing polymers [16–19] where the T_2 is known to be inversely proportional to degree of cross linking. The relaxation times are obtained to study the behavior of EPDM rubber in bonded and debonded cases. Carr-Purcell-Meiboom-Gill (CPGM) pulse sequences were used for determining the T_2 relaxation times. The T_2 relaxation data obtained for the cylindrical shell with EPDM rubber in the two zones (A & B) are shown below in Table 1.

Observations on Relaxation Times

From the relaxation time data it is observed that the T_2 of the debond zone rubber is slightly higher than the bonded case. In case of debonded zone the T_2 relaxation corresponding to only free rubber whereas in case of bonded rubber includes combination of bonded nature of EPDM which is relatively stiffer and intact with the cylindrical shell. The present study is done for the case of created debond in the sample. However, studies to be carried out on the naturally debonded case to get more information on the behavior of EPDM characteristics to assess the bond quality of rubber layers. Further extensive work needs to be carried out to arrive at detailed correlations with elastic properties of rubber. Hence, the preliminary relaxation studies using NMR infers useful information in characterization of materials and evaluating the bond quality in the critical applications.

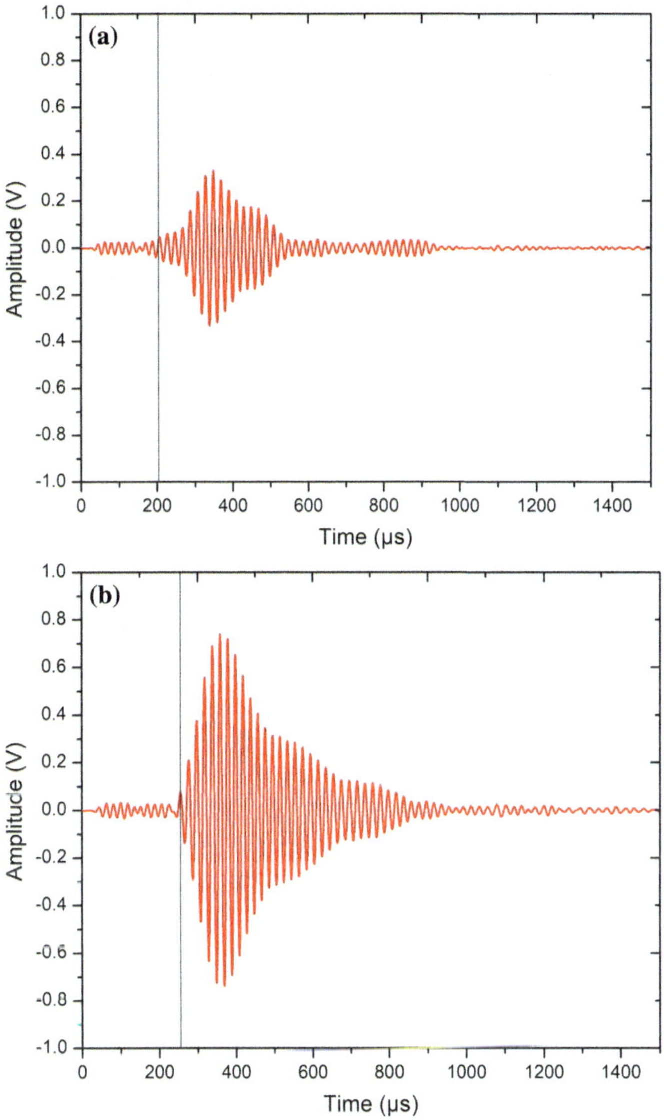

Fig. 11 **a** AUS data of good zone in cylindrical sample. **b** AUS data of debond zone in cylindrical sample

6 Conclusions

The low frequency NMR studies on GFRP with EPDM samples and Rocasin rubber is found to be very useful in estimation of rubber thickness, debonds/air gaps present at the interface and also the approximate extent of the debond with an accuracy of

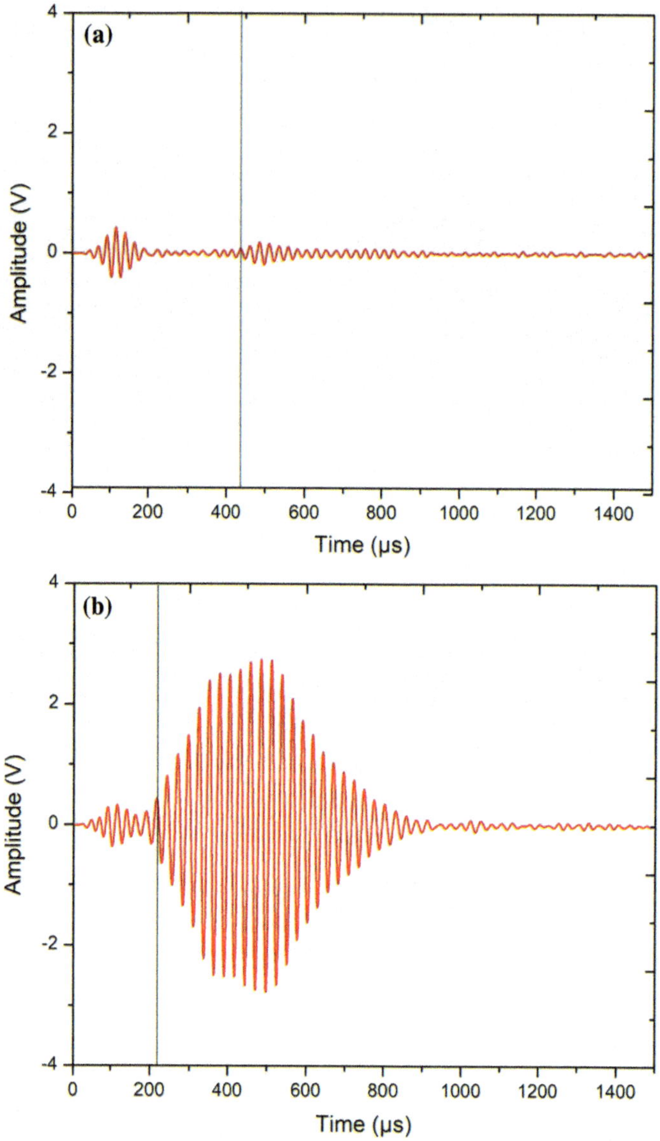

Fig. 12 **a** AUS data of good zone of flat sample. **b** AUS data of debond zone of flat sample

the order of 100 μm. The curvature effects are also well captured by the NMR data of composite cylinder with EPDM rubber sample. The comparison of NMR data with AUS and DR data indicated that in addition to identification of the defects, quantification of rubber thickness and extent of debonds are possible to investigate

Fig. 13 DR image of debond zone

Table 1 Transverse relaxation times (T_2) of EPDM

Sample nomenclature	$T_{2\ short}$ (ms)
E Glass shell with EPDM Rubber—cylinder zone-A (no debond)	5.1
E Glass shell with rubber—cylinder zone-B (debond)	9.0

using single sided NMR. It is a very useful tool in field applications due to its portability, non-invasive detection using single sided access. The bond quality can be correlated with the NMR signature which helps in detection of the interface soundness. Comparison of EPDM and Rocasin signatures indicate the capability of NMR in characterization of different materials and can be correlated to the chemical and elastic nature of materials. The preliminary Relaxation studies show characteristic nature of bonded and debonded regions of EPDM rubber. However, it needs further investigations on the relaxation studies in corroborating the results with the material characteristics.

Acknowledgements The authors are thankful to the Director, DRDL and Director, ASL for the motivation, support and encouragement in carrying out the present work.

References

1. Kumar M (2017) Buragohain: composite structures: design, mechanics, analysis, manufacturing, and testing, 1st edn. Boca Raton, CRC Press. https://doi.org/10.1201/9781315268057

2. Kim JK, Mai YW, Mai YW (1998) Engineered interfaces in fiber reinforced composites. Elsevier Science
3. Adams RD, Cawley P (1998) A review of defect types and non-destructive testing techniques for composite and bonded joints. NDT Int 21(4):208–222. https://doi.org/10.1016/0308-9126(88)90333-1
4. Blümich B (2019) Essential NMR: for scientists and engineers, 1st edn. Springer- Berlin Heidelberg
5. Sanjaya KS, Srinivas K, Narasimha Rao R, Buragohain MK, Chaitanya S (2021) Detection of planar defects in multi layered gfrp composite structures using low field nuclear magnetic resonance. Mater Eval 79(9):897–904. https://doi.org/10.32548/2021.me-04201
6. Mallick PK (2007) Fiber-reinforced composites: materials, manufacturing, and design, 3rd edn, Boca Raton, CRC Press. https://doi.org/10.1201/9781420005981
7. Blümich B (2003) NMR imaging of materials. Oxford University Press. https://doi.org/10.1093/acprof:oso/9780198526766.001.0001
8. Blümich B, Perlo J, Casanova F (2008) Mobile single-sided NMR. In: Progress in nuclear magnetic resonance spectroscopy, vol 52, no 4, pp 197–269. https://doi.org/10.1016/j.pnmrs.2007.10.002
9. Mauler J, Casanova F, Blümich B (2013) Shaping the sensitive volume of a single sided NMR-sensor to profile Cylindrical samples with high resolution. Appl Magn Reson 44:1325–1334. https://doi.org/10.1007/s00723-013-0484-4
10. Mareeswaran S, Sasikumar T (2017) The acousto-ultrasonic technique: a review. Int J Mech Eng Technol (IJMET) 8(6):418–434
11. Raju PK (1997) Acousto-ultrasonic technique for non-destructive evaluation of composites and structures. J Acoustic Soc Am 102(5). https://doi.org/10.1121/1.420199
12. Sahoo SK, Srinivas K, Rao RN, Buragohain MK (2020) NDE of glass fiber reinforced composite structures using single sided solid-state proton NMR. Mater Today Proceed 21(2):1239–1243. https://doi.org/10.1016/j.matpr.2020.01.111
13. Srinivas K, Durga Rani N, Murthy BVSR (2014) Single sided NMR for NDE of GFRP-rubber interface, national seminar and exhibition on non-destructive evaluation, NDE 2014, 4–6 December, Pune, India
14. Halmshaw R (1995) Industrial Radiology: Theory and Practice, 2nd edn. Springer, Netherlands
15. Halmen N, Kugler C, Hochrein T, Heidemeyer P, Bastian M (2017) Determination of the degree of cross-linking and curing with single-sided NMR. In: Proceedings sensor, pp 561–566. https://doi.org/10.5162/sensor2017./p1.2
16. Sahoo SK, Rao RN, Kuchipudi S, Buragohain MK (2020) Application of single sided NMR and acousto-ultrasonic methods for NDE of composites structures. Adv Appl Mech Eng Lect Notes Mech Eng. https://doi.org/10.1007/978-981-18-1201-8_61
17. Sahoo SK, Narasimha Rao R, Srinivas K, Buragohain MK, Chaitanya S (2020) A novel NDE approach towards evaluating adhesive bonded interfaces. Mater Today Proceed 26(2):1191–1197. https://doi.org/10.1016/j.matpr.2020.02.240
18. Sahoo SK, Rao RN, Kuchipudi S, Prasad YLVD, Buragohain MK (2019) NDE of GFRP composite laminates using single sided NMR and acousto-ultrasonic methods. Non-destruct Test Eval 17:28–31
19. Busse F, Rehorn C, Küppers M, Ruiz N, Stege H, Blümich B (2020) NMR relaxometry of oil paint binders. Magn Resonan Chemistr 58(9):830–839. https://doi.org/10.1002/mrc.5020

A Novel Approach to Predict the Cured Mechanical Properties of HTPB-Based Composite Solid Propellant

I. Kingstone Lesley Jabez, Urmila Das, and Nakka Sudarshan

1 Introduction

Hydroxyl Terminated Polybutadiene (HTPB) with hydroxyl value (OH value) in the range of 40–50 mg KOH/g is preferred for Composite Solid Propellant (CSP) of ballistic missiles and launch vehicles [1]. Variation (subtle) in molecular weight and its distribution, which is not reflected in terms of change in OH value, leads to variation in achieved mechanical properties of cured CSP despite keeping R value (NCO/OH ratio) within desirable range [2]. Batch to batch variation of other ingredients and climatic conditions too contribute for the uncertainty in the achieved mechanical properties of CSP. Thus, extensive formulation trials are warranted to overcome the uncertainty, which leads to cost and time overrun. Authors have carried out lab level experiments and devised Gumstock based scheme to overcome the uncertainty in the achieved mechanical properties of CSP. For few R values (NCO/OH ratio), on which Gumstock level experiments were carried out, authors have prepared 20 kg batches of actual CSP itself. Results and findings of the same are discussed in this paper.

I. Kingstone Lesley Jabez (✉) · U. Das · N. Sudarshan
ASL DRDO, Hyderabad, Telengana 500058, India
e-mail: lesleyjabezik.asl@gov.in

M. R. Manohara Babu et al. (eds.), *Recent Advances in Composites Engineering and Research*, Springer Proceedings in Materials 39,
https://doi.org/10.1007/978-981-99-8807-5_26

2 Experimental Details

2.1 Formulations Used for Study and Details of Gumstock-Level Experiments

HTPB (binder), Ambilink or Adduct (which is prepared by dissolving 1 parts by weight of cross-linker Trimethylol propane (TMP) in two parts by the weight of chain extender 1,4-butanediol (BDO)) and curative Toluene Diisocyanate (TDI), which were responsible for the cross-linking and formation of cured network of CSP were only used for the experimental trials. All the above ingredients of same batch, meeting QA requirements, were used for the study. Formulations used for the study are given in Table 1. Quantity of Ambilink and total quantity of HTPB and TDI in each formulation were same as those of established CSP formulation, wherein, HTPB + TDI amounts to 10.78% and Ambilink is 0.12%. Calculated R value (NCO/OH ratio) of formulations under study are given in Table 2. X-value in Table 2 is nothing but the quantity of HTPB in the formulation.

Batch size of each formulation was 500 g. Ingredients were mixed in glass beaker using glass rod as per the formulation given in Table 1. Thereafter, the contents were cast as slabs of 6 mm thickness. Thus, eight numbers (one for each formulation and one extra slab for each of Formulation-3 and 5) of approximately 200 mm × 200 mm slabs of 6 mm thickness were obtained. Six numbers of slabs (one from each formulation), after ambient keeping till flow had stopped, were cured at 50 °C for 8 days. Remaining 2 slabs (one from each of formulation-3 and 5) were stored in cold chamber at 20 °C for 8 days. Cured slabs were named as **Gumstock**. Dumbbells were cut from Gumstock, cured at 50 °C for 8 days and Gumstock/slabs kept at 20 °C

Table 1 Formulations used for study

Ingredients	Formulation					
	1	2	3	4	5	6
HTPB	10.01	10.04	10.05	10.07	10.09	10.10
Ambilink	0.12	0.12	0.12	0.12	0.12	0.12
TDI	0.77	0.74	0.73	0.71	0.69	0.68

Table 2 Calculated R values of formulations

Formulation	*X* value	*R* value (~NCO:OH)
1	10.01	0.88
2	10.04	0.84
3	10.05	0.83
4	10.07	0.81
5	10.09	0.78
6	10.10	0.77

for 8 days and tested as per ASTM D638 maintaining cross head speed of 50 mm/min [3]. UTM of Teenius make was used for the tensile testing. Samples representative Gumstock of each formulation were subjected to DSC under nitrogen environment using Differential Scanning Calorimeter of Discovery make, till degradation [4–7]. TGA of samples representative of Gumstock of each formulation was carried out till degradation using TGA of TA Instruments Q600 SDT make. Heating rate of 10 °C/ minute was maintained during DSC and TGA runs.

2.2 Details of 20 kg Batches and 700 kg Batch of CSP Carried Out at Selected X/R Values

Actual Composite Solid Propellant (CSP) of batch size 20 kg was prepared by mixing the ingredients in 20 L planetary mixer of Ross make. Three 20 kg batches of CSP were prepared, first one at X value of 10.05 (R value of 0.83), second batch at X value of 10.07 (R value of 0.81) and the last batch at X value of 10.09 (R value of 0.78). Baseline formulation used for above 20 kg batches of CSP is given in Table 3.

Same batches/lots of ingredients were used in all the three 20 kg batches of CSP. HTPB, TDI and TMP and BDO of ambilink are from the same source as used in Gumstock level formulations of Table 1. These 20 kg batches were intended to understand the effect of other ingredients especially aluminium and Ammonium Percholorate (AP) on the mechanical properties of cured CSP. One more objective of 20 kg batches was to find correlation between gumstock-level properties and CSP-level properties. Aluminium powder of particle size 15 + 3 μm was used in all the three 20 kg batches. AP was used in coarse and fine form. AP coarse was of mean diameter 280 ± 20 μm and AP fine was of mean diameter 40 ± 2 μm in all the three 20 kg batches. AP Fine to coarse ratio of 1:1.9 was maintained in all the three batches. After curing, mechanical properties were evaluated of cured CSP in respect of each 20 kg batch, as per ASTM D638 (cross head speed of 50 mm/min).

Table 3 Baseline formulation for 20 kg batches of CSP

Ingredients	Weight %		
	$X = 10.05$ $R = 0.83$	$X = 10.07$ $R = 0.81$	$X = 10.09$ $R = 0.78$
HTPB	10.05	10.07	10.09
TDI	0.73	0.71	0.69
Ambilink	0.12	0.12	0.12
Plasticizer, antioxidant, Burn-rate modifier	3.145	3.145	3.145
Aluminium (Metal Fuel)	18.0	18.0	18.0
Ammonium perchlorate (Oxidizer) + Anticaking agent	67.955	67.955	67.955
Total	100	100	100

In order to understand how far the Gumstock level findings hold good for the production batch of CSP, one 700 kg batch of CSP was prepared at X value of 10.07 (R value of 0.81), using the same lots of ingredients and same particle size and Fine to coarse ratio of AP. CSP of 700 kg batch too was evaluated for mechanical properties.

3 Results and Discussion

3.1 Gumstock-Level Experiments

Mechanical Properties: Results of mechanical properties of Gumstock pertaining to various formulation of Table 1 of Sect. 2.1 are given in Table 4. It may be noted that each formulation of Table 1 of Sect. 2.1 is of different R value as evident from Table 2 of Sect. 2.1. Therefore, by plotting the achieved tensile strength of various formulations against respective R value, Tensile strength is represented as a function of R value in Fig. 1. Similarly, % Elongation and E Modulus are represented as function of R value in Figs. 2 and 3, respectively.

As evident from Table 4 and Fig. 1, achieved tensile strength of Gumstock is within 1.68–3.26 kg-f/cm^2. Even though, tensile strength corresponding to few R values does not explain the effect of R value on the achieved tensile strength of Gumstock, from Fig. 1, it can be concluded that tensile strength tends to increase with increase in R value and vice versa. As evident from Table 4 and Fig. 2, % Elongation is quite sensitive to change in R value. % Elongation of Gumstock decreases with increase in R value. As evident from Table 4 and Fig. 3, E Modulus of Gumstock increases with increase in R value. These observations are well in line with those of cured CSP reported in literature [8, 9]. Thus, in terms of change in mechanical properties with respect to change in R value, Gumstock behaves similar to cured CSP.

It is surprising to note that Gumstock pertaining to Formulation-3 ($X = 10.05$ and $R = 0.83$) and Formulation-5 ($X = 10.09$ and $R = 0.78$), despite being kept at 20 °C

Table 4 Achieved mechanical properties of Gumstock pertaining to various formulations (cured at 50 °C for 8 days)

Properties	1	2	3	4	5	6
	$R = 0.88$	$R = 0.84$	$R = 0.83$	$R = 0.81$	$R = 0.78$	$R = 0.77$
Tensile strength (kg-f/cm^2)	3.26 (3.06–3.36)	2.76 (2.60–2.98)	2.29 (2.12–2.41)	2.84 (2.48–3.50)	1.68 (1.58–1.82)	2.02 (1.65–2.35)
% Elongation	218.9 (198–225)	291.7 (266–313)	359.1 (346–388)	460.4 (388–588)	550.0 (526–571)	592.0 (487–662)
E Modulus (kg-f/cm^2)	2.57 (2.49–2.69)	1.56 (1.37–1.75)	0.969 (0.79–1.09)	0.771 (0.58–0.91)	0.337 (0.32–0.34)	0.34 (0.31–0.39)

Value in parenthesis indicates minimum and maximum of test results

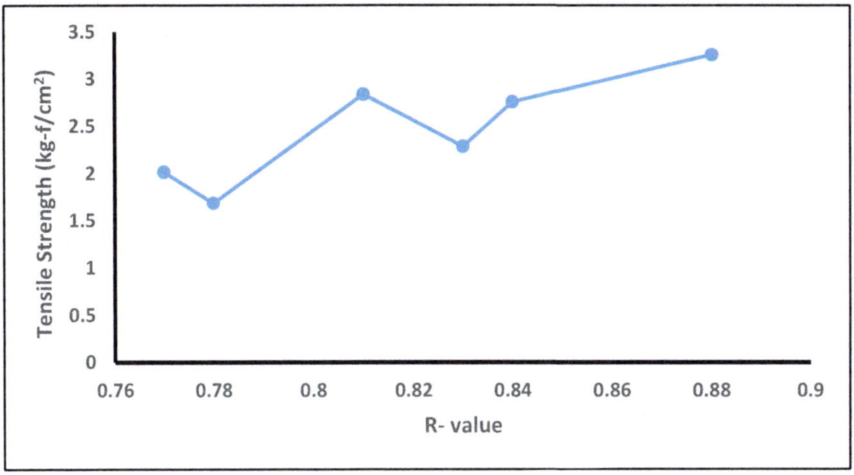

Fig. 1 Tensile strength of Gumstock as a function of *R* value

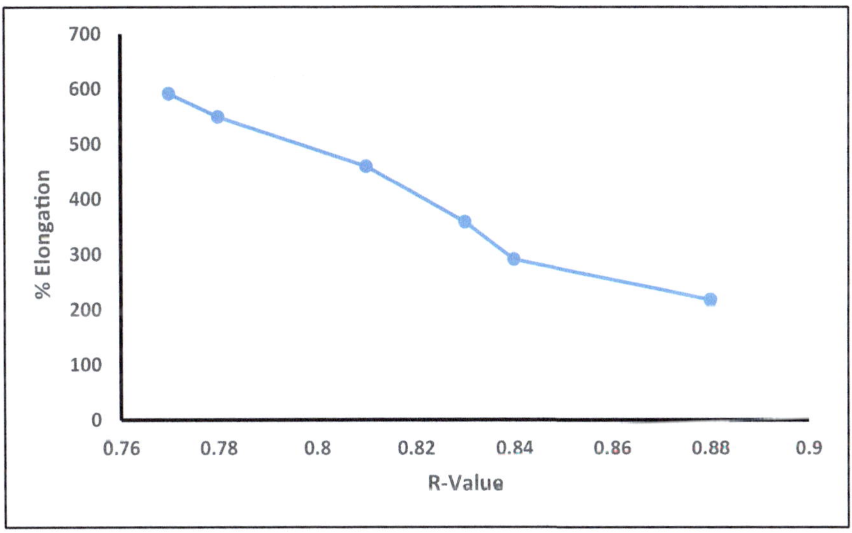

Fig. 2 % Elongation of Gumstock as a function of *R* value

for 8 days, exhibited mechanical properties well comparable to those of respective Gumstocks cured at 50 °C for 8 days. Details of comparison are given in Table 5. Table 5 confirms that polyurethane formation, that is, crosslinking reaction among HTPB, TMP, BDO and TDI can reach completion even at 20 °C [10].

Thermal Analysis by DSC and TGA: DSC curves of Gumstock samples pertaining to Formulation-1 ($R = 0.88$), Formulation-2 ($R = 0.84$), Formulation-3 ($R = 0.83$),

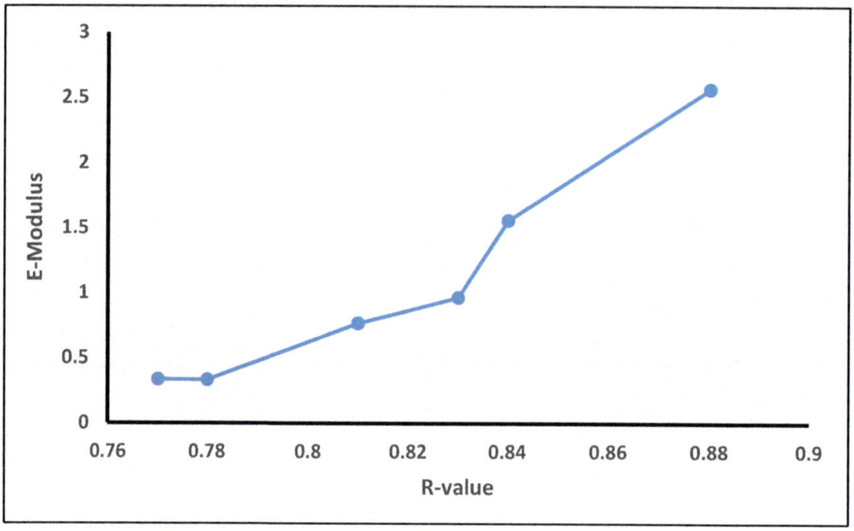

Fig. 3 E Modulus of Gumstock as a function of *R* value

Table 5 Comparison of mechanical properties of Gumstock cured at 50 °C/8 days and 20 °C/8 days

Properties	Formulation-3 $R = 0.83$		Formulation-5 $R = 0.78$	
	@50 °C/8 days	@20 °C/8 days	@50 °C/8 days	@20 °C/8 days
Tensile strength (kg-f/cm^2)	2.29 (2.12–2.41)	2.83 (2.65–3.00)	1.68 (1.58–1.82)	1.79 (1.45–2.04)
% Elongation	359.1 (346–388)	345.8 (322–364)	550.0 (526–571)	481.4 (393–549)
E Modulus (kg-f/cm^2)	0.969 (0.79–1.09)	1.233 (1.19–1.26)	0.337 (0.32–0.34)	0.46 (0.41–0.53)

Formulation-4 ($R = 0.81$), Formulation-5 ($R = 0.78$) and Formulation-6 ($R = 0.77$) are given in Fig. 4. As evident from Fig. 4, DSC curves are comparable for all the formulations. Further, degradation occurs at about 300 °C (abrupt upward shift in DSC curve) in respect of Gumstocks pertaining to all the formulations. It is thereby confirmed that there is no remarkable difference in cured network of Gumstock due to change in *R* value (NCO/OH ratio).

Above findings are further confirmed by TGA curves of Gumstock pertaining to various formulations, which are given in Fig. 5. Figure 5 confirms that thermal stability of cured network of Gumstock pertaining to all the formulations with different *R* values are well comparable. As evident from Fig. 5, irrespective of *R* value, degradation starts at 300 °C for the Gumstock of each formulation and it is very abrupt at about 400 °C for all the formulations. Thus findings of DSC have been corroborated by TGA in respect of Gumstock.

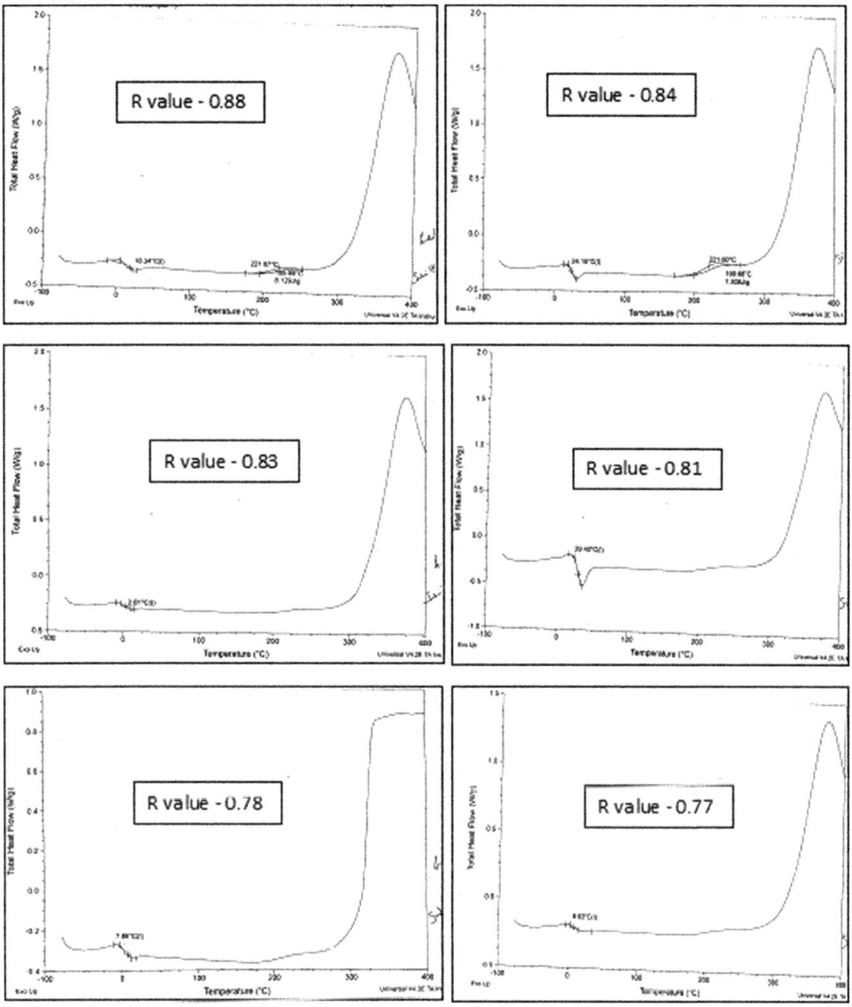

Fig. 4 DSC curves of Gumstocks of various formulations (various *R* values)

3.2 Composite Solid Propellant (CSP) Batches

CSP of 20 kg Batches in Comparison with Respective Gumstock: Mechanical properties evaluated of cured CSP of 20 kg batches (of *X* values 10.05, 10.07 and 10.09 and their respective *R* values are 0.83, 0.81 and 0.78) pertaining to Formulations given in Table 3 of Sect. 2.2 are given in Table 6, in comparison with those of respective Gumstocks (pertaining to Formulation-3, 4 and 5 of Table 1 of Sect. 2.1). Tensile strength has been depicted as a function of R value in Fig. 6 for CSP and corresponding Gumstock comparatively for the R values given in Table 6. Similarly,

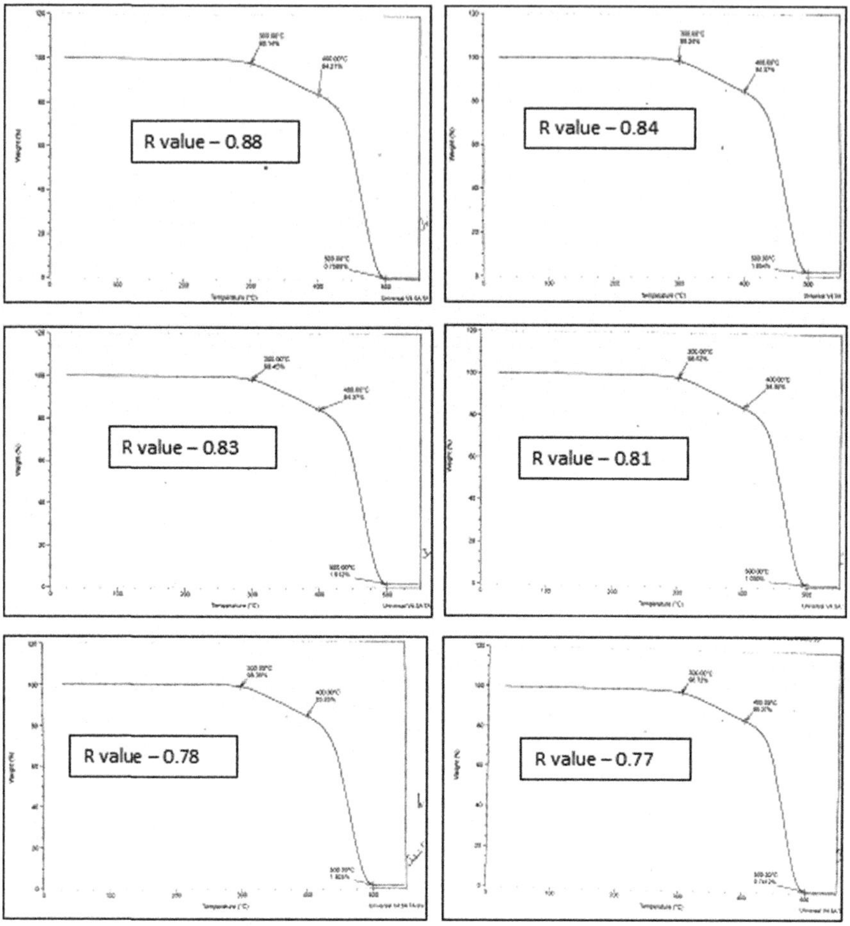

Fig. 5 TGA curves of Gumstocks of various formulations (various *R* values)

% Elongation and E Modulus have been depicted as a function of R value in Figs. 7 and 8, respectively, for CSP and corresponding Gumstock comparatively for the R values given in Table 6.

As evident from Table 6 and Figs. 7 and 8, Gumstock exhibits the behaviour of CSP as far as the change in % Elongation and change in E Modulus with respect to *R* value are concerned**.** As the % Elongation of CSP decreases with increase in *R* value, so does the % Elongation of the Gumstock. As the E Modulus of CSP increases with increase in *R* value, so does the E Modulus of the Gumstock too. As the % Elongation of Gumstock (which ranges from 359.1–550) and that of CSP (which ranges from 32.45–43.16) are plotted against R values in the same figure, rate of decrease in % Elongation with respect to R value does not seem to be same in Fig. 7.

Table 6 Mechanical properties of CSP of 20 kg batches in comparison with those of respective Gumstocks

X/R values	Tensile Strength (kg-f/cm^2)		% Elongation		E-Modulus (kg-f/cm^2)	
	Gumstock	CSP	Gumstock	CSP	Gumstock	CSP
10.05/ 0.83	2.29 (2.12–2.41)	10.25 (9.88–10.43)	359.1 (346–388)	32.45 (30.23–34.89)	0.969 (0.79–1.09)	53.2 (50.86–56.70)
10.07/ 0.81	2.84 (2.48–3.50)	8.70 (8.54–8.94)	460.4 (388–588)	35.93 (34.89–37.78)	0.771 (0.58–0.91)	40.99 (38.83–43.47)
10.09/ 0.78	1.68 (1.58–1.82)	6.81 (6.71–7.03)	550.0 (526–571)	43.16 (41.33–44.22)	0.337 (0.32–0.34)	29.21 (27.44–31.61)

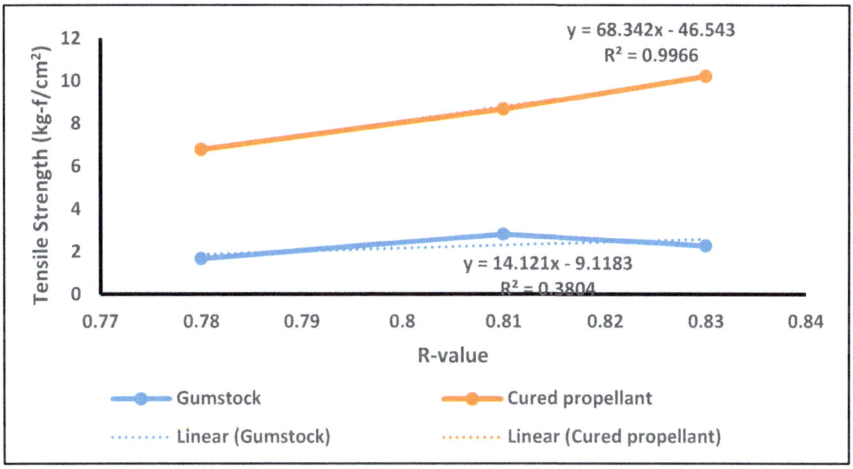

Fig. 6 CSP versus Gumstock wrt Tensile strength as a function of *R* value

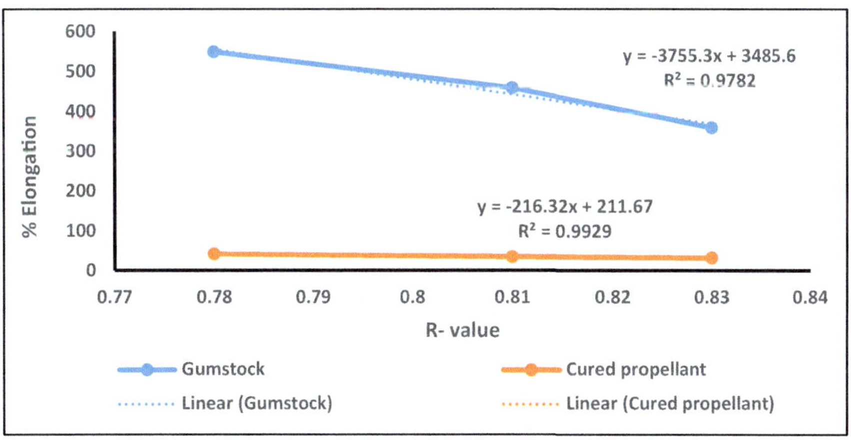

Fig. 7 CSP versus Gumstock wrt % Elongation as a function of *R* value

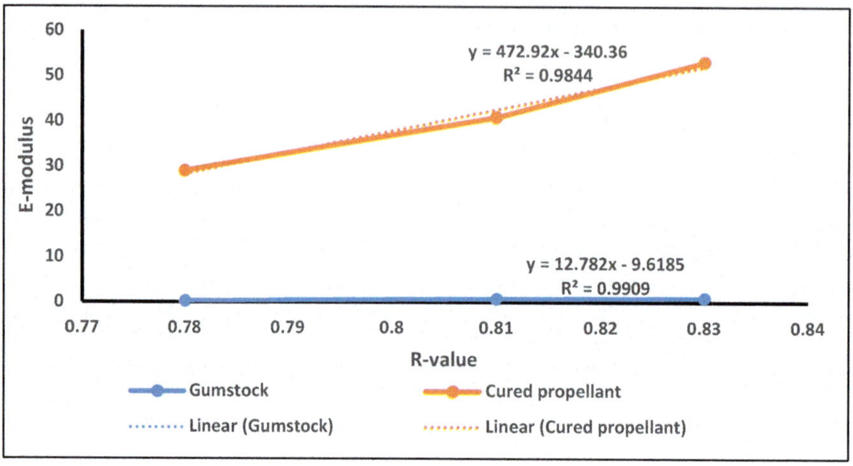

Fig. 8 CSP versus Gumstock wrt E Modulus as a function of *R* value

Similar reason holds good for the seemingly different rate of increase in E Modulus with respect to *R* value in Fig. 8. But, indeed, the decreasing trend of % Elongation with respect to *R* value and the increasing trend of E Modulus with respect to *R* value are the same for both CSP and corresponding Gumstock. Thus, Gumstock proves to be effective and simple tool to overcome the uncertainty in the mechanical properties of cured CSP as through Gumstock level properties, most prospective *X* value and *R* value can be predicted, thereby number of formulation trials at CSP level can be minimized. This leads to huge saving of time and cost.

CSP of Production Scale (700 kg Batch) at $X = 10.07$ and $R = 0.81$ in Comparison with Corresponding 20 kg Batch and Gumstock: Achieved mechanical properties of CSP in 700 kg Production Batch carried out on the above X and R values are given in Table 7 in comparison with those of corresponding 20 kg batch and Gumstock.

As evident from Table 7, except for % Elongation, no appreciable scale up effect has been seen in 700 kg production batch as the achieved Mechanical Properties

Table 7 Mechanical properties of CSP of 700 kg batch in comparison with those of corresponding 20 kg batch and Gumstock

Properties	*X* value = 10.07 and *R* value = 0.81		
	Gumstock	20 kg batch	700 kg batch
Tensile strength (kg-f/cm^2)	2.84 (2.48–3.50)	8.70 (8.54–8.94)	8.21 (8.01–8.50)
% Elongation	460.4 (388–588)	35.93 (34.89–37.78)	39.02 (36.44–41.11)
E Modulus (kg-f/cm^2)	0.771 (0.58–0.91)	40.99 (38.83–43.47)	40.43 (37.61–43.37)

are well comparable for both 20 kg batch and 700 kg batch both produced using Planetary Mixers (Ross Make) of capacities 20 L and 900 L respectively. Table 7 confirms that as Gumstock has proven to be effective and simple tool to overcome uncertainty in the achieved mechanical properties of CSP of 20 kg batch, so also it will be an equally effective tool to overcome uncertainty in case of CSP produced in Production batch of size 700 kg.

In general, E Modulus of 35–40 kg-f/cm^2 and % Elongation of 35–40 are preferable for the CSP of large Rocket Motors. Therefore, Table 7 confirms that Gumstock level % Elongation must be around 460 and Gumstock level E Modulus must be around 0.77 kg-f/cm^2 in order to achieve required mechanical properties in the production batch for the type of formulation under discussion in this paper with the given particle size and fine to coarse ratio of AP oxidizer. Thus, for the formulation for 20 kg batches and 700 kg batch of CSP discussed in this paper, Gumstock can be prepared at different X values while keeping R value around 0.80 and that X value which gives % Elongation around 460 and E Modulus around 0.77 kg-f/cm^2 can be found out. That X value and the corresponding formulation as given in Table 3 of Sect. 2.2, may be confirmed by running 20 kg trial of CSP before clearing the formulation for production of CSP for the actual Rocket Motor. This will not only get rid of uncertainty in the large Rocket Motor CSP but also will reduce number of formulation trials required to finalize the formulation for casting of Propellant in large Rocket Motors.

4 Conclusion

Authors have successfully introduced and demonstrated Gumstock as an effective tool to finalize the propellant formulation for large Rocket Motors. Further, they have experimentally proven that the behaviour of change in mechanical properties especially % Elongation and E Modulus with respect to R value (NCO/OH ratio) is the same for both Gumstock and CSP. Based on Gumstock experiments, authors have proven that cross-linking reaction reaches completion even at 20 °C within 8 days. Methodology based on Gumstock level properties, has been successfully devised by the authors to finalize the formulation for casting of propellant in large Rocket Motors. Methodology, thus promulgated, even though, is specific to formulation of CSP with oxidizer of given particle size and fine to coarse ratio, it paves way to establish Gumstock level acceptance criteria by generating data and correlating with production scale batch for CSP of other formulations too. This, in turn, will reduce the number of formulation trials thereby save time and cost and avert surprises in the achieved mechanical properties of Rocket Motor Propellant.

References

1. Klager K (1984) Polyurethanes the most versatile binder for solid composite propellants. In AIAA Paper No. 84-1239, AIAA/SAE/ASME joint propulsion conference 20, 11 June 1984. Cincinatti, OH.
2. Kingstone Lesley Jabez I, Das U, Manivannan R et al (2019): Influence of HTPB prepolymer on achieved properties of composite solid propellant. High Perform Polym 31(9–10):1162–1172
3. ASTM Standard D638-14. Standard test method for tensile properties of plastics
4. Restrepo-Zapata NC, Osswald TA et al (2015) Vulcanization of EPDM rubber compounds with and without blowing agents: identification of reaction events and TTT-diagram using DSC data. Polym Eng Sci 55:2073–2088
5. Stelescu MD, Airinei A, Grigoras C et al (2010) Use of differential scanning calorimetry (DSC) in the characterization of EPDM/PP Blends. Int J Thermophys 31:2264–2274
6. TA 389, Characterization of the degree of cure of thermosetting resins by DSC
7. Kingstone Lesley Jabez I, Das U, Anne SB et al (2021) Evaluation of ethylene propylene diene terpolymer based insulation against nitrile rubber based insulation for large composite rocket motor casing. Polym Adv Technol 32:111–122
8. Manjari R, Joseph C, Pandureng LP et al (1993) Structure- property relationship of HTPB-based propellants. I. Effect of hydroxyl value of HTPB resin. J Appl Polym Sci 48:271–278
9. Manjari R, Somasundaran UI, Joseph VC et al (1993) Structure- property relationship of HTPB-based propellants. II. Formulation tailoring for better mechanical properties. J Appl Polym Sci 48:279–289
10. Kingstone Lesley Jabez I, Das U, Anne SB et al (2021) Cure behavior of ethylene propylene diene terpolymer insulation for large composite rocket motor casing. Polym Compos 42:3996–4004

Basalt Fabric as a Superior Reinforcement for Low-Velocity Impact Protection

Saiyam Dobhal, T. Sreekantha Reddy, Ch. Subrahmanyam, and P. Rama Subba Reddy

1 Introduction

An armor is a protective equipment wrapped around a body to shield it from any incoming threats in the form of kinetic or chemical energy. It is practically a boon when it comes to providing safeguard against any incoming threat, but we still need to ensure it doesn't turn into a curse. Since the inception of protective equipment, there have been attempts at making them better and lighter. The problem with currently available traditional armor materials is limited energy absorption and heaviness as a result weight of armor increases steeply to counter the latest threats. Increase of armor weight reduces combat efficiency of the fighting systems. Therefore, it is advised to have a material with higher energy absorption. Gradually, towards the end of twenty-first century polymer composites (mostly thermosets) started gaining popularity as matrix choice for armor materials [1]. Through their use one could prepare light-weight armor which would not only have high efficiency but would also be protected from moisture and dirt without compromising on the field mobility. Composites are used in vehicle armor as add on armor and in structures as part of structural armor. During the operations, the vehicle shall be experiencing both low velocity and ballistic impact loads. The criteria and mechanics however, are different for low and high velocity impact scenarios. Therefore, a thorough analysis on behavior of armor composites under low-velocity impact is extremely useful to asses a laminate's performance.

S. Dobhal · Ch. Subrahmanyam
Indian Institute of Technology Hyderabad, Hyderabad, Telangana 500085, India

T. Sreekantha Reddy · P. Rama Subba Reddy (✉)
Defence Metallurgical Research Laboratory, DRDO, Hyderabad, Telangana 500058, India
e-mail: rsreddy.dmrl@gov.in

M. R. Manohara Babu et al. (eds.), *Recent Advances in Composites Engineering and Research*, Springer Proceedings in Materials 39,
https://doi.org/10.1007/978-981-99-8807-5_27

1.1 Current Progress

Currently, epoxy and phenolic resin systems with a wide variety of woven and non-woven fibers are used to prepare high-performance laminates [2]. Glass, Aramid, carbon and ultra-high molecular polyethylene fibers are widely used as reinforcements for preparing ballistic grade composite laminates. Glass is available in a variety of grades such as E-Glass, S-Glass, R-Glass, etc. with each grade bestowing the fiber a unique characteristic. Glass fibers are strong, cheap and easily available in the worldwide market which makes them immensely popular as a reinforcement choice [3]. However, they come with their own set of shortcomings. Glass fibers are made from Silica which in turn comes from sand, a natural (nonrenewable) resource [4]. Moreover, they are coated with a sizing layer which makes them chemically compatible with different resins but uses up a lot of highly reactive chemicals [5]. Similarly, Aramids, which are well known for their flexibility and high tensile strength, have a tedious production process and cost higher than their counterparts [6]. Carbon fabric, arguably one of the strongest reinforcement options with high specific strength and modulus. UHMWPE is expensive and less stiff, though it is an excellent energy absorber, it is not suitable for structural armor due to the lower stiffness, hence it is restricted as spall liner and backing to ceramic armors. However, all these high-performance polymer fibers exert a heavy toll on the environment and are among the costliest fabric materials [7, 8]. Thus, there is a need to develop laminates based on green alternatives which do not subside in terms of strength either.

1.2 Basalt-Reinforced Epoxy Laminates

In recent years, a lot of attention is being shifted toward basalt fiber. Basalt is a form of rock obtained by cooling down of lava. The rock is ground, crushed and thereby melted to be drawn as fibers. Basalt is a green material which is abundantly present in the earth's crust [9]. It is an environmentally friendly green alternative which has a mild carbon footprint as compared to most commercially available fibers. The properties of basalt, E-Glass and aramid are listed in Table 1 [10–14].

Table 1 Comparison of different properties of widely used reinforcements [10–14]

Reinforcement	Density (g/cm^3)	Modulus (GPa)	Tensile strength (MPa)
E Glass	2	70	2050
S Glass	2.5	80	4000
Kevlar	1.4	100	2700
UHMWPE (Dyneema)	0.97	25	2500
Basalt	2.5	80	3000

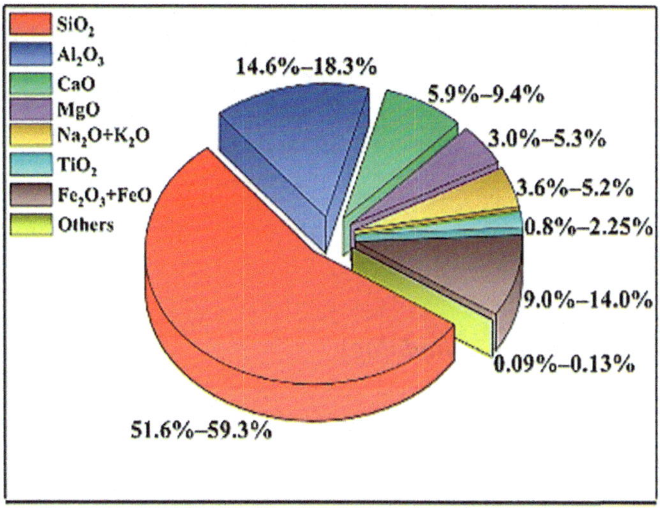

Fig. 1 Composition of basalt fiber

In today's scenario it is essential to prepare, use and promote materials which are environmentally benign and safe. Glass fibers are a major health hazard while working and disposing off scrap fiber is always an issue. Kevlar production utilizes and wastes a lot of solvent and carbon fiber production releases a lot of harmful gases in the environment. Primarily composed of Silica (51–60%), Alumina (14–18%), Iron oxide (9–14%) and Lime (6–9%), basalt fabric, could be a solution towards the search of a safe and strong material owing to abundant availability of raw material and a primitive production process [14] (Fig. 1).

Mechanically, basalt performs at par with S-Glass and Carbon fibers in several aspects while being far superior than Kevlar and E-glass in most aspects. The laminates prepared from these fabrics has potential to use as structural composite as well as backing composite for steel and ceramic armour. It is worth to study and understand the behaviour of basalt fiber-based laminates under mechanical and impact loads. Therefore, the main objective of the present work is to study the low velocity impact behaviour of Basalt/Epoxy composite laminates and compare their performance and damage modes with Glass and aramid fiber reinforced laminates.

Table 2 Fiber specifications

Fiber	Weave	Thickness	GSM
E-Glass	Plain woven	0.4	450
Kevlar	Plain woven	0.5	500
Basalt	Plain woven	0.48	450
S-Glass	Plain woven	0.6	690

2 Experimental

2.1 *Materials and Methodology*

Commercially available epoxy resin (EP 1556) and its hardener 1972 were used as matrix to prepare laminates. Woven E-Glass (450 gsm), S-Glass (700 gsm), Kevlar-749(500 gsm) and Basalt(450 gsm) fabrics were procured from local sources and used as reinforcements. It was tried as much as possible to take fabrics with similar characteristics for comparison. S-Glass which has slightly superior properties was also kept to draw a solid conclusion as to which fabric performs better. The specifications of fibers used is summarized in Table 2.

Composite laminates were prepared through hand layup followed by hot compression moulding method. Resin content of 25 ± 1%, fiber content at 75 ± 1% and thickness of 5 ± 0.1 mm was maintained for all the combinations. Typically, the resin was mixed with hardener in 100:60 ratio and then the mix resin was applied to the fabrics carefully using hand lay-up method. The fabric was allowed to be properly impregnated and then stack of resin coated fabrics (plys) are placed inside metallic mould in a hot press. The temperature was slowly increased up to 80 °C and 40 bar pressure was applied for it to cure at specified time of 3 h (Fig. 2).

2.2 *Characterization*

The cured laminates were cut into specimens of required dimensions using a diamond blade cutting machine. The dimensions for samples were determined according to the ASTM standards. The standards and sample dimensions are mentioned in Table 3.

Physical Properties

The prepared laminates were checked for areal density, density, hardness and resin content as per the ASTM test methods. The hardness values were recorded using Barcole Hardness tester. For resin content, ASTM D2584 was followed. A 10 × 10 × 10 mm portion of the laminate was weighed and placed inside a silica crucible which was then kept inside a muffle furnace and heated at 600 °C for 4 h.

Mechanical property characterization is significant as a laminate undergoes several types of loads during its lifetime and during impact. Especially, during

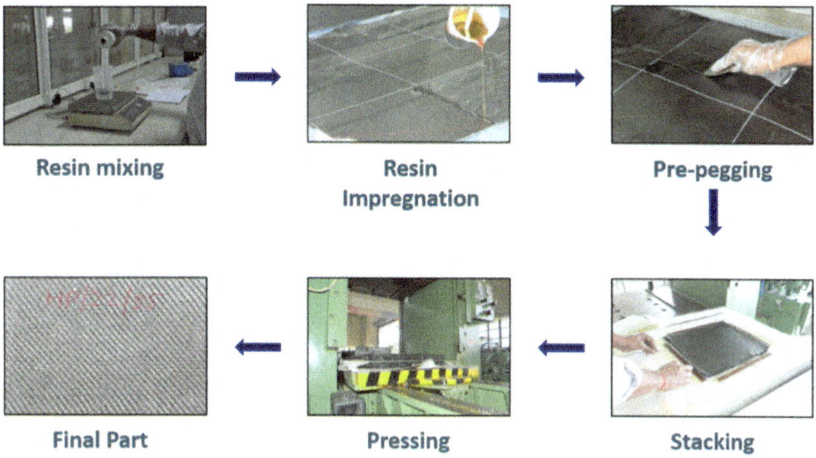

Fig. 2 Preparation of laminates

Table 3 ASTM standards and sample specifications

Test	ASTM	Sample dimensions (mm)
Tensile test	D3039	250 × 15 × 5
Flexural test	D790	100 × 10 × 5
Short beam strength	D2344	50 × 10 × 5
Compression	D695	20 × 10 × 5

low-velocity impact, compressive stresses are generated on the strike-face while the back-face has to withstand tensile stresses [15]. Hence, a thorough investigation of mechanical response is essential in determining the best material choice.

2.3 *Low-Velocity Impact Tests*

Low-velocity impact test was carried out using an Instrumented drop-weight impact tester (Instron CEAST-9350) as shown in Fig. 3. The samples were placed over a platform inside the test chamber where the indenter strikes at the center of the sample from a specified height. The striker was equipped with force transducer of 45 kN capacity to measure the force exerted by the test specimen on the impactor during the impact. The impact energy was varied from 50 to 250 J to cover velocities ranging from 4.34 to 10 m/s which constitute the ideal low-velocity impact range according to Cantwell et al. [16]. The energy corresponding to every impact velocity has been listed below in Table 4. Impactor mass 5.277 kg is equipped with an indenter of conical shape which measures around 3 cm in length with a 16 mm diameter tip. The drop tower has an anti-rebound mechanism in place to prevent multiple hits on

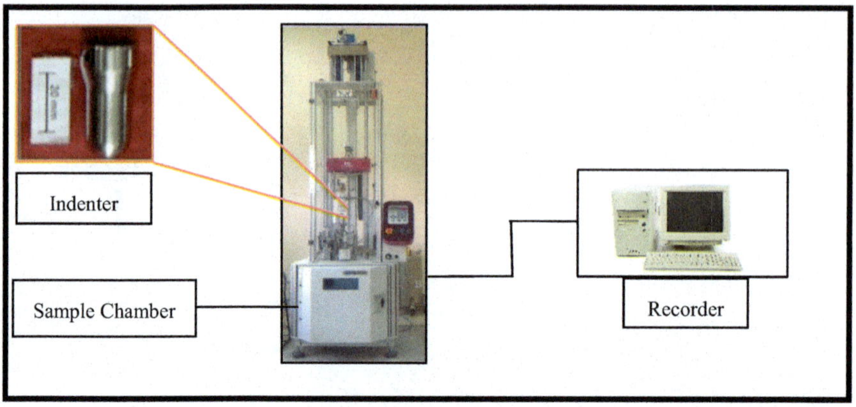

Fig. 3 Drop tower assembly for low-velocity impact testing

Table 4 Low velocity test parameters (refer in text)

Impact energy (J)	Impact velocity (m/s)
50	4.3
100	6.0
150	7.5
200	8.7
250	9.7

the test sample. Composite samples were subjected to total of five different impact energies in increasing order of magnitude to determine if the impact range is above or below the perforation threshold and analyzed for their behavior.

2.4 *Failure Analysis*

The post impacted samples were subsequently examined for different damage modes through infrared (IR) thermography followed by stereomicroscopy analysis. The samples were first analyzed through IR thermograph (FLIR EcoTherm), which is an excellent non-destructive tool to analyze the extent of damage in thin laminate capable of pointing out damages sustained on the surface as well as in-between layers [17]. The sample is subjected to a high intensity thermal flash (with an energy of 9.6 kJ) that passes through the sample for 10 ms while the Infrared camera records the image. Thereafter, the laminates subjected to 250 J impact energy, which sustained maximum damage were chosen and cut horizontally along the point of impact and placed under a stereomicroscope (M165C, Leica Micro Systems) to examine the failure mode of fibers and at point of impact and through thickness for understanding the damage modes and impact absorption mechanics.

3 Results and Discussion

Physical properties of the laminates are given in Table 5. It is observed that, the measured resin content and fiber content are in the range of 75 ± 2% for all the laminates. Barcole hardness which indicates cure state of laminate, it is found to be 40 for Glass and basalt laminates, while for Kevlar it is found to be 20 which is slightly lesser as compared to the other laminates though matrix system is similar. This may be due to the presence of aramid fiber which are softer than the other fibers. Among all the laminates, Kevlar laminate shows lower density and areal density, probably due to the low density of fibers. Basalt fiber-based laminate shows areal density of 11–12 kg/m^2 which is almost close to the glass fiber-based laminates.

The mechanical properties of are summarized in Fig. 4. For each test, 5 samples were tested and the value closest to average was chosen to be displayed. As per our observations, S-Glass reinforced laminates showed highest tensile strength while Basalt reinforced laminates showed higher compressive and flexural strength. Hence, Laminates prepared with Basalt as a reinforcement show excellent mechanical properties mostly on par with S-Glass, which is a mechanically superior and costlier material with higher GSM (700) as compared to basalt (450). Hence, Basalt can be considered as a robust material capable of withstanding a wide spectrum of loads encountered in structural applications. As far as failure modes during tensile test are concerned, Basalt and E-Glass show brittle failure while Kevlar and S-Glass fail predominantly due to delamination.

3.1 Low-Velocity Impact

Force–Displacement

A force–displacement curve is recorded in an instrumented drop tower as a response to the incoming impact. The low velocity impact performance of all laminates from 50 to 250 J can be described using force–displacement curves. A closed loop usually indicates a rebound or partial penetration while an open curve is obtained on complete perforation of the specimen [18].

Table 5 Physical properties of reinforced laminates

Laminate reinforcement	Fiber content (%)	Resin content (%)	Barcole hardness	Laminate density (Kg/m^3)	Areal density (Kg/m^2)
E-Glass	74 ± 2	26 ± 2	30	1.74	10
S-Glass	76 ± 1	24 ± 1	40	1.78	12
Kevlar	75 ± 2	25 ± 2	20	1.11	6
Basalt	75 ± 1	25 ± 1	40	1.76	11

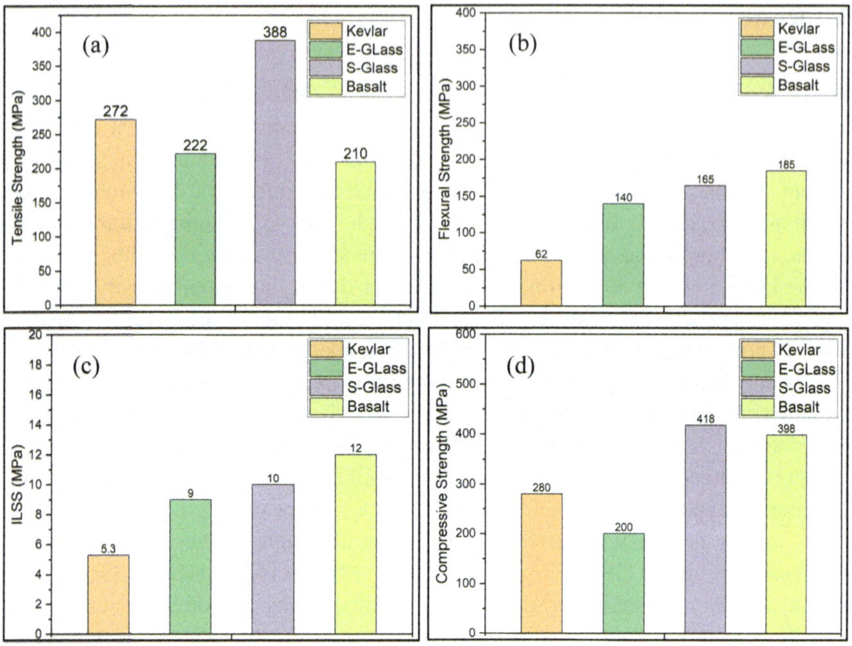

Fig. 4 **a** Tensile, **b** flexural, **c** ILSS and **d** compressive strengths of laminates

A force–displacement curve can also be seen as a testament to the toughness of the laminate. A larger curve means higher energy absorption and dissipation.

Energy-Time

The energy transfer and dissipation over the course of an impact can also be interpreted using energy-time curve [19]. The maximum value indicates the maximum impact energy absorbed by the laminate whereas, the flat portion gives the final absorbed energy. The difference between these two can be used to determine the rebound energy or elastic energy of the laminate (Fig. 5).

It can be seen from the force displacement curves that Basalt provides a better response to low velocity impact as compared to any other chosen reinforcement. The closed loop with high enclosed area even at higher impact energies indicate a strong network capable of effectively absorbing and dispersing incoming trauma. It can also be seen from the force displacement curves that E-Glass gets perforated at 150 J and while others do not show complete perforation even at 250 J energy. For all other reinforcements, impactor gets rebounded at every energy level.

Also, interestingly enough, basalt shows a smaller number of oscillations in the force displacement curve at all energy levels as compared to any other fabric. Moreover, the Energy-time curve also justifies the observation as rebound energy is highest for basalt at 250 J.

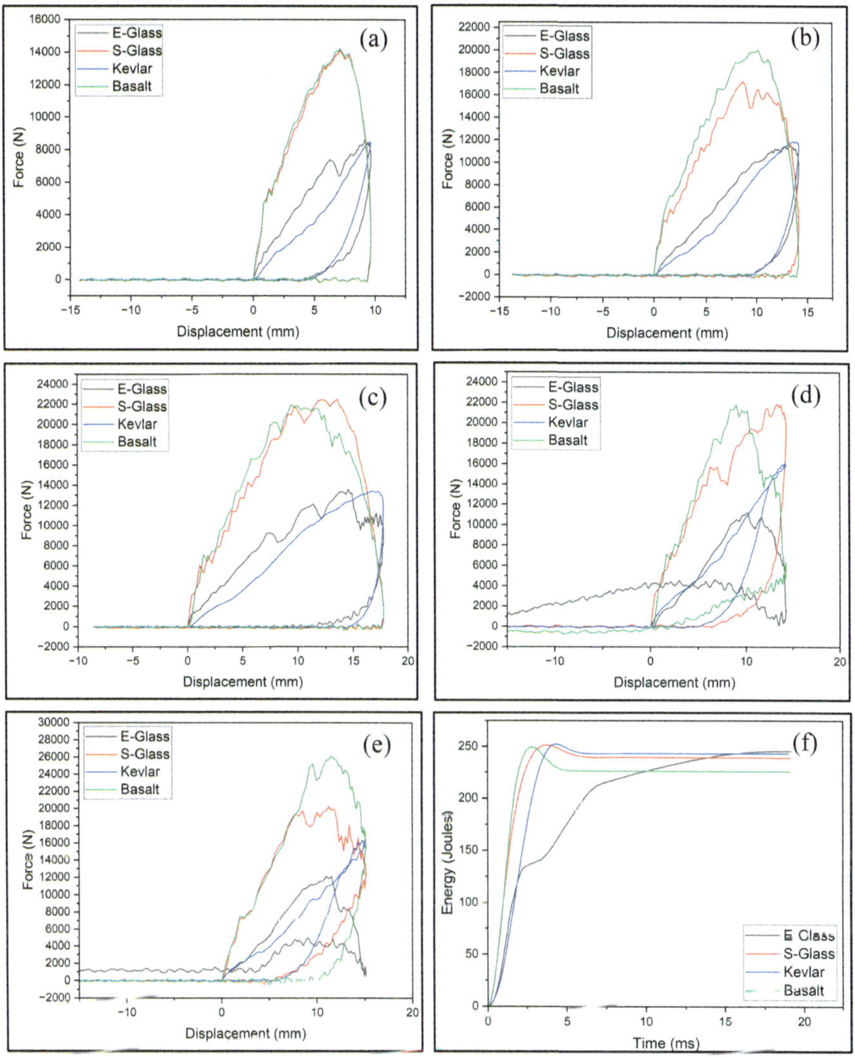

Fig. 5 Force–displacement curves for impact at **a** 50 J, **b** 100 J, **c** 150 J, **d** 200 J, **e** 250 J and energy-time response (**f**) of laminates subjected to 250 J impact

The reason why basalt stands out can also be understood by looking at the graphs in Fig. 6. As expected, Kevlar, owing to its high tensile strength and tenacity, absorbs most energy as compared to any other reinforcement at any energy level. But it is also the one that shown highest deformation post-impact. Which means, in any scenario where it is used in front of a body that it is supposed to shield, the deformation could be so severe (reaching to 22 mm for 250 J) that it may actually transfer the shock to the body component during an event of impact. While S-Glass and Basalt show decent levels of energy absorption with almost 50% less deformation. This arises due to difference in inherent nature of the materials and their failure modes.

Energy Profile Diagram

The energy profile of a laminate is obtained when absorbed energy is plotted against the energy of impact [20]. It is used to denote whether a laminate shows complete perforation at given energy. Looking at Fig. 7, it can be clearly observed that only for E-Glass reinforced laminates, the absorbed energy is equal to the impact energy at 200 and 250 J. This indicates that only E-Glass reinforced laminates show complete perforation after 150 J while others are able to withstand the impact as their perforation threshold tends to be higher than the energy values imparted in the current study.

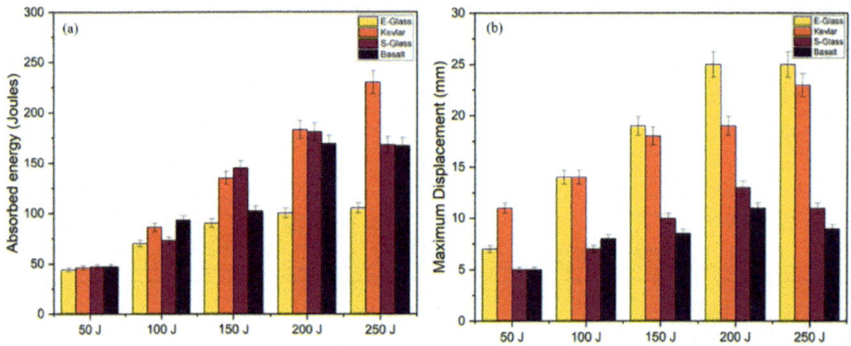

Fig. 6 **a** Max energy absorbed and **b** maximum displacement sustained by the laminates at different impact energies

Fig. 7 Energy profile diagram of different laminates

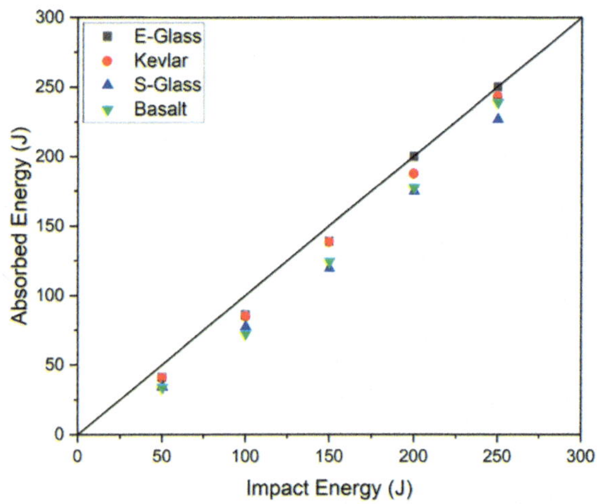

Table 6 Predicted perforation energies from curve fitting of Fig. 10

Laminate reinforcement	Perforation threshold (J)
E-Glass	230
Kevlar	285
Basalt	280
S-Glass	335

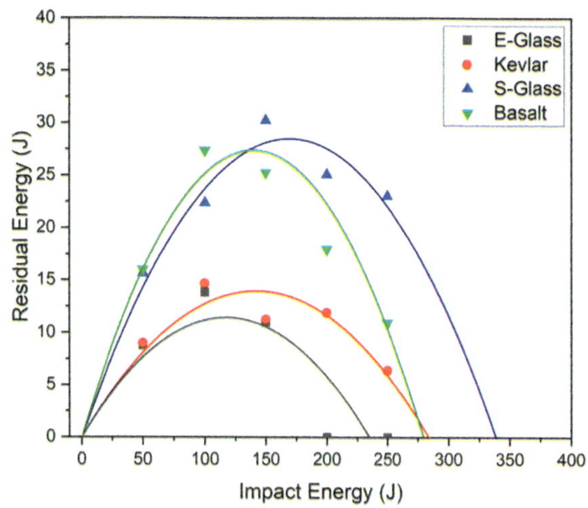

Fig. 8 Perforation energy threshold diagrams of different laminates with 2nd order polynomial fit

Perforation Energy Threshold

Since, majority of the samples are left unperforated, a perforation energy threshold diagram that is plotted between Residual energy (E_r) and Impact energy can provide useful insight by predicting the energy at which indenter will perforate the laminate [20]. The experimental data is fitted using a 2nd order polynomial equation who's non-zero roots will be the perforation energy threshold. Here, as expected, E-Glass has the lowest predicted threshold value of 175 J followed by Kevlar and Basalt at approximately 280 J while S-Glass turns out to be the one with the highest perforation threshold at 335 J. The values are summarized in Table 6 (Fig. 8).

3.2 Stereomicroscopy

The results show that the despite having the same matrix, the samples fail differently under same load. Thus, the failure mode of the laminate under low-velocity impact depends predominantly upon the nature of fabric. E-Glass, S-Glass and Basalt due to their ceramic nature show brittle failure while Kevlar owing to its elastic nature and excellent compressibility disperses energy via sustaining compressive stresses

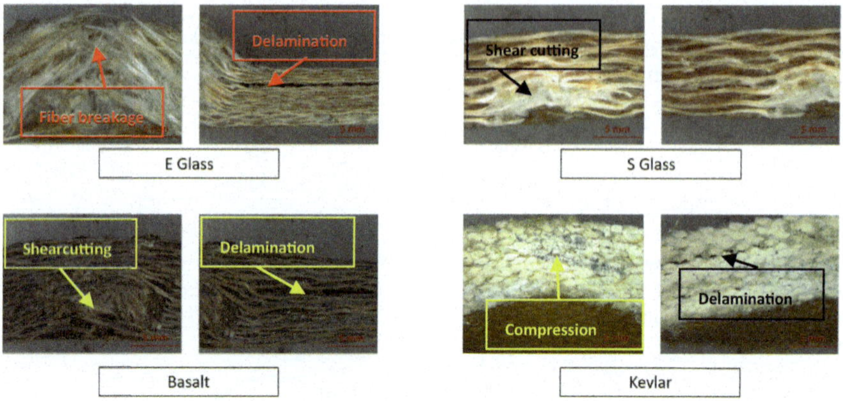

Fig. 9 Stereomicroscopic analysis of damaged samples

and delamination. Kevlar being a long chain polymer, absorbs shock via dispersing the shockwave over the entire matrix while sustaining high deformation in the fibers. Basalt and S-Glass on the other hand are brittle materials that show brittle failure at the point of impact and disperse the shockwave via fiber cracking and breakage. The Extent of damage is highest for E-Glass which shows fiber breakage and pullout due to complete perforation. Basalt and S Glass show partial perforation while the impactor fails to perforate Kevlar. Even though Kevlar does not allow the indenter to pass through, the area of damage, as seen through visual examination and extent of delamination surrounding the point of impact is much higher for Kevlar as compared to other reinforcements (Fig. 9).

3.3 IR Thermography

Thermographic analysis of samples shows that the damage area increases as the impact energy is increased. E-Glass which sustains the most damage, shows the highest damage area followed by Kevlar which absorbs the highest energy. Through thermography images, the energy dispersion mechanism of a material can also be seen. For brittle materials a distinct star shape pattern is visible (Fig. 10) on the strike-face which corresponds to the brittle fracture of matrix component.

The damage areas in the back-face are mostly in shape of distorted circle. The calculated areas are summarized in below in Table 7.

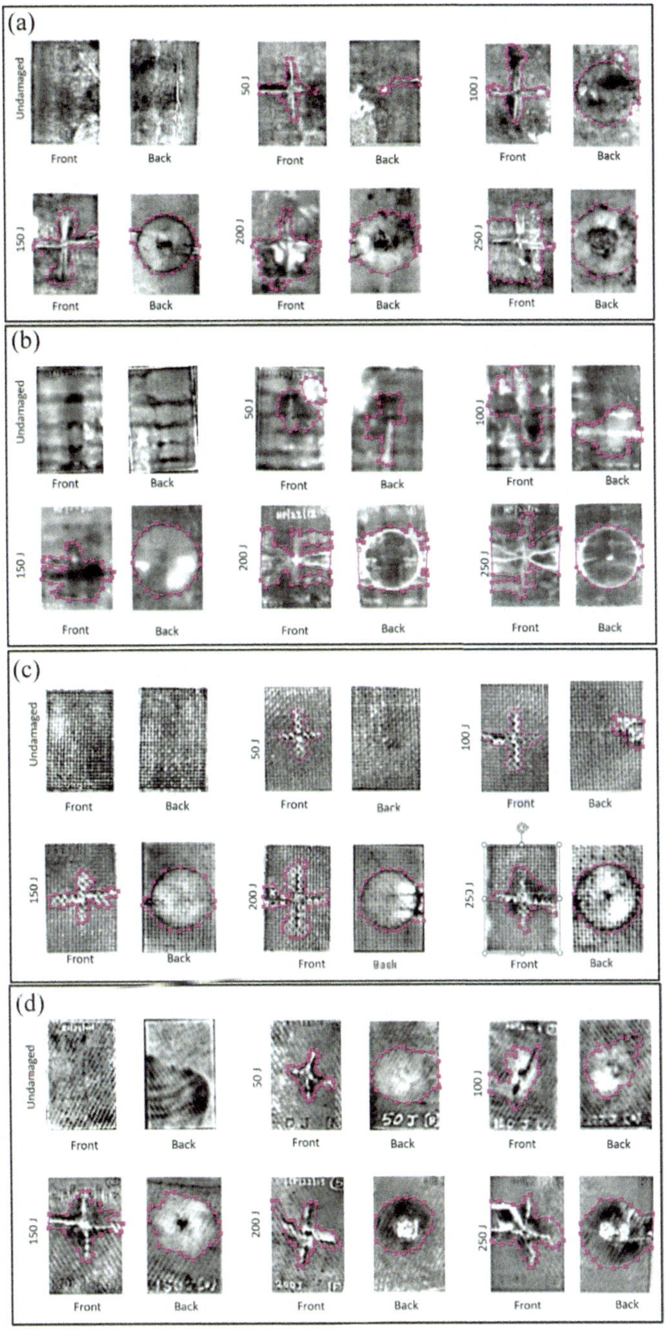

Fig. 10 IR thermography images of damaged **a** E-Glass, **b** Kevlar, **c** S-Glass and **d** Basalt laminates

Table 7 Damage areas of laminates subjected to low-velocity impact

Sample	50 J	100 J	150 J	200 J	250 J
E-Glass	F – 2100 R – 574	F– 3300 R- 7026	F – 4560 R- 6281	F – 5386 R- 6294	F – 7928 R- 7900
Kevlar	F – 3400 R- 2600	F – 4780 R- 3200	F – 3762 R- 6085	F – 4188 R- 6088	F – 6633 R- 6737
Basalt	F – 1496 R- 5112	F – 3093 R- 3858	F – 3318 R- 5076	F – 2619 R- 5951	F – 4630 R- 6079
S-Glass	F – 2100 R – 500	F – 2700 R- 1296	F – 2543 R- 5972	F – 4704 R- 6382	F – 2543 R- 5972

4 Conclusion

Basalt reinforced laminates show outstanding properties as compared to E-Glass, S-Glass and Kevlar reinforced epoxy laminates.

(i). Mechanical tests show that basalt reinforced laminates perform better than their counterparts in terms of flexural, ILSS and compressive strength while also showing good tensile strength values.
(ii). Low velocity studies reveal that they absorb high amount of energy with low deformation. The energy-time plots show that the rebound energy is highest for basalt reinforced laminates.
(iii). Perforation threshold of 5 mm thick basalt laminates was found to be 275 J. Which is an improvement over E glass and comparable to Kevlar.
(iv). Stereo-microscopic analysis confirms that basalt/epoxy laminates disperse energy via brittle failure which leads to shearing of fibers in top layer.
(v). They display an all-round performance dominating all structural and dynamic aspects. These outcomes are essential in cementing the position of basalt reinforced epoxy laminates as low-cost, light-weight materials for protective equipment.

These findings are also useful for formulating future studies on high-velocity ballistic impact performance for vehicle and personnel protection armours.

References

1. Tabiei A, Nilakantan G (2008) Ballistic impact of dry woven fabric composites: a review. Appl Mech Rev 61(1)
2. Hani AR, Roslan A, Mariatti J, Maziah M (2012) Body armor technology: a review of materials, construction techniques and enhancement of ballistic energy absorption. Adv Mater Res 488:806–812
3. Grujicic M, Pandurangan B, Zecevic U, Koudela KL, Cheeseman BA (2007) Ballistic performance of alumina/S-2 glass-reinforced polymer-matrix composite hybrid lightweight armor against armor piercing (AP) and non-AP projectiles. Multidiscip Model Mater Struct

4. Bender JR, Hadley JG (1994) Glass fiber manufacturing and fiber safety: the producer's perspective. Environ Health Perspect 102(suppl 5):37–40
5. Liu Z, Zhang L, Yu E, Ying Z, Zhang Y, Liu X, Eli W (2015) Modification of glass fiber surface and glass fiber reinforced polymer composites challenges and opportunities: from organic chemistry perspective. Curr Org Chem 19(11):991–1010
6. Tanner D, Fitzgerald JA, Phillips BR (1989) The Kevlar story—an advanced materials case study. Angew Chem, Int Ed Engl 28(5):649–654
7. Hassan MF, Sabri MA, Fazal H, Hafeez A, Shezad N, Hussain M (2020) Recent trends in activated carbon fibers production from various precursors and applications—a comparative review. J Anal Appl Pyrol 145:104715
8. Patel K, Chikkali SH, Sivaram S (2020) Ultrahigh molecular weight polyethylene: catalysis, structure, properties, processing and applications. Prog Polym Sci 109:101290
9. Jamshaid H, Mishra R (2016) A green material from rock: basalt fiber–a review. J Textile Inst 107(7):923–937
10. Alagar M, Thanikai Velan TV, Ashok Kumar A (2000) Mechanical properties of e-glass fiber reinforced siliconized epoxy polymer composites. Polym Compos 21(5):739–744
11. Yeon J, Chowdhury SC, Gillespie JW Jr (2022) Mechanical properties and damage analysis of S-glass: a reactive molecular dynamics study. Compos B Eng 234:109706
12. Hindeleh AM, Halim NA, Ziq KA (1984) Solid-state morphology and mechanical properties of Kevlar 29 fiber. J Macromol Sci Part B Phys 23(3):289–309
13. Spiegelberg S, Kozak A, Braithwaite G (2016) Characterization of physical, chemical, and mechanical properties of UHMWPE. In: UHMWPE biomaterials handbook. William Andrew Publishing, pp 531–552
14. Deák T, Czigány T (2009) Chemical composition and mechanical properties of basalt and glass fibers: a comparison. Text Res J 79(7):645–651
15. Richardson MOW, Wisheart MJ (1996) Review of low-velocity impact properties of composite materials. Compos A Appl Sci Manuf 27(12):1123–1131
16. Cantwell WJ, Morton J (1989) Comparison of the low and high velocity impact response of CFRP. Composites 20(6):545–551
17. Inagaki T, Ishii T, Iwamoto T (1999) On the NDT and E for the diagnosis of defects using infrared thermography. NDT E Int 32(5):247–257
18. Reddy PRS, Reddy TS, Madhu V, Gogia AK, Rao KV (2015) Behavior of e-glass composite laminates under ballistic impact. Mater Des 84:79–86
19. Reddy TS, Reddy PRS, Madhu V (2017) Response of e glass/epoxy and Dyneema® composite laminates subjected to low and high velocity impact. Procedia Eng 173:278–285
20. Taraghi I, Fereidoon A, Taheri-Behrooz F (2014) Low-velocity impact response of woven Kevlar/epoxy laminated composites reinforced with multi-walled carbon nanotubes at ambient and low temperatures. Mater Des 53:152–158

Superior Ballistic Performance of Ti–Al_3Ti Based Metal–Intermetallic Composites Against 7.62 AP Projectiles

G. Sukumar, K. Muralidharan, P. Ponguru Senthil, P. Prakasa Rao, G. Balaji, S. G. Savio, and B. Bhav Singh

1 Introduction

Over the last few decades, there has been a tremendous focus on the weight reduction of armoured vehicles. Weight reduction improves the vehicle's survivability, manoeuvrability, and deployability in rugged terrains such as mountains. Amongst various strategies, introducing advanced materials with superior ballistic performance is one of the ways that enable weight reduction. Traditionally, metallic armour materials such as rolled homogenous armour steels (medium carbon low alloy steels) and aluminium alloys (5083, 2519, 7017) are widely used in the structure of armoured vehicles [1]. By comparison, ceramics (such as Al_2O_3, B_4C, and TiB_2) and composites (such as E-glass–epoxy and S-glass–epoxy) are used in add-on armour applications [1].

In addition to traditionally used metallic alloys, metal matrix composites have also been studied for armour applications. Jo et al. [2] studied the dynamic deformation and ballistic impact response of 7075 Al alloy reinforced with either B_4C or SiC or a combination of B_4C and SiC particles. Amongst all the composites, the hybrid composites (with B_4C and SiC reinforcements) showed the highest dynamic compressive strength (more than 1.5 GPa) and good elongation (about 10%). Under dynamic loading, Al matrix deformation, ceramic particle cracking, and ceramic/Al matrix interface debonding were the primary energy dissipating mechanisms in all three composites. The ballistic performance of the hybrid composite was higher than the 7075 alloy reinforced with either SiC or B_4C. Sudhakar et al. [3] studied the

G. Sukumar (✉) · K. Muralidharan · P. Ponguru Senthil · P. Prakasa Rao · G. Balaji · S. G. Savio · B. Bhav Singh
Armour Design and Development Division, Defence Metallurgical Research Laboratory, Hyderabad, India
e-mail: gsukumar.dmrl@gov.in

M. R. Manohara Babu et al. (eds.), *Recent Advances in Composites Engineering and Research*, Springer Proceedings in Materials 39,
https://doi.org/10.1007/978-981-99-8807-5_28

ballistic performance of surface composite developed on 7075 Al alloy through friction stir processing using different sizes of B_4C ceramic particles. Further, functionally graded composites based on Ti–TiB_2 have also shown promising ballistic performance [4]. Various new materials (bio-mimicking materials) inspired by biological systems have recently been studied to understand the protection mechanisms operating in these materials [5–7].

Earlier studies on mollusc (such as abalone and conch) shells have shown that they have optimized mechanical properties due to their hierarchical structure (consisting of nano-, micro-, and mesoscale features). This hierarchical structure is derived from primary constituents such as calcium carbonate and a series of organic binders [5–7]. $CaCO_3$ is the main constituent in the mollusk shell, and it has strength and toughness values of 30 MPa and 1 MPa-$m^{1/2}$, respectively. However, the mollusk shell has a much higher tensile strength (100–300 MPa) and toughness values (3–7 MPa-$m^{1/2}$) due to its hierarchical structure [5–7]. There have been attempts to mimic the laminate design found in mollusk shells to obtain unique mechanical properties. Inspired by these biological structures, recently, Ti–Al_3Ti metal–intermetallic laminate (MIL) composites have been produced by a simple and cost-effective method called reactive foil sintering [8]. Limited ballistic studies on Ti–Al_3Ti MIL composites against laboratory-type projectiles have shown that Ti–Al_3Ti MIL composites are promising for armour applications [5, 9–11]. However, studies on the ballistic impact response of Ti–Al_3Ti MIL composites against actual ammunitions used in war scenarios are lacking in the literature. In this regard, the present study is undertaken to synthesize and evaluate the ballistic performance of Ti–Al_3Ti composites against 7.62 mm armour-piercing (AP) projectiles. Further, failure mechanisms involved during ballistic impact are also studied.

2 Experimental Set-Up

Reactive foil sintering method [8] has been utilized for producing Ti-Al_3Ti (MIL) composite in this study. In the reactive foil sintering method, Ti–Al_3Ti (MIL) composite is produced from elemental metal foils of pure Ti and pure Al using a controlled reaction (at high temperature and pressure) carried out in the open-air environment. The Ti and Al foil thicknesses were selected so that the reaction of the Al with the nearby Ti layer completely consumes the Al layer [8]. After the reaction, the final composite consists of Ti and Al_3Ti layers. The processing details and experimental set-up used in the present study are shown in Fig. 1a and b, respectively. The processing parameters for the present work are chosen based on earlier work [8]. The initial foil thicknesses of Al and Ti were 200 μm and 130 μm, respectively. MIL composites with various final thicknesses (MIL-1 to MIL-4) were produced by varying the initial numbers of Al and Ti foils.

The phases in the final MIL composite formed after reaction sintering were studied using X-ray diffraction patterns obtained using Cobalt Kα_1 radiation with a wavelength of 0.1789 nm. The room temperature compressive strength of the Ti–Al_3Ti

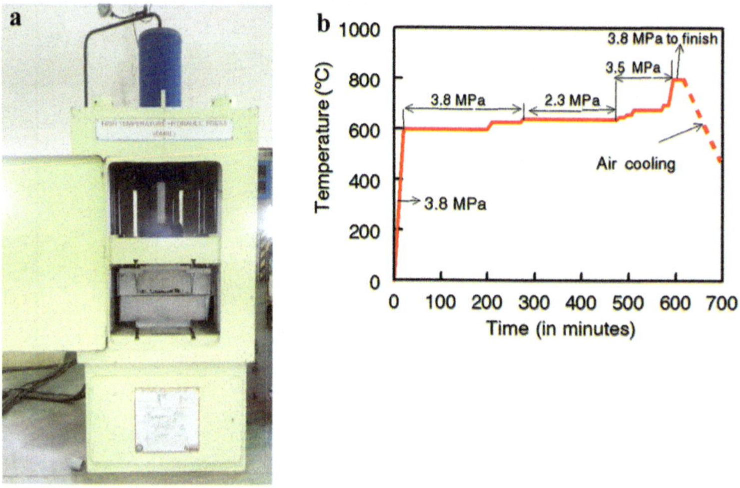

Fig. 1 **a** High-temperature hydraulic press used in present work, **b** Typical processing schedule used in present work

MIL composites (in the direction perpendicular to laminate orientation) was evaluated using a hydraulic test system (Make: Dak systems Inc.). The dimensions of the compression sample were 9 mm (l) × 9 mm (b) × 9 mm (w). The hardness profile across different layers in MIL composite was measured using a Vickers hardness tester at a load of 1 kg. Further, the Ti–Al_3Ti MIL composites were evaluated against 7.62 AP projectiles using a universal gun at a small arms range. The ballistic test was carried out with standard DOP test configuration [12]. Figure 2 shows the schematic of the ballistic set-up used in the present work [12]. The details of 7.62 AP projectiles are given in Ref. [12]. The nominal velocity of projectiles was 830 ± 10 m/s. The steel confined Ti–Al_3Ti MIL composite was clamped together with 50-mm-thick 6061 Al alloy (in T6 condition) backing at the firing stand. The ballistic performance of MIL composites was determined by measuring the residual depth of penetration in the back 6061 Al alloy. In addition, reference depth of penetration in standalone Al6061 alloy backing was also measured against 7.62 AP impacted at a similar velocity. The typical configuration of the target and associated depth of penetration (DOP) measurements are shown in Fig. 3. Differential mass efficiency [13, 14] was used as a parameter to compare the ballistic performance of different materials studied.

Differential mass efficiency is given as follows:

$$\Delta e_c = \frac{\rho_b \times (P_0 - P_b)}{\rho_{\mathrm{MIL}} \times t}$$

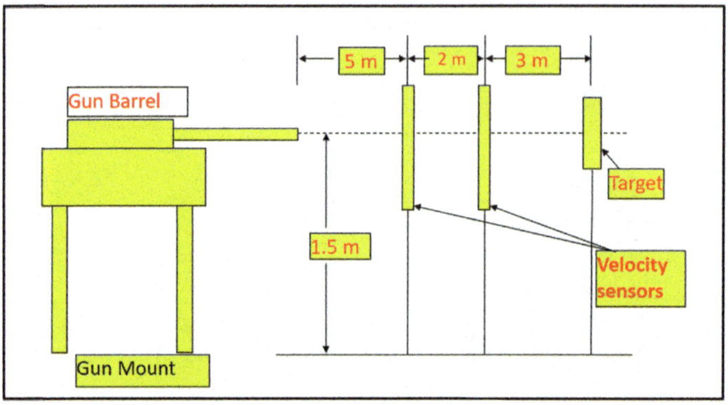

Fig. 2 Schematic of ballistic test set-up used in present work

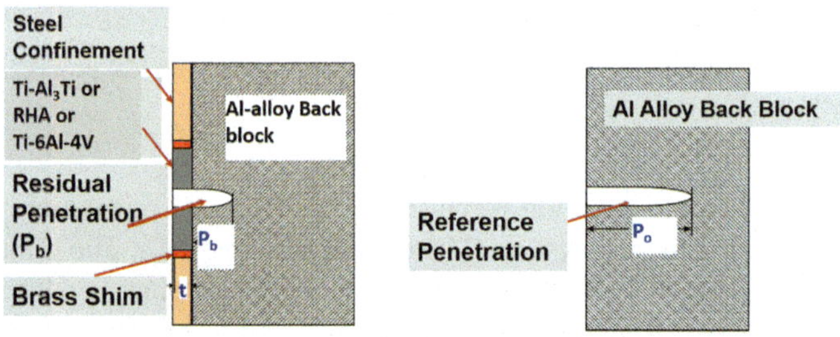

Fig. 3 Typical DOP test configuration of the target

Here, ρ_b is the density of 6061 Al alloy backing, P_0 is the reference depth of penetration, P_b is the residual depth of penetration, ρ_{MIL} is the density of the MIL composite, and t is the thickness of the MIL composite.

Further, the ballistic performance of MIL composites was compared with conventionally used RHA steel (quenched and tempered at 650 °C) and Ti–6Al–4V alloy (in mill annealed condition). For comparative ballistic evaluation, 5-mm-thick RHA steel and 11-mm-thick Ti–6Al–4V steel plates were tested in the same ballistic test configuration that is used for evaluating MIL composites. Typical tensile properties of RHA steel, Ti–6Al–4V, and 6061 Al alloy backing are given in Table 1.

Table 1 Typical tensile properties of RHA steel, Ti–6Al–4V, and 6061 Al alloy backing

S. No.	Material	Yield strength (MPa)	Ultimate tensile strength (MPa)	% Elongation
1	6061 Al alloy (T6)	250	290	15
2	RHA steel	970	1050	16
3	Ti–6Al–4V	900	960	12

After the ballistic test, the crater cross-sections of the Ti–Al_3Ti MIL composite were studied using optical metallography to understand the failure mechanisms activated during ballistic impact.

3 Results and Discussion

The typical microstructure of the Ti–Al_3Ti MIL composite is shown in Fig. 4a. The microstructure consists of alternating layers of Ti and Al_3Ti with different colour contrast. Figure 4b shows the XRD pattern of the Ti–Al_3Ti MIL composite.

It is confirmed from XRD patterns that the final MIL composites (formed after reaction sintering) consist of Ti and Al_3Ti intermetallic compound. No other phase formation is observed in the XRD patterns. The formation of Al_3Ti is favoured over other phases in Ti–Al alloy systems such as TiAl, Ti_3Al, $TiAl_2$, and $TiAl_5$ due to its thermodynamic stability and faster kinetics of formation [15]. The thicknesses of the Al_3Ti intermetallic compound and the remaining Ti layer formed after the controlled reaction are 243 ± 8 μm and 51 ± 9 μm, respectively. The Ti and Al_3Ti layer volume was 21% and 79%, respectively. It can be seen that Al_3Ti has hardness values in the range of 400–550 VHN, and the residual Ti had hardness values of about 160–170 VHN (Fig. 5). Hence, the resulting Ti–Al_3Ti MIL composite after reaction sintering has alternate hard and soft layers mimicking the structure found in the mollusk shell (Figs. 4a and 5).

Figure 6 shows the engineering stress–strain curves of the Ti–Al_3Ti composites after compression testing in a direction perpendicular to the laminate orientation (through thickness direction).

The compressive strength of the Ti–Al_3Ti MIL composite was 939 ± 5 MPa. The compressive strength of the Ti–Al_3Ti MIL composite is close to the quasi-static tensile yield strength of RHA steel and Ti–6Al–4V (Table 1). Figure 7 shows the

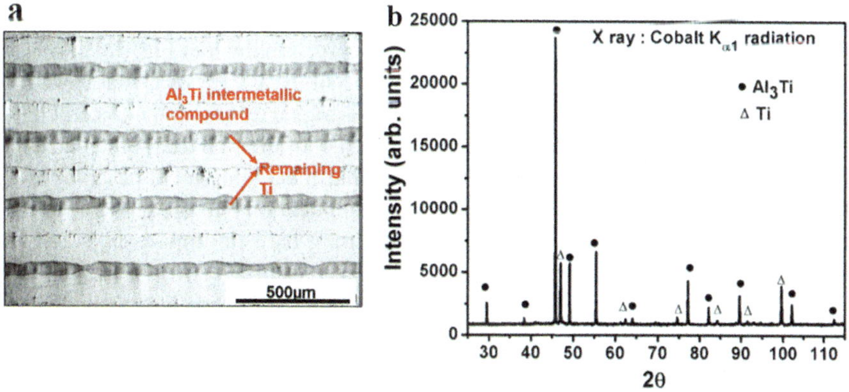

Fig. 4 **a** Microstructure of Ti–Al_3Ti MIL composite, **b** XRD pattern of the Ti–Al_3Ti MIL composite

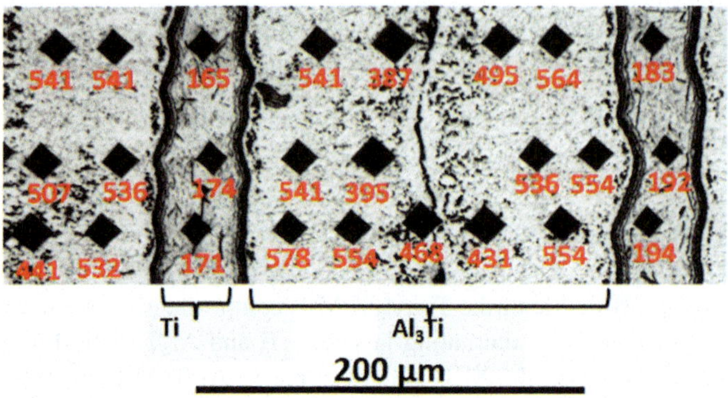

Fig. 5 Hardness profile across different locations in Ti–Al_3Ti MIL composite (number represents the Vickers hardness values in VHN)

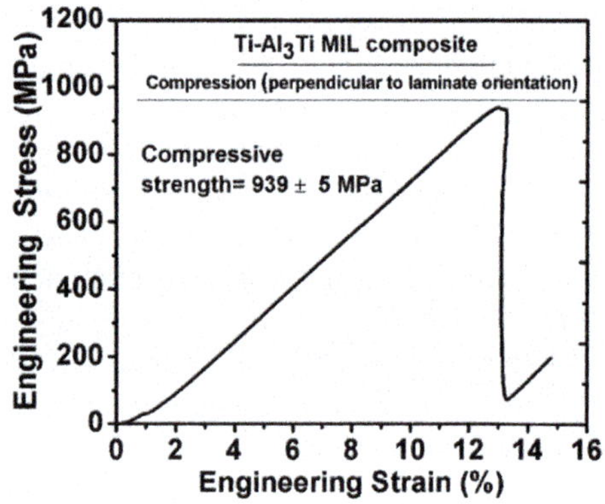

Fig. 6 Engineering stress–strain curve of Ti–Al_3Ti MIL composite

front face, rear face, and crater cross-section of a typical MIL composite (MIL-4 in Table 2) after ballistic testing against 7.62 AP projectiles. The penetration channel consists of a front spall and ductile hole growth regions. Delamination of the composite layers is observed near the exit region as shown in Fig. 7c.

The complete ballistic results of the present study are given in Table 2. It is observed that the MIL composites have about 30% higher differential mass efficiency in comparison with RHA steel and about 10% higher differential mass efficiency than that of Ti–6Al–4V. The present study shows that metal–intermetallic composites have the potential to be used in future armour applications. In order to understand the energy dissipating mechanisms activated during ballistic impact, post-ballistic microstructural analysis was carried out.

Figure 8 shows the optical micrograph of the crater cross-section of MIL composite (MIL-4) after 7.62 AP projectile impact. The crater cross-section contains a front spall region and a ductile hole growth region (Fig. 8). In the front spall region, both the Ti and Al_3Ti layers were failed and removed from the plate due to high stresses generated during initial impact. It has been reported that crack first

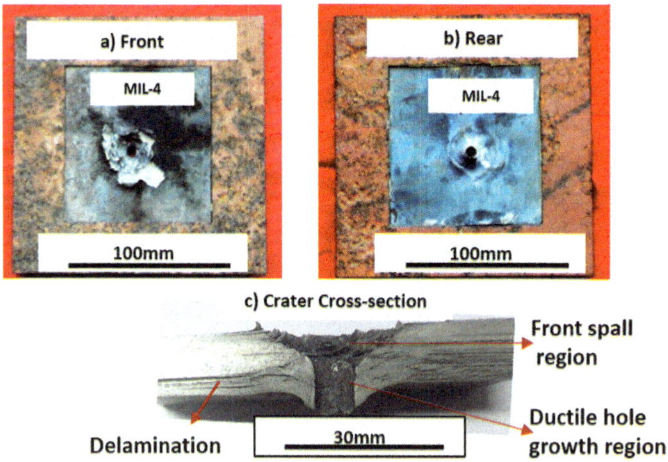

Fig. 7 **a** Front face, **b** rear face, and **c** crater cross-section of the Ti–Al_3Ti MIL composite

starts in brittle Al_3Ti intermetallic during ballistic loading [10, 16, 17]. Many transverse brittle cracks were seen along Al_3Ti intermetallic layer. Through thickness, shear cracks propagating through both Ti and Al_3Ti layers were observed near the penetration channel. In addition, delamination of individual Ti and Al_3Ti layers was observed near the exit region (Fig. 8). In many locations, bending and cracking of ductile Ti layers were observed (Fig. 8). The post-ballistic microstructural analysis shows that the presence of alternate soft Ti and hard Al_3Ti layer activates multiple energy absorption mechanisms (such as shear cracks, transverse cracks, delamination, bending, and cracking) which result in the improved ballistic performance of Ti–Al_3Ti MIL when compared to RHA steel and Ti–6Al–4V.

4 Conclusions

- In this work, Ti–Al_3Ti MIL composites were successfully prepared using the reactive foil sintering method. The thicknesses of the final Al_3Ti intermetallic compound and the remaining Ti layer after the controlled reaction are 243 ± 8 μm and 51 ± 9 μm, respectively. The volume fractions of Ti and Al_3Ti layer are 21% and 79%, respectively.
- In the Ti–Al_3Ti MIL composite, Al_3Ti intermetallic layer has hardness values in the range of 400–550 VHN, and the residual Ti layer shows hardness values of about 160–170 VHN. The compressive strength of the Ti–Al_3Ti MIL composite is 939 ± 5 MPa
- Against 7.62 AP projectiles, the Ti–Al_3Ti MIL composites have about 30% higher differential mass efficiency than RHA steel and about 10% higher differential mass efficiency than that of Ti–6Al–4V.

Table 2 Ballistic performance of Ti–Al_3Ti composites, RHA steel, and Ti–6Al–4V against 7.62 AP projectiles

Sample ID	Velocity	The thickness of the target and configuration	Density (g/cm^3)	DOP in back 6061Al alloy (mm)	Differential mass efficiency
MIL-1	830 ± 10 m/s	10.4 mm MIL-1 + 50 mm 6061 Al alloy	3.6	23.6	2.18
MIL-2		11.5 mm MIL-2 + 50 mm 6061 Al alloy	3.5	26.4	1.81
MIL-3		11.5 mm MIL-3 + 50 mm 6061 Al alloy	3.5	23	2.04
MIL-4		9 mm MIL-4 + 50 mm 6061 Al alloy	3.6	29	2.05
RHA steel		5 mm RHA + 50 mm 6061 Al alloy	7.8	33 ± 1	1.4
Ti-6Al-4 V		11 mm Ti–6Al–4V + 50 mm 6061 Al alloy	4.5	23 ± 2	1.64
6061 (reference)		50 mm 6061 Al alloy + 50 mm 6061 Al alloy	2.73	53 ± 1	1

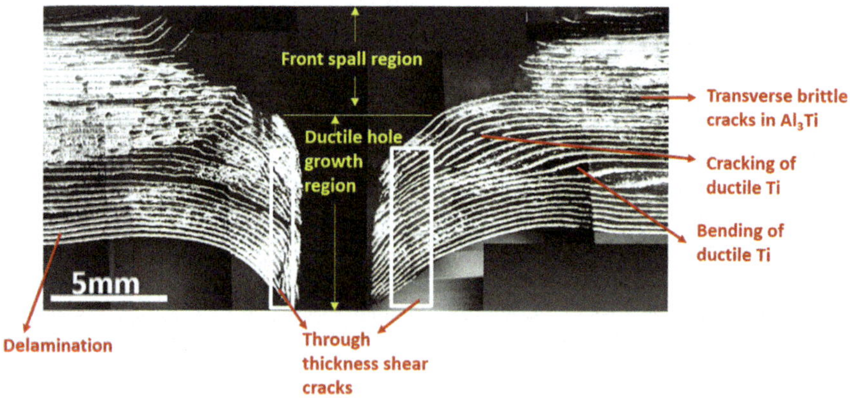

Fig. 8 Optical micrograph of the Ti–Al_3Ti metal–intermetallic laminate (MIL-4)

- The superior ballistic performance of Ti–Al_3Ti MIL composite can be attributed to the laminate structure with alternate soft and hard layers, which activates multiple energy-absorbing mechanisms during ballistic impact. Transverse brittle cracks in Al_3Ti, through-thickness shear cracks, delamination of Ti and Al_3Ti layers, and bending and cracking of Ti are the energy absorption mechanisms activated during ballistic impact against 7.62-mm AP projectiles.

Acknowledgements The authors would like to thank Director, DMRL, Hyderabad, for giving permission to publish this work. The authors would like to acknowledge the funding from DRDO, Govt of India for this work.

References

1. Crouch I (2016) The science of armour materials. Woodhead Publishing
2. Jo MC, Choi JH, Yoo J, Lee D, Shin S, Jo I, Lee SK, Lee S (2019) Novel dynamic compressive and ballistic properties in 7075-T6 Al-matrix hybrid composite reinforced with SiC and B4C particulates. Compos B Eng 174. https://doi.org/10.1016/j.compositesb.2019.107041.
3. Sudhakar I, Madhu V, Madhusudhan Reddy G, Srinivasa Rao K (2015) Enhancement of wear and ballistic resistance of armour grade AA7075 aluminium alloy using friction stir processing. Defence Technol 11:10–17. https://doi.org/10.1016/j.dt.2014.08.003
4. Gupta N, Bhanu Prasad VV, Madhu V, Basu B (2012) Ballistic studies on TiB_2-Ti functionally graded armor ceramics. Def Sci J 62:382–389. https://doi.org/10.14429/dsj.62.2666
5. Vecchio KS (2005) Synthetic multifunctional metallic-intermetallic laminate composites. JOM 57:25–31. https://doi.org/10.1007/s11837-005-0229-4
6. Menig R, Meyers MH, Meyers MA, Vecchio KS (2000) Quasi-static and dynamic mechanical response of haliotis rufescens (abalone) shells. Acta Mater 48:2383–2398
7. Menig R, Meyers MH, Meyers MA, Vecchio KS (2001) Quasi-static and dynamic mechanical response of Strombus gigas (conch) shells. Acta Mater 297:203–211
8. Harach DJ, Vecchio KS (2001) Microstructure evolution in metal-intermetallic laminate (MIL) composites synthesized by reactive foil sintering in air. Metall Mater Trans A 32:1493–1505. https://doi.org/10.1007/s11661-001-0237-0
9. Cao Y, Zhu S, Guo C, Vecchio KS, Jiang F (2015) Numerical investigation of the ballistic performance of metal-intermetallic laminate composites. Appl Compos Mater 22:437–456. https://doi.org/10.1007/s10443-014-9416-1
10. Grujicic M, Snipes JS, Ramaswami S (2016) Penetration resistance and ballistic-impact behavior of Ti/$TiAl_3$ metal/intermetallic laminated composites (MILCs): a computational investigation. AIMS Mater Sci 3:686–721. https://doi.org/10.3934/matersci.2016.3.686
11. Yuan M, Yao Y, Han F, Xin L, Shen X (2020) Effects of ductile Al on the anti-penetration performance of the Ti-Al_3Ti laminated composites. Results Phys 18. https://doi.org/10.1016/j.rinp.2020.103308
12. Savio SG, Ramanjaneyulu K, Madhu V, Bhat TB (2011) An experimental study on ballistic performance of boron carbide tiles. Int J Impact Eng 38:535–541. https://doi.org/10.1016/j.ijimpeng.2011.01.006
13. Rosenberg Z, Bless SJ, Yeshurun Y, Okajima K (1988) A new definition of ballistic efficiency of brittle materials based on the use of thick backing plates. Impact Loading Dyn Behavior Mater 1:491–498
14. Bless SJ, Rosenberg Z, Yoon B (1987) Hypervelocity penetration of ceramics. Int J Impact Eng 5:165–171. https://doi.org/10.1016/0734-743X(87)90036-4

15. Peng LM, Li H, Wang JH (2005) Processing and mechanical behavior of laminated titanium-titanium tri-aluminide (Ti-Al_3Ti) composites. Mater Sci Eng, A 406:309–318. https://doi.org/10.1016/j.msea.2005.06.067
16. Yuan MN, Li L, Wang ZJ (2018) Study of the microstructure modulation and phase formation of Ti–Al_3Ti laminated composites. Vacuum 157:481–486. https://doi.org/10.1016/j.vacuum.2018.09.002
17. Xueling F, Meini Y, Qiang Q (2018) Failure mechanisms of Ti-Al_3Ti metal-intermetallic laminate composites under high-speed impact. Rare Metal Mater Eng 47:2615–2620

Synthesis and Structural Properties of Flexible Ferroelectric Polymer–Ceramics Nanocomposite Films for Multifunctional Applications

Venkata Sreenivas Puli, S. Narendra Babu, and N. V. Prasad

1 Introduction

Dielectric materials with high dielectric constant and low dielectric loss are of prime importance in multifunctional applications. Perovskite materials with a ferroelectric nature are in the first place to occupy in functionality. Usually, ceramic materials will have a high dielectric constant with a low dielectric breakdown strength, which is limiting their usage in electrical energy storage capacitor devices. Polymers are also gaining more and more attention due to low cost, high dielectric breakdown strength, and low dielectric loss. However, the only drawback with polymers is their low dielectric constant. Great efforts are underway to surpass these issues and develop materials with high dielectric constant, low dielectric loss, and high dielectric breakdown strength. In this direction, various combinations of materials such as glass–ceramics and polymer–ceramic nanocomposites are in search. Much research has been in progress on ceramic–polymer nanocomposites, especially for piezoelectric, pyroelectric transducers, and energy storage capacitor applications [1, 2] Therefore, changing the volume fraction or weight fraction of the constituents in the polymer–ceramic nanocomposites is receiving much more attention for desired properties.

In our previous work [3–5], we have reported how the dielectric behavior of ceramic [$BaTiO_3$-BTO, (Ba,Sr)TiO_3-BST, (Ba,Ca)TiO_3-BCT, Ba(Zr,Ti)O_3-BZT, (x)($Ba_{0.70}Ca_{0.30}$)-(1-x) Ba($Zr_{0.20}Ti_{0.80}$)O_3]-polymer [poly(methyl methacrylate)-PMMA,PVDF, poly(glycidyl methacrylate)-PGMA)] nanocomposites change by

V. S. Puli (✉)
Smart Nanomaterials Solutions, Casselberry, Orlando, FL 32707, USA
e-mail: pvsri123@gmail.com

S. Narendra Babu · N. V. Prasad
Department of Physics, Materials Research Laboratory, Osmania University, Hyderabad, Telangana 500007, India

M. R. Manohara Babu et al. (eds.), *Recent Advances in Composites Engineering and Research*, Springer Proceedings in Materials 39,
https://doi.org/10.1007/978-981-99-8807-5_29

changing the volume/weight fraction of ceramic filler. These encouraging results motivated our current research efforts to study the combined effect of flexible polymer–ceramic nanocomposites for energy storage applications to achieve desired dielectric properties and will be presented elsewhere. Herein, we report the synthesis and structural properties of PVDF–BCZT flexible polymer–ceramic nanocomposite films for capacitor applications.

2 Experimental Procedures

2.1 *Synthesis of BCZT NPs*

BCZT NPs were synthesized using the solgel synthesis technique as reported elsewhere [4].

2.2 *Fabrication of PVDF and PVDF–BCZT Polymer–Ceramic Nanocomposite Films*

Commercial PVDF powder of molecular weight 400,000 Mw obtained from Sigma-Aldrich, Inc., MO, USA, was used as received. PVDF polymer solution was prepared by dissolving polymer powder in dimethylformamide (DMF) solvent. Once a clear and stable solution was obtained, different volume percentages of OH functionalized BCZT NPs (10, 20, and 30 vol%) were gradually added to it. Sonication technique was used for thorough mixing, and then, polymer–ceramic nanocomposite solution was stirred using a magnetic stirrer at 170 °C for 2 h on a hot plate. Both PVDF and PVDF–BCZT polymer–ceramic nanocomposite films were prepared using the solution casting method at on glass slide at 170 °C, and then, peeled off free-standing films were hot-pressed at 180 °C for 5 min. The schematic diagram representing polymer–ceramic nanocomposite films formation is shown in Fig. 1.

2.3 *Structural Characterization*

Structural characterization of as prepared BZCT NPs, PVDF, and PVDF–BCZT films was done using the X-ray diffraction (XRD) technique. Rigaku X-ray diffractometer with Cu K_{α} radiation operating at 40 kV and 40 mA was used for obtaining XRD patterns. The XRD data were recorded at a step size ~0.02° and at a scan rate of 1°/min, for the entire angular range of 10°–80°. Renishaw in Via™ confocal Raman microscope was employed to obtain Raman spectra of all the samples under study at

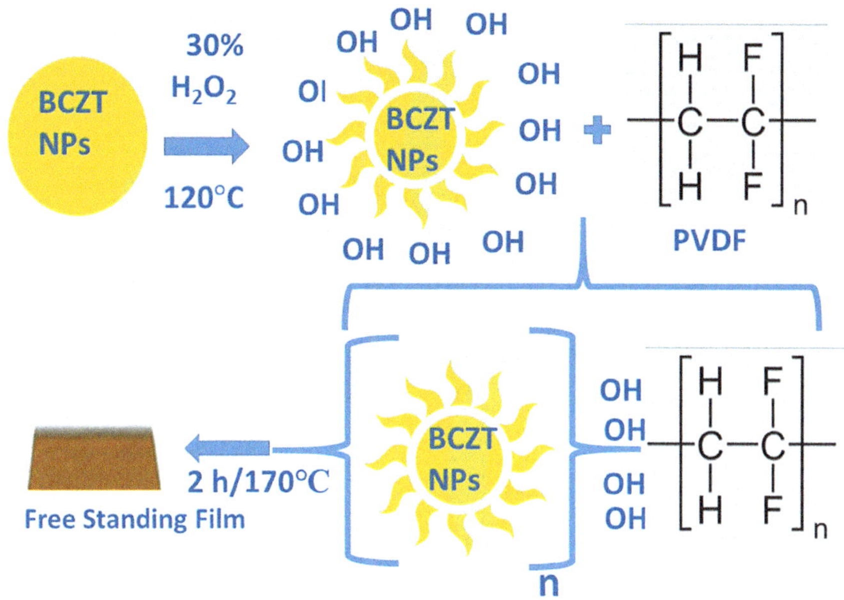

Fig. 1 Schematic diagram representing the fabrication of PVDF–BCZT at 10, 20, and 30 vol% polymer–ceramic nanocomposites films

room temperature. Transmission electron microscopy (TEM) images of BCZT NPs were collected with a FEI G2 F30 Tecnai TEM operated at 200 kV.

3 Results and Discussion

3.1 X-Ray Diffraction

The XRD patterns of BCZT NPs, PVDF, and PVDF – BCZT polymer–ceramic films were shown in the 2θ range from 20° to 80° in Fig. 2a, b, c, respectively.

Room temperature diffraction patterns of BCZT NPs were as shown in Fig. 2a with the angle 2θ ranging from 10° to 80° with a scan rate of 1° per minute. BCZT NPs were phase indexed to single-phase tetragonal perovskite crystal structure with space group (*P4mm*) at room temperature. XRD diffraction patterns of PVDF film were indexed according to α, β, and γ phases reported in the literature [1, 6–8]. Figure 2b shows that the semi-crystalline diffraction peaks corresponding to 2θ were found at around 17.66°, 18.39°, 19.33°, 26.62°, 32.2°, 33.15°, 35.9°, 38.8° and were indexed to (100), (020), (110), (021), (130), (200), (002), (111). As shown in the XRD patterns, PVDF films have shown intense peaks at about 17.66°,18.39°,

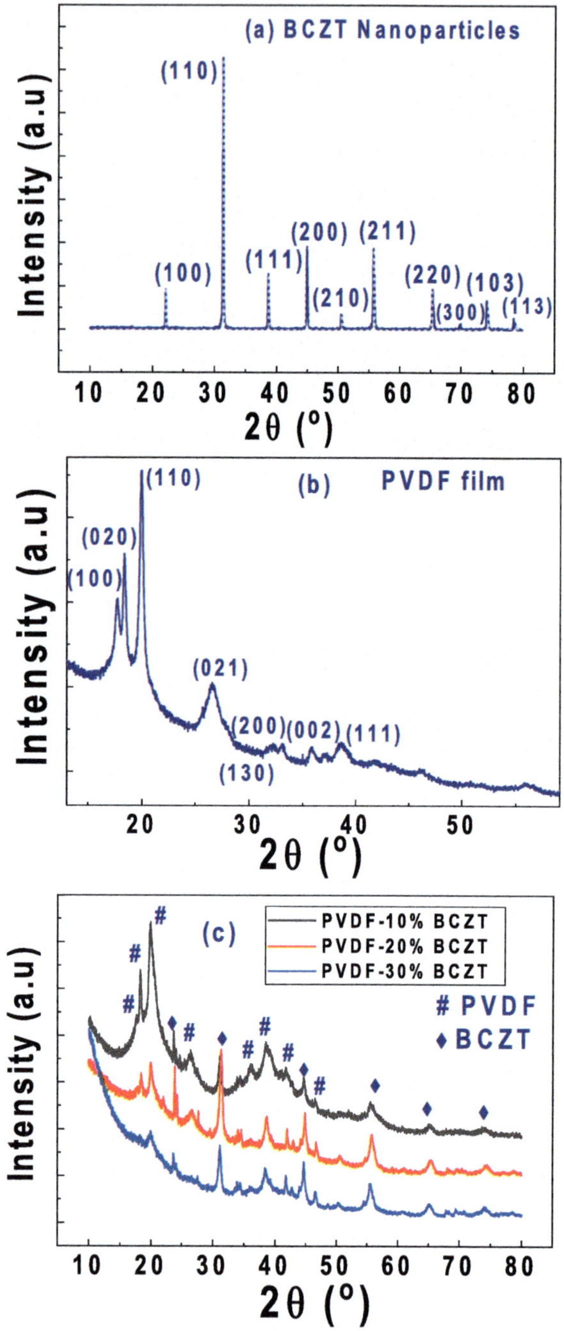

Fig. 2 **a–c** XRD patterns of BCZT NPs, PVDF film, and PVDF–BCZT at 10, 20, and 30 vol% polymer–ceramic nanocomposites films

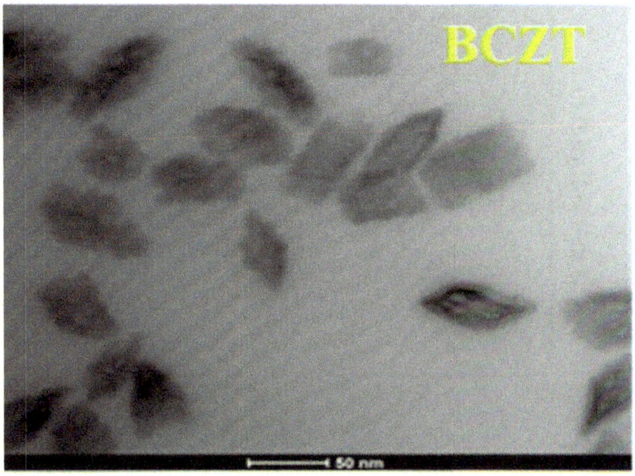

Fig. 3 Transmission electron microscopy images of BCZT NPs

and 26.62° corresponding to monoclinic α-phase and peak at around 19.33° corresponding to (110) reflections of the orthorhombic β-phase. Remaining four weak peaks at 32.2°, 33.15°, 35.9°, 38.8° corresponding to (130), (200), (002), and (111) reflections were ascribed to the monoclinic α-phase, respectively [6–8]. XRD patterns of polymer–ceramic nanocomposite films (Fig. 2c) confirm the coexistence of mixed peaks corresponding semi-crystalline PVDF and tetragonal perovskite BCZT. It is clearly visible from XRD patterns that the peaks were shifted to higher angle in the measured region due to the intermixing of polymer and ceramic.

3.2 Transmission Electron Microscopy of BCZT NPs

TEM image of BCZT NPs synthesized using the solgel synthesis technique is shown in Fig. 3. Here, one can see the BCZT particles are found as individual nanosize particles without any agglomerates. The size of these NPs is in the range of ~50 nm [4]. TEM image also revealed that the BCZT NPs are of diamond shape.

3.3 Raman Spectroscopy

The Raman spectra of pristine BZCT and PVDF–BCZT polymer–ceramic films are shown in Fig. 4a, b. Pristine polymer PVDF films have shown interesting Raman dispersed polar β-phase modes as shown in Fig. 4b. Most significant CF_2 symmetric stretching mode corresponding to ferroelectric phase has been witnessed around 800–815 cm^{-1} in Fig. 4b. As shown in Fig. 4b-inset, the low frequency region (100–800 cm^{-1}) of the spectrum shows perovskite Raman modes.

The Raman spectra of BCZT nanoparticles corresponding to tetragonal features of perovskite structure were reported elsewhere [9]. These modes were in complete

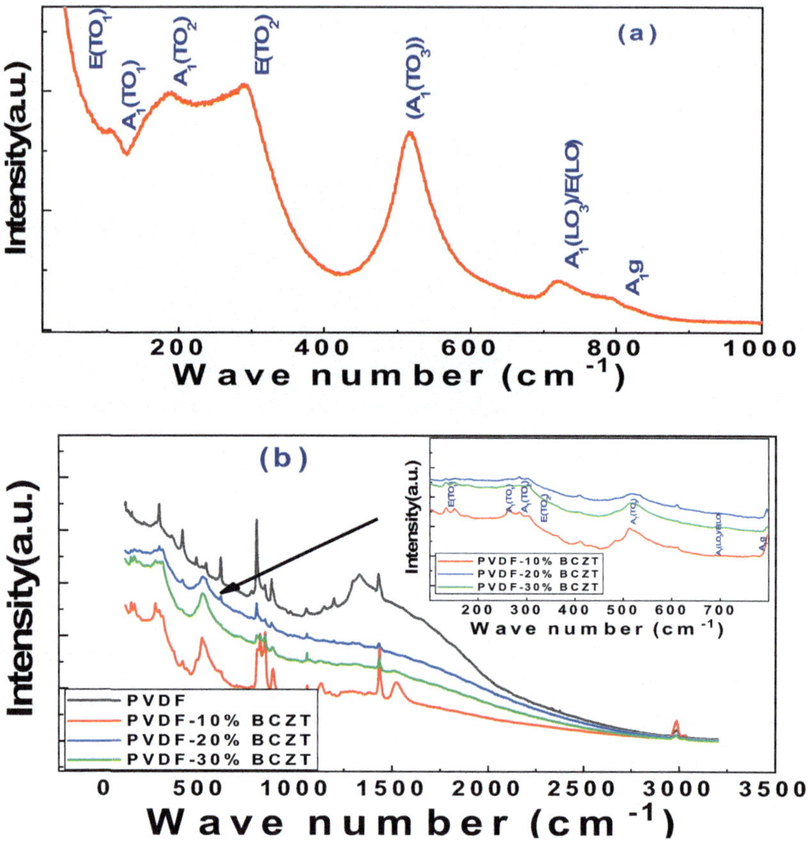

Fig. 4 **a**, **b** Room temperature Raman spectra of BCZT NPs and PVDF–BCZT at 10, 20, and 30 vol% polymer–ceramic nanocomposites films

agreement with conventional BTO ceramics at room temperature [10]. Raman modes corresponding to tetragonal phase were found, respectively: $E(TO_1)$ ~ 37 cm^{-1},$A_1(TO_1)$ at 105 cm^{-1}, $A_1(TO_2)$ at 185 cm^{-1}, $B_1/E(TO_2)$ at 294 cm^{-1}, $A_1(TO_3)$ at 517 cm^{-1}, and [E(LO), A_1(LO)] at 715 cm^{-1}, A_{1g} at 795 cm^{-1}. Particularly, the $B_1/E(TO_2)$ mode at 294 cm^{-1} is the signature peak for tetragonal perovskite crystalline structure in BTO-based materials and is due to the existence of disordered Ti^{4+} with TiO_6 octahedral in the lattice. In addition to perovskite phase modes, corresponding mode for polymer PVDF was also clearly noticed in the polymer–ceramic nanocomposites in the wave number region from 800 to 3500 cm^{-1}.

4 Conclusions

In summary, BCZT NPs were successfully synthesized using a solgel synthesis technique. PVDF–BCZT polymer–ceramic nanocomposite films were grown using solution casting method, and then, free-standing films were hot-pressed at 180 °C. XRD studies confirmed tetragonal perovskite structure of BCZT NPs. TEM images revealed that the synthesized BCZT NPs in diamond shape with size are around 50 nm. Raman analysis demonstrated that the mode corresponding to tetragonal perovskite structure of BCZT NPs is due to the existence of disordered Ti^{4+} with TiO_6 octahedral, also polar β-phase modes corresponding to pristine PVDF in polymer–ceramic nanocomposite films. Polymer–ceramic composite films prepared in this technique are may be useful in electrical energy storage capacitor applications.

References

1. Muralidhar C, Pillai PKC (1988) Dielectric behavior of barium titante-polymervinylidene fluoride composites. J Mater Sci 23:1071–1076
2. Muralidhar C, Pillai PKC (1986) Pyroelectric behavior in barium titanate/polyvinylidene fluoride composites. IEEE Trans Electr Insulation EI-21(3)
3. Ejaz M, Puli VS, Elupula R, Adireddy S, Riggs BC, Chrisey DB, Grayson SM (2015) Core-shell structured poly(glycidyl methacrylate)/$BaTiO_3$ nanocomposites prepared by surface-initiated atom transfer radical polymerization: a novel material for high energy density dielectric storage. J Polym Sci, Part A: Polym Chem 53:719–728
4. Puli VS, Ejaz M, Elupula R, Kothakonda M, Adireddy S, Katiyar RS, Grayson SM, Chrisey DB (2016) Core-shell like structured barium zirconium titanate-barium calcium titanate–poly(methyl methacrylate) nanocomposites for dielectric energy storage capacitors. Polymers 105:35–42
5. Adireddy S, Puli VS, Lou TJ, Elupula R, Sklare SC, Riggs BC, Chrisey DB (2015) Polymer-ceramic nanocomposites for high energy density applications. J Sol-Gel Sci Technol 73:641–646
6. Cai X, Lei T, Sund D, Lin L (2017) A critical analysis of the α, β and γ phases in poly(vinylidene fluoride) using FTIR. RSC Adv 7:15382–15389
7. Okada D, Kaneko H, Kato K, Furumi S, Takeguchi M, Yamamoto Y (2015) Colloidal crystallization and ionic liquid induced partial β-phase transformation of poly(vinylidene fluoride) nanoparticles. Macromolecules 48:2570–2575
8. Xia W, Wang Z, Xing J, Cao C, Xu Z (2016) The dependence of dielectric and ferroelectric properties on crystal phase structures of the hydrogenized P(VDF-TrFE) films with different thermal processing. IEEE Trans Ultrasonics Ferroelectrics Frequency Control 63(10):1674–1680
9. Puli VS, Kumar A, Chrisey DB, Tomozawa M, Scott JF, Katiyar RS (2011) Barium zirconate-titanate/barium calcium-titanate ceramics via sol–gel process: novel high-energy-density capacitors. J Phys D: Appl Phys 44:395–403
10. Puli VS, Li P, Adireddy S, Chrisey DB (2015) Crystal structure, dielectric, ferroelectric and energy storage properties of La-doped $BaTiO_3$ semiconducting ceramics. J Adv Dielectr 5(3):1550027

Synthesis and Structural Studies of Polymer-Ceramic Nanocomposite Thick Films for Energy Storage

S. Narendra Babu, K. Srinivas, and Venkata Sreenivas (Narayanabhatla) Puli

1 Introduction

There has been an increasing interest in electrical energy storage capacitors because of their scientific and technological importance in the broad spectrum of applications, such as electric vehicles, spacecraft, photovoltaic power generation, and medical devices [1]. Hence, compact, low-cost energy storage capacitors with high power density and high energy density are in high demand for next-generation energy needs. In this direction, a lot of research work is ongoing with respective materials such as ceramics, polymers, glasses, polymer nanoceramics, and glass-ceramics [2–4].

High dielectric constant ceramics are of prime importance in electronic device applications. However, the main drawback of these materials is low dielectric loss. Ceramics with porous microstructure are detrimental to the device's performance. Polymer materials have gained importance due to high dielectric breakdown strength for capacitor applications; however, more important drawback with polymers is their low dielectric constant values. Hence, either ceramics or polymers alone cannot reach the goal of achieving high power density and high energy density due to their respective drawbacks.

Hence, a search for alternatives is very much needed. Among the various materials of choice for energy storage, polymer-ceramic nanocomposites are gaining more importance due to their combined effect of high dielectric ceramics with low dielectric

S. Narendra Babu (✉)
Materials Research Laboratory, Department of Physics, Osmania University, Hyderabad, Telangana 500007, India
e-mail: narendraphysics@gmail.com; narensimha@osmania.ac.in

K. Srinivas
Advanced Systems Laboratory, DRDO, Kanchanbagh P.O., Hyderabad, Telangana 500058, India

V. S. Puli
Smart Nanomaterials Solutions, Casselberry, Orlando, FL 32707, USA

M. R. Manohara Babu et al. (eds.), *Recent Advances in Composites Engineering and Research*, Springer Proceedings in Materials 39,
https://doi.org/10.1007/978-981-99-8807-5_30

loss polymers with high dielectric breakdown strength. Due to the aforementioned reasons, herein, we report the synthesis and structural properties of P(VDF-TrFE-CTFE)-$Ba_{(1-x)}La_xTiO_3$ polymer-ceramic nanocomposites for energy storage.

2 Experimental Procedures

2.1 Materials

Barium nitrate ($Ba(NO_3)_2$)(99.95%), lanthanum (III) acetate hydrate $La(CH_3CO_2)_3{\cdot}xH_2O$ (99.99%), sodium hydroxide NaOH(98%), 1-butanol anhydrous (99.9%), titanium(IV) n-butoxide-$Ti(OCH_2CH_2CH_2CH_3)_4$ (98%), oleic acid (99.8%, anhydrous), 1-decanol (98%), N,N-Dimethylformamide DMF (99.8%) and toluene (99.8%, anhydrous), high pure precursors were purchased from Sigma-Aldrich, MO, USA, and pure grade ethanol (200 Proof, absolute) procured from Pharmco-Aaper, USA, and used without any further purification.

2.2 *Synthesis of Ceramic $Ba_{(1-x)}La_xTiO_3$ (x = 0.003,0.006,0.010) NPs*

Quasi monodispersed ceramic $Ba_{(1-x)}La_xTiO_3$ NPs, where $x = 0.003,0.006,0.010$ designated as $Ba_{0.997}La_{0.003}TiO_3$-BLT-1, $Ba_{0.994}La_{0.006}TiO_3$-BLT-2, and $Ba_{0.990}La_{0.01}TiO_3$-BLT-3 were synthesized using a modified solvothermal synthesis route reported in detail elsewhere [5, 6].

Three different stoichiometric BLT-1, BLT-2, and BLT-3 reaction solutions were prepared individually at room temperature by mixing oleic acid,1-butanol,1-decanol, $Ti(OCH_2CH_2CH_2CH_3)_4$, and aqueous solution of $Ba(NO_3)_2$ and/or $La(CH_3CO_2)_3{\cdot}xH_2O$. The prepared solutions viz.BLT-1, BLT-2, and BLT-3 were then added to NaOH solution (3.5 g in 17.5 mL milli-Q water). These final solutions, after through mixing for 10 min using a magnetic stirrer, were transferred to Teflon-lined stainless steel autoclaves (Parr, model 4748, 125 mL). These autoclaves were maintained at 150 °C for 24 h and then cooled to room temperature. White milky suspensions obtained were precipitated into absolute ethanol and separated using a centrifuge. These suspensions were further purified in three successive cycles by dispersing them in toluene and then precipitated with ethanol and then centrifuged for 10 min at 6000 rotations per minute. A total of about 1.25 g of BLT ceramic NPs were obtained. The hydroxylation of BLT NPs is done according to the synthetic protocol reported by Li et al. [7].

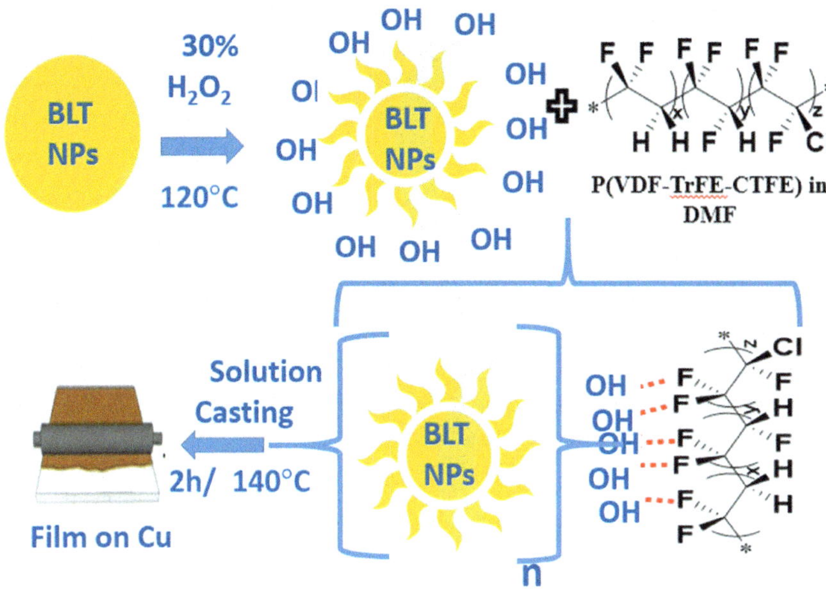

Fig. 1 Schematic representing the development of P(VDF-TrFE-CTFE)-BLT polymer-ceramic nanocomposites films

2.3 Preparation of Polymer and Polymer-Ceramic Nanocomposite Films

Both P(VDF-TrFE-CTFE) and P(VDF-TrFE-CTFE)-BLT polymer-ceramic nanocomposite films were obtained using the solution casting method. Pristine polymer and polymer-ceramic nanocomposite solutions were prepared by dissolving polymer powders in DMF solvent. These clear stable solutions were thoroughly mixed using the sonication technique and then stirred using a magnetic stirrer at 120 °C for 2 h. Later, they were solution casted on Cu substrate at 140 °C and baked for 2 h on the hot plate. The schematic diagram of the combination of P(VDF-TrFE-CTFE) and BLT NPs for the formation of polymer-ceramic nanocomposites is shown in Fig. 1.

2.4 Structural Characterization

X-ray diffraction (XRD) measurement was carried out using Rigaku X-ray diffractometer. The wavelength of the X-ray was 1.543 Å (Cu K_α radiation, 40 kV and 40 mA). The XRD data were recorded at a step size of 0.02° and step scan of 1°/min. Room temperature Raman spectra were obtained using Renishaw inVia™ confocal

Raman microscope. Transmission electron (TEM) microscopy data of BLT NPs were collected with a FEI G2 F30 Tecnai TEM operated at 200 kV.

3 Results and Discussion

The as-synthesized BLT NPs were characterized using XRD, Raman spectroscopy, and TEM, and the results on XRD and Raman spectroscopy reported elsewhere [5]. TEM images of washed BLT NPs, viz.BLT-1, BLT-2, and BLT-3, synthesized using the solvothermal synthesis technique are shown in Fig. 2a–c. The size of the NPs was in the range of ~ 10–15 nm. TEM images revealed BLT NPs shapes are bimodal (cubic and spherical). All BLT compositions have shown nano-sized individual particles without any agglomerates.

The XRD patterns of the polymer film P(VDF-TrFE-CTFE) and polymer-ceramic nanocomposite P(VDF-TrFE-CTFE)-BLT are shown in the 2θ range from 20° to 80° (Fig. 3a, b). BLT NPs XRD patterns were indexed to the cubic perovskite phase. The particle size of BLT NPs is determined using Scherer's equation from XRD line broadening patterns. The particle size of BLT NPs was around ~ 15 nm. These values are in complete agreement with TEM measurements. Indexed XRD patterns were matched with cubic $BaTiO_3$ (JCPDS data No. 31-0174) with one single peak at 45° corresponding to (002) which is also visible from XRD patterns presented in polymer-ceramic nanocomposites as shown in Fig. 3. XRD patterns of P(VDF-TrFE-CTFE) films have shown a prominent characteristic peak, at 2θ ~ 18.72° and 19.06°. Especially, the (110) and (021) diffraction peaks almost disappear. Polymer-ceramic nanocomposites were indexed to the polycrystalline cubic perovskite phase. In addition to these peaks, substrate peaks at around 43° ~ (111), 50.2° ~ (200), and 73.9° ~ (220) corresponding to copper (Cu) were also noticed.

The Raman spectra of pristine polymer P(VDF-TrFE-CTFE) and polymer-ceramic nanocomposites films are shown in Fig. 4a, b. Pristine polymer P(VDF-TrFE-CTFE) films have shown dispersed polar β-phase modes which are shown in Fig. 4a.

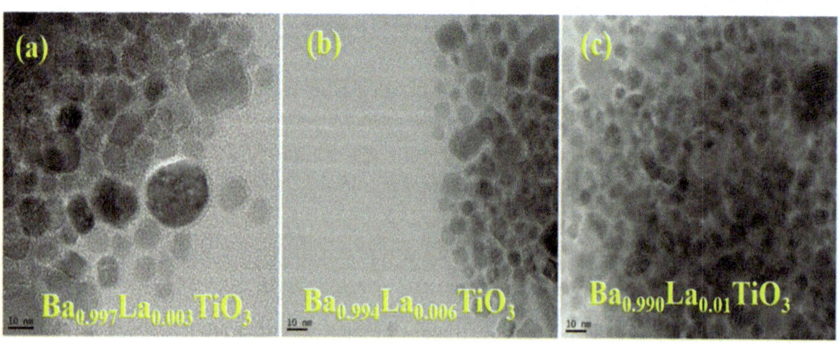

Fig. 2 Transmission electron microscopy (TEM) images BLT-1, BLT-2, and BLT-3 NPs

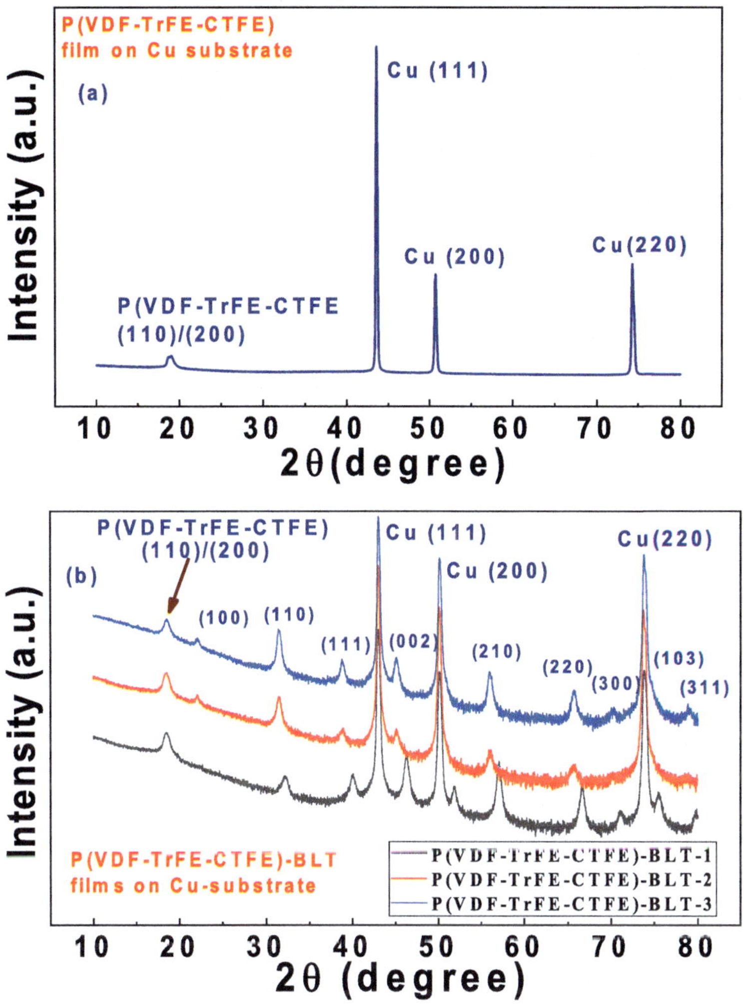

Fig. 3 **a**, **b** XRD patterns of P(VDF-TrFE-CTFE) polymer film and P(VDF-TrFE-CTFE)-BLT polymer-ceramic nanocomposite films

Most significant CF_2 symmetric stretching mode at 800–815 cm^{-1} corresponding to the ferroelectric phase has been observed.

As shown in Fig. 4b, the low-frequency region (100–800 cm^{-1}) of the spectrum shows perovskite Raman modes. The Raman spectra of BLT nanoparticles corresponding to tetragonal features of perovskite structure were reported elsewhere [5]. Raman modes corresponding to tetragonal phase were found respectively: $A_1(TO_1)$ at 185 cm^{-1}, $A_1(TO_2)$ at 260 cm^{-1}, $B_1/E(TO_2)$ at 307 cm^{-1}, $A_1(TO)$ at 521 cm^{-1}, [E(LO), $A_1(LO)$] at 720 cm^{-1}, and $A_1(g)$ at 800 cm^{-1}. Particularly the $B_1/E(TO_2)$ mode at 307 cm^{-1} is the signature peak for tetragonal perovskite crystalline structure

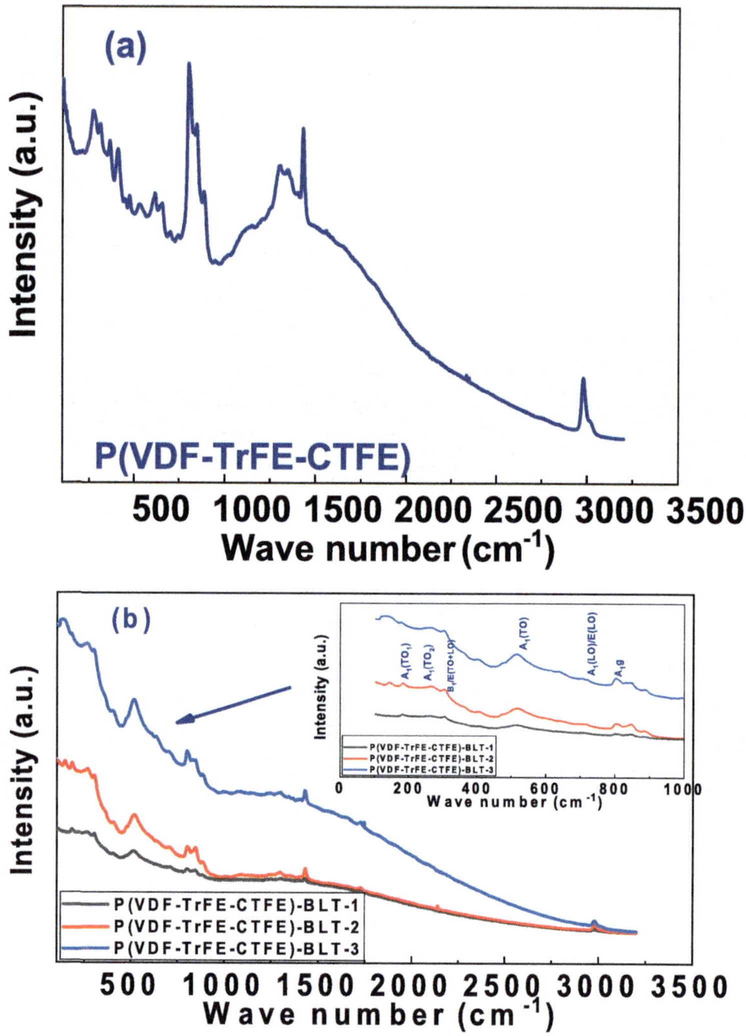

Fig. 4 **a**, **b** Room temperature Raman spectra of P(VDF-TrFE-CTFE) polymer film and P(VDF-TrFE-CTFE)-BLT polymer-ceramic nanocomposites films

in BTO-based materials. In addition to perovskite phase modes, corresponding modes for polymer (PVDF-TrFE-CTFE) were also clearly noticed in the polymer-ceramic nanocomposites.

4 Conclusions

In summary, size and shape-controlled BLT nanoparticles were successfully synthesized using a modified solvothermal synthesis technique. We have demonstrated the highly dispersed BLT NPs in P(VDF-TrFE-CTFE). Ferroelectric P(VDF-TrFE-CTFE)-BLT polymer-ceramic nanocomposite films were grown on Cu substrate using the solution casting method. The highly crystalline perovskite cubic phase of BLT NPs present in polymer-ceramic nanocomposite films is confirmed with XRD. Size and shape-controlled BLT NPs were successfully presented in transmission electron microscopy. Raman spectroscopy results revealed that the mode corresponding to the tetragonal perovskite phase BLT NPs due to the existence of disordered Ti^{4+} with TiO_6 octahedral, along with polar β-phase modes was observed, respectively, in pristine P(VDF-TrFE-CTFE) in polymer-ceramic nanocomposite films. Polymer-ceramic composite films fabricated in the present study are may be useful in ferroelectric energy storage capacitor applications.

References

1. Whittingham MS (2008) Materials challenges facing electrical energy storage. MRS Bull 33:411–419
2. Puli VS, Pradhan DK, Chrisey DB, Tomozawa M, Sharma GL, Scott JF, Katiyar RS (2013) Structure, dielectric, ferroelectric, and energy density properties of $(1-x)$ BZT–xBCT ceramic capacitors for energy storage applications. J Mater Sci 48(5):2151–2157
3. Ejaz M, Puli VS, Elupula R, Adireddy S, Riggs BC, Chrisey DB, Grayson SM (2015) Core-shell structured poly(glycidyl methacrylate)/$BaTiO_3$ nanocomposites prepared by surface-initiated atom transfer radical polymerization: a novel material for high energy density dielectric storage. J Polym Sci Part A: Polym Chem 53:719–728
4. Zhang W, Xue S, Liu S, Wang J, Shen B, Zhai J (2014) Structure and dielectric properties of $Ba_xSr_{1-x}TiO_3$-based glass ceramics for energy storage. J Alloys Compd 617:740–745
5. Puli VS, Adireddy S, Elupula R, Molugu S, Shipman J, Chrisey DB (2017) Synthesis and structural properties of $Ba_{(1-x)}La_xTiO_3$ perovskite nanoparticles fabricated by solvothermal synthesis route, AIP Conf Proc 1837:020001
6. Adireddy S, Puli VS, Lou TJ, Elupula R, Sklare SC, Riggs BC, Chrisey DB (2015) Polymer-ceramic nanocomposites for high energy density applications. J Sol-Gel Sci Technol 73:641–646
7. Li CC, Chang SJ, Lee JT, Liao WS (2010) Efficient hydroxylation of $BaTiO_3$ nanoparticles by using hydrogen peroxide. Colloids Surf A: Physicochemical Eng Aspects 361(1–3):143–149

Development and Extraction Process of Calotropis Gigantea Natural Fibres

Matlapudi Madhavi, V. Nikil Murthy, and Haritha Nukaraju

1 Introduction

"In this study, an effort was made to use CG fibre as a reinforcement in the form of twisted fibre. Specific chemical treatment was performed to increase the properties of CG fibres and to reduce hydrophilic and hygroscopic benefit using NaOH [1]". 'To lower the hydrophilicity of the fibres while keeping their characteristics, the proper concentration of NaOH in the treatment solution must be determined [1]'. 'Bast fibres were isolated from the bark of the Calotropis gigantea plant and part of them were cleaned with an alkaline solution (5 wt. percent). Calotropis fibres with untreatable physico-chemical and alkali-containing qualities were investigated for their chemical composition and therapeutic potential. The chemical study revealed that alkaline treatment eliminated the majority of non-cellulose compounds, which was validated by FTIR analysis'. 'The results of X-ray exposure showed that alkali fabrics had a higher crystallinity index than untreated fibres, which was consistent with the results of mechanical tests. [2]'. 'Strands are things like hair that is a continuous strand or long strands. Man-made natural (NF) fibre and synthetic fibre are the two sorts. NFs are a group of substances that include hair from plants and animals. Some can be twisted into threads or strings. They can be used to safeguard structures. The natural fibres found in vegetables contain cellulose, a glucose polymer bound to lignin with the changing steps of other common substances. Synthetic fibres are a key component of synthetic materials like petrochemicals, but other varieties, as modular rayon and lyocell, are obtained from ordinary cellulose [3]'. 'Bio composites are compounds that are made up of a decaying grid and also decaying NFs

M. Madhavi · V. Nikil Murthy · H. Nukaraju (✉)
Maturi Venkata Subba Rao Engineering College, Mechanical Engineering, Hyderabad, Telangana 501510, India
e-mail: haritha_mech@mvsrec.edu.in

M. R. Manohara Babu et al. (eds.), *Recent Advances in Composites Engineering and Research*, Springer Proceedings in Materials 39,
https://doi.org/10.1007/978-981-99-8807-5_31

as sponsors. Bio composites have attracted extraordinary interest mainly because of their biodegradability; besides the strength they impart [4]'. Natural fibres (NFs) have been used by Asians extensively for a long time in making ropes and bags. 'NFs are widely used in automotive and construction materials'. 'Major vehicle makers have switched to the use of plant filaments for some car components [D. B. Dittenber, et al.]'. '90% of the world's jute is produced in India and Bangladesh and 75% of kenaf generation is due to India and China [D Saravana Bavan, et al.]'. India has a record of wide usage of other natural fibres like silk, cotton, sisal, banana, and coir, and their use is on the increase. India is the world's second-biggest maker and buyer of filaments, materials, after China [Q. R. Jiang, et al.]. **Calotropis Gigantea** is one fibre that is not adequately exploited despite its availability in abundance. The renewability and cheap availability make **Calotropis Gigantea** economically and environmentally viable alternative source of natural fibres. The broad objective of this development effort is to exploit the abundantly available raw material source for extracting valuable fibres in a manner that is cost-effective, fast, and consistent in terms of quality, and to study its potential as reinforcement in polymer composites, for possible use in multiple applications.

In this work, details of a mechanical device developed for the extraction of this fibre is presented. Also, a resin-based composite is made from fibres produced from the machine, which is tested for strength and the results are presented. It is hoped that automation of the extraction process will help in the production of adequate quantity to make the fibres viable contenders for use in many modern applications like automobiles, domestic appliances, and the building industry.

2 Machine Development for Calotropis Gigantea Extraction Mechanization

The conventional manual techniques used for extraction of Calotropis Gigantea fibres take lengthy intervals for soaking; Peeling is time-taking and slow, making it unremunerated from a business aspect. Also, the water used for soaking the stems, except systematically disposed off can cause pollutants of water or soil. The extraction procedure involves the subsequent steps:

i. Cut branches of Calotropis **Gigantea** for stems of 150–200 mm length and 8–10 mm diameter
ii. Soak in water for about 7 days to make it soft for easy removal of fibre from **Calotropis Gigantea** stems, while changing the water regularly to prevent bacterial contamination.
iii. Peel the outer layer from the stem
iv. Rinse slowly and thoroughly to wash away slurry waste from the fibre
v. Dry the fibre at room temperature.

The machine is developed to automate the following:

i. Feed pre-soaked **Calotropis Gigantea** stems into the machine, by a suitable mechanism
ii. Apply mechanical action for a surface to apply pressure on the stem and create a relative motion to it to and directing a water jet spray to help to remove the fibre, for about 5 min.
iii. Collect water with fibres, and subject it to centrifugal motion for sufficient time, for separating the biomass, to get clean and untangled fibres.

3 Schematic Representation of the Machine

The process material incorporated in the fibre extraction machine is depicted in Fig. 1 schematic diagram. The extraction processes of fibre are mentioned in the order:

(i) Feeding of pre-soaked **Calotropis Gigantea** stems to the mechanical fibre extraction units wherein, the fibres are subjected to the combination of lateral and compression forces created between the top plate including the rotatable knurled array of rods and the bottom plate having the rectangular slots and the circular holes which results into a fibre slurry.
(ii) Passing of fibre slurry to a fibre extraction unit which is subjected to the frictional force of the water jet sprayed from the spray unit arranged to the top of the mechanical fibre extraction unit.
(iii) Separating the fibres in fibre extraction unit using centrifugal motion by a to-and-fro rotational movement at a certain speed and time, preferably executing 180o rotation in 30 s. and time ranging 1–6 min.
(iv) Treating and recycling of wastewater generated during fibre extraction unit in wastewater collecting unit.

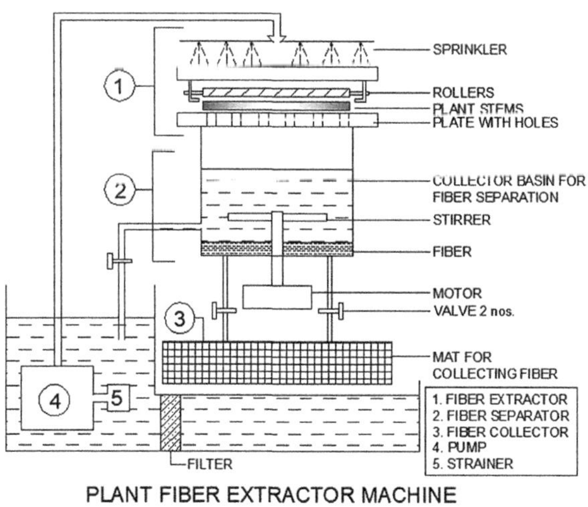

Fig. 1 Schematic diagram of plant fibre extractor machine

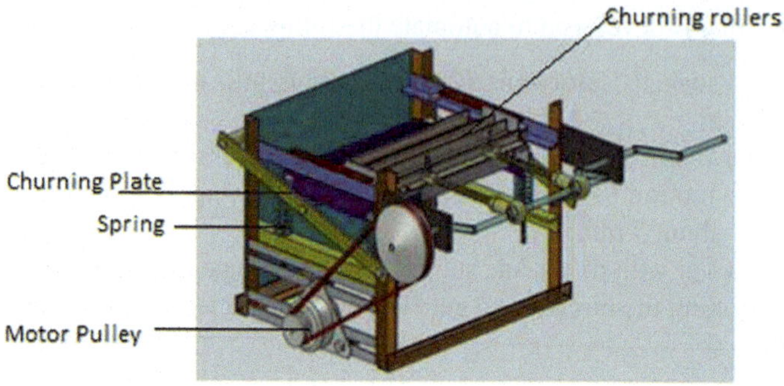

Fig. 2 Fibre extractor

4 Fibre Extraction Machine Subsystems

The major subsystems of the machine developed for the extraction of **Calotropis Gigantea** fibres are: (1) Fibre Extractor: Fig. 2 shows mechanical fibre extraction unit for defibration from **Calotropis Gigantea** stems into fibre slurry. Calotropis Gigantea stems are pre-soaked and fed onto a plate at the top of the machine. A knurled cluster of rollers runs to and fro on the stems in a perpendicular orientation. The roller set is also pulled down by springs onto the base plate to grip the stems while rolling action takes place. There is a relative motion between stem and roller. The base plate on which the stems roll has circular holes, whose edges also assist peeling. Thus the fibres are exposed to the mix of horizontal and vertical forces, which together cause fibre peeling action.

A water shower is also sprayed on top of the rollers, which aids fibre separation from stems and drains the water slurry including fibres.

The remaining parts described below are depicted in the 3D model in Fig. 3.

Fibre Separation Unit: It is used for separating fibers from the fibre slurry, using a fluid disruption mechanism via to- and-fro oscillation, of 180° in steps of 0.5 min for 4 min.

Fibre collector and Wastewater treatment Unit: This is a collection unit for treating and recycling water, and is used for removing the debris and colloidal matter. Wastewater Treatment Unit includes the filters meant for pre-treatment, and made-up of any of the materials like polypropylene, nylon, glass fibres as well as composites or their combinations thereof. The pore sizes of filters are in the range of 10 μm, for removing the debris/colloidal matter.

Pump: The pump is used to pump water from the water collection unit (bathtub) to the shower. The water is continuously added to the extractor to spray the stems at the top.

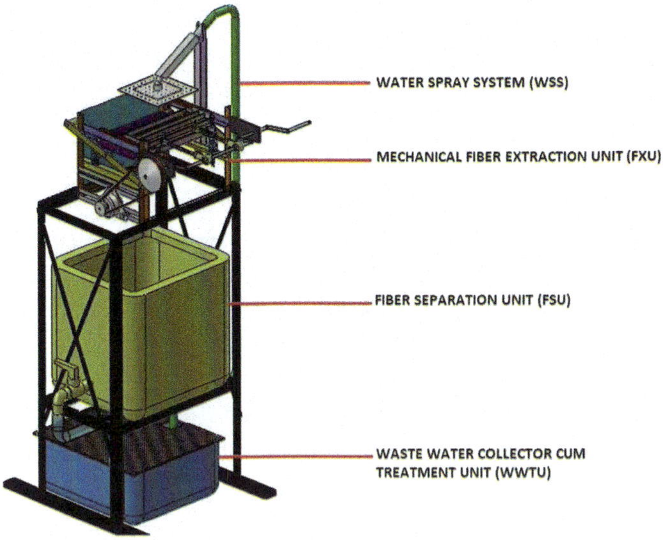

Fig. 3 Plant fibre extraction machine (3D model)

5 Features of the Machine Developed

Figure 4 depicts the developed Plant Fibre Extraction Machine. It consists of four main components or units, namely, Fibre Extraction, Fibre Separation, Waste Water Treatment, and Spray system. The output unit controls by free hand and mechanization process.

The to-and-fro motion of the rollers can be produced test manually by rotating a handle on one side of the crankshaft. In this manual mode, peeling 10–12 stems of 120–150 mm length takes 10–12 min. In machine mode, the shaft is rotated by an electric motor using pulleys for speed reduction, and the time taken is about 6–8 min.

In either case, about 3 L of water is fed through the sprayer unit into the separator unit for churning purposes. **2 kg of fibre can be extracted in a day. During the extraction procedure, the machine can put 12 stems into each cluster. Approximately 50 gm of fibre may be extracted from 240 stems of 120 to 150 mm length.**

Manual mode: 50 gm of fibre may take more than 1 h to extract fibres manually

The short fibres were extracted from the **Calotropis Gigantea** stems after soaking them in water for 5 days.

Both fresh or dried stems can be used for extraction of Calotropis fibre. Figure 5 depicts photographs of the Calotropis Gigantea, stems, and extracted fibres.

Fig. 4 Prototype of fibre extraction machine

Fig. 5 Photographs of **a Calotropis Gigantea** plant, **b** stems, **c** extracted fibres

Fig. 6 Wax Polishing, applying resin with brush and mixture of fibres and resin

6 Preparation of Composite and Test Results

In a 60:40 volume proportion, short Calotropis Gigantea fibres are amalgamate with Epoxy.

For its properties, the specimens are made according to the ASTMD256standard procedure. A 4 mm thick and 150 mm × 150 mm thick layer is developed on the mould employing a hand-laying process (Fig. 6). The laminate is cured for 8–10 h in room temperature for complete curing (Fig. 7).

7 Mechanical Properties

The aforementioned laminates are used to produce test specimens that meet ASTM, D638 Standard.

Fig. 7 Cured laminate of 150 mm × 150 mm

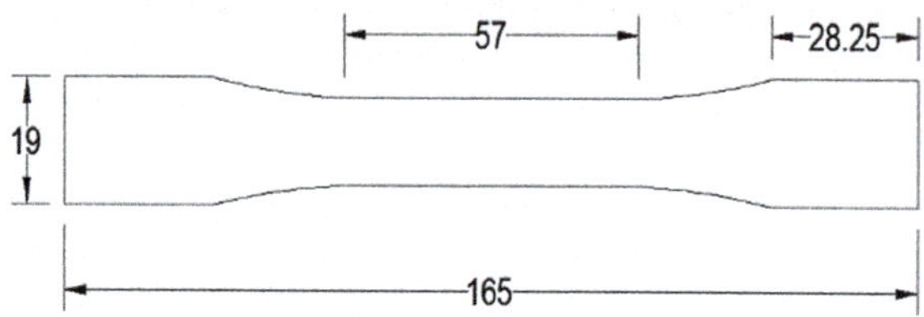

Fig. 8 Tensile test specimen (dimensions in mm)

Thermogravimetric analysis, tensile strength testing are performed. Figures 8, 9, and 10 depict tensile test specimens, specimens during test, and specimens after the test, respectively.

Results of the tests are presented in Table 1. Figure 11 shows TGA data of the sample. Figure 12 shows stress strain curve of the sample.

The tensile strength of Calotropis Gigantea fibre-reinforced epoxy is 14.68 MPa, according to the test findings it has very little tensile strength compared to other natural composites, % elongation is 5.52% and reaching plastic zone at a fast rate, in Thermogravimetric analysis (TGA) it is observed that **Calotropis Gigantea** is able to withstand temperature of 358° C. Further increase in temperature tends to weight loss. The extension with load is linear, with the little plastic zone.

Fig. 9 Water jet machine used to cut specimen and testing of specimens using UTM

Fig. 10 Tested specimens

Table 1 Test parameter versus test value

S. No.	Test parameter	Test value
1	Tensile strength	14.68 MPa
2	% Elongation	5.52%
3	Modulus of elasticity	235.5 MPa
4	TGA analysis	Resin content: 88.4%
		Fiber content: 11.6%

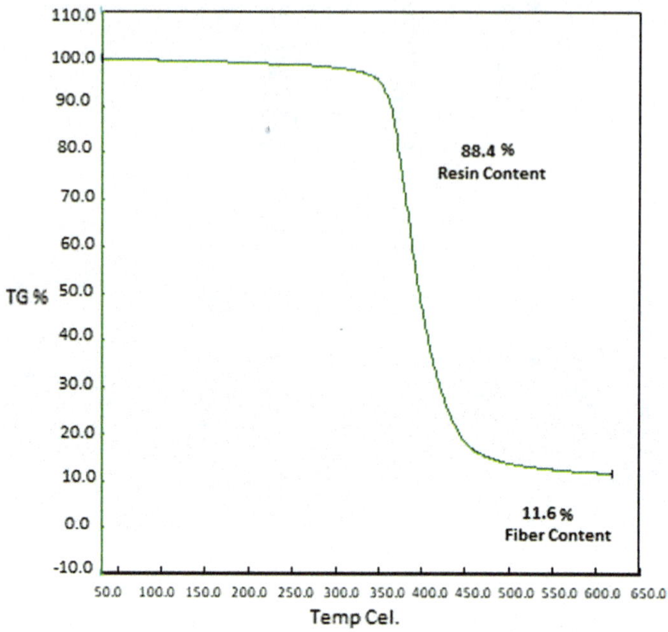

Fig. 11 TGA graph

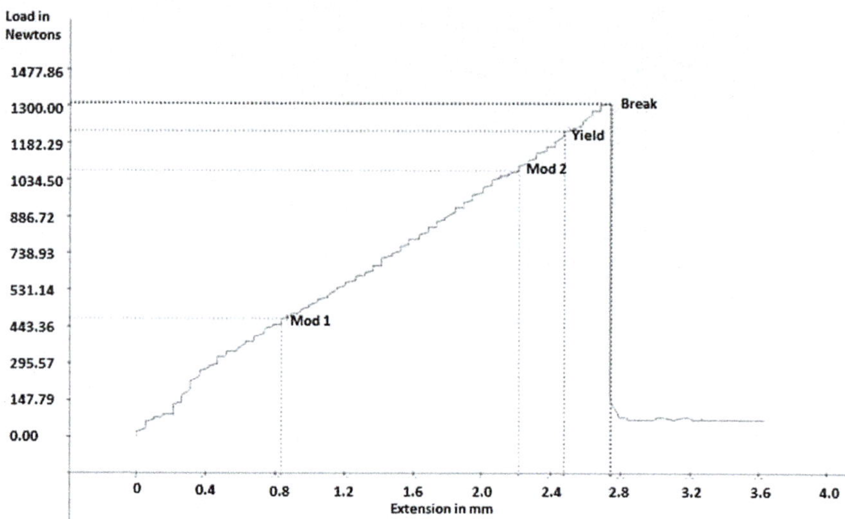

Fig. 12 Stress–strain graph

8 Conclusion

Calotropis Gigantea is a plant that can be found in abundance throughout the country. The goal of the research is to design a machine that can extract fibres from plant stems. The machine's functional aspects are discussed in full in the paper. The prototype developed can produce about 2 kg of dried fibre per day. Mechanical features such as power intensity, expansion percentage, stiffness mode, and TGA analysis are all measured by various samples. The fibres produced by the machine are utilized to build epoxy epoxy composites that meet the customer's requirements. The characterization implies that **Calotropis Gigantea**/epoxy has very less tensile strength compared to other natural composites. This combination has a better thermal stability than natural combinations, which is a benefit. These materials have the potential to be used as thermal heaters in applications with low temperatures.

Acknowledgements Our Sincere thanks to IEI (India) for sponsoring an amount of Rs.45,000/- under PG project entitled 'Automated Machine for Bio-degradable Natural Fibre Extraction and Conditioning', in June 2018, Project ID: R5/2/PG/2017–18/RDPG2017008.

Patent publication details.

Published a patent on title 'Process for Extraction of **Calotropis Gigantea** Fibres for Preparing Composite Structures used in Multiple Application'. with Patent No. 201741040315, CBR NO. 34856

References

1. Renugadevia K, Devanb PK, Chandra Sekhara Reddya M et al (2020) Optimization of surface treatment in Calotropis Gigantea (CG)-fibre yarn by simple techniques and characterization of CG fibre yarn reinforced laminate. J Mater Res Technol 9(6):12187–12200
2. Ramasamya R, Obi Reddyb K, Varada Rajuluc A (2017) Extraction and characterization of Calotropis gigantea bast fibers as novel reinforcement for composites materials. J Nat Fibers, ISSN: 1544-0478 (Print) 1544-046X (Online) Journal homepage: http://www.tandfonline.com/loi/wjnf20. Pickering KL, AruanEfendy MG et al (2016) A review of recent developments in natural fibre composites and their mechanical performance. Compos Part A: Appl Sci Manuf 83:98–112
3. Shah DU, Porter D, VollrathCan F (2014) Silk become an effective reinforcing fibre? A property comparison with flax and glass reinforced composites. Compos Sci Technol, 101:173–183
4. Jiang QR, Yang YQ (2012) Biocompatible natural silk fibers from *Argema mittrei*. J Biobased Mater Bioenergy 6(5):558–563

Effect of Coagent Concentration on Crosslinking, Physico-Mechanical Properties of the Peroxide-Cured Hydrogenated Acrylonitrile Butadiene Rubber (HNBR)

G. Rajesh, Ivaturi Srikanth, P. M. Asseref, and Ch. Subrahmanyam

1 Introduction

Hydrogenated acrylonitrile butadiene rubber (HNBR) is a specialty elastomer that recently found applications ranging from seals, belts, and hoses [1]. The main advantage of HNBR is resistance to amine corrosion inhibitors and hydrogen sulfide found in oil fields [2–4]. HNBR is commercially prepared by catalytic hydrogenation of acrylonitrile butadiene rubber (NBR), which results in a highly saturated backbone with fewer residual double bonds (0.9–5.5 wt%). HNBR is effectively a tetrapolymer of ethylene, propylene(P), acrylonitrile(A), and butadiene(B) rubber, as shown in Fig. 1 [5–7]. The resultant rubber has excellent mechanical properties, high-temperature resistance and resists degradation to oxygen and ozone. HNBR grades with low unsaturation (0.59%) are exclusively cured by peroxides [8, 9]. The challenges faced during peroxide curing of HNBR include poor scorch time, slow cure rate and inferior mechanical properties [10]. Literature reports are available on using various coagents and peroxides for curing various saturated elastomers

G. Rajesh (✉) · I. Srikanth · P. M. Asseref
Advanced Systems Laboratory, DRDO, Dr.A.P.J. Abdul Kalam Missile Complex, Hyderabad 500058, India
e-mail: rajeshg.asl@gov.in

I. Srikanth
e-mail: isrikanth.asl@gov.in

P. M. Asseref
e-mail: asserefpm.asl@gov.in

G. Rajesh · Ch. Subrahmanyam
Department of Chemistry, Indian Institute of Technology (IITH), Kandi, Sangareddy 502284, India
e-mail: csubbu@chy.iith.ac.in

M. R. Manohara Babu et al. (eds.), *Recent Advances in Composites Engineering and Research*, Springer Proceedings in Materials 39,
https://doi.org/10.1007/978-981-99-8807-5_32

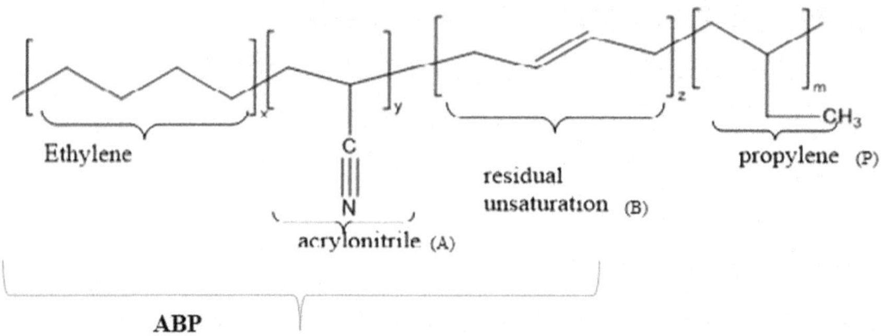

Fig. 1 Structure of HNBR

like EPM and HNBR rubbers [11–15]. Reports are available on measurements of mechanical and cure characteristics on HBNR cured by peroxides [16–23]. Most studies are carried out on HNBR/blends and filled with various fillers like carbon black, silica, graphite, carbon nanotubes, etc., [24–31]. There are no reports, best of my knowledge, discussing coagent's effect on crosslink density, physico-mechanical and thermal properties of the peroxide-cured HNBR rubber. The paper describes the impact of addition of triallyl isocyanurate (TAIC) coagent with different dicumyl peroxide (DCP) concentrations on crosslink density measured by cure characteristics and heat of formation values. The effect of coagents on the physico-mechanical characteristics of HNBR rubber is studied, and possible reasons are described in the present study.

2 Experimental

2.1 Materials

The following are details of ingredients, peroxide and cogent used in the studies with source and properties given in Table 1.

Figures 2 and 3 describe the structures of DCP and TAIC used in the formulations.

2.2 Formulations and Sample Preparations

As listed in Table 2, the formulations were selected for all batches where the concentration of coagent 1,3,5 triallyl isocyanurate was kept constant. Table 3 describes the variation of peroxide concentration and designation used for various batches. The rubber and ingredients were mixed in a laboratory-size two-roll mill, 6″ × 13″ (M/s Plus One Machinefabrik, Belgaum, India), with roller speed 1:1.15 at 40 °C. The

Table 1 List of ingredients

Ingredients	Specifications	Source
HNBR (Therban 3907)	Mooney viscosity ML (1 + 4) 100 °C: 71 MU, Acrylonitrile 39.2 wt. %, Residual bonds 0.5%	Arlanxeo performance elastomers, Switzerland
Zinc oxide	White powder, particle size 4μ	Synochem India Ltd.
Stearic acid	Acid value 204.5 mg KOH/g	TGV SRAAC Ltd.India
Wingtack 98	Hydrocarbon resin, Yellow granules with softening point 98 ± 2 °C	Cray Valley, USA
Dicumyl peroxide (DCP)	Appearance: powder Active oxygen: Assay: 98%	Arkema Peroxides India Pvt. Ltd.
Vulcofac TAIC-70	1,3,5 triallyl isocyanurate Appearance: White powder	Safic-Alcan, UK

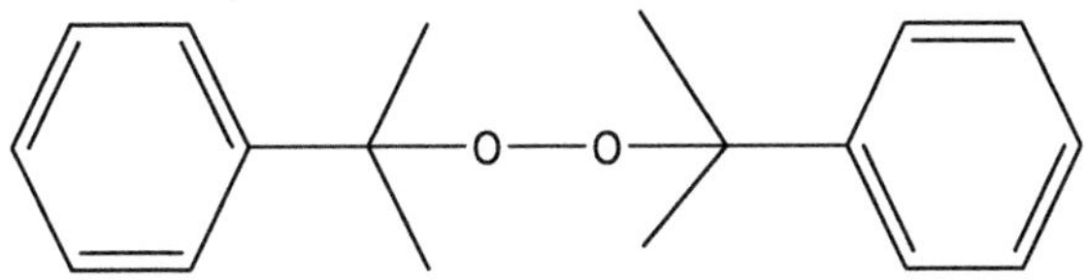

Fig. 2 Dicumyl peroxide

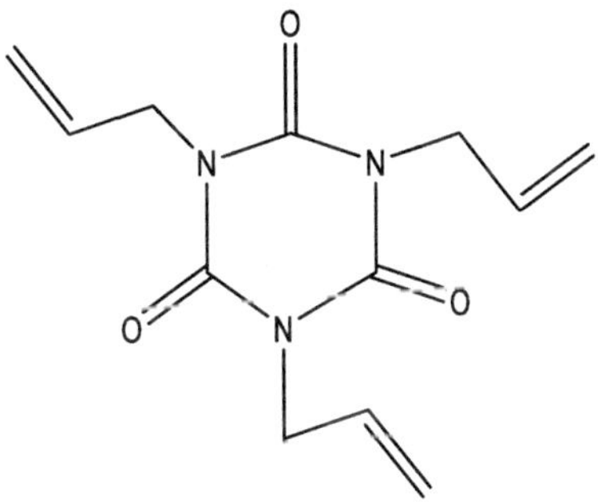

Fig. 3 Triallyl isocyanurate

compounded rubber was then sheeted out and kept at room temperature for 16 h before being used for molding.

Table 2 Basic formulation

Ingredients	PPhr
Therban 3907	100
Zinc Oxide	2
Stearic Acid	1
Wingtack 98	5
Vulcofac TAIC-70	3

Table 3 Formulations of HNBR compound with the variation of peroxide

	Designation (B = HNBR 39%ACN), D- DCP						
	BD1	BD3	BD5	BD7	BD9	BD11	BD13
The concentration of DCP (PPhr)	1	3	5	7	9	11	13

2.3 *Rheological Measurements*

The cure characteristics were measured at 170 °C for 30 min using Moving Die Rheometer (MDR) 3000 from M/s MonTech, Germany, with test frequency 1.67 Hz and strain 7%, respectively. Scorch time (ts1) was determined as the time to achieve a torque rise of 1 unit above minimum torque. Cure time was defined as reaching a 95% complete cure state (tc95). Torque difference between the maximum (S′ max) and minimum storage torque ((S′ min) was used as an indication of crosslink density.

2.4 *Post-curing Cycle*

The post-curing was evaluated using the uncured specimens of typical peroxide-filled HNBR BD5. Storage shear Modules (G′) values were determined by following a regular programming cycle (Test1) as shown in Table 4 in MDR. It is well known that G′ relates to several crosslinks formed between the polymer chains [32]. The temperature for post-curing is determined based on the plot of G′ versus temperature obtained from rheograph. A fresh uncured sample was subjected to programming cycle (Test 2) to evaluate the time required for post-curing of HNBR.

Table 4 Post-curing conditions

S. No.	Test conditions	Cycle
1	• Frequency: 1.57 Hz, Strain: 3.58 • Initialization: 100 °C	• Isothermal: 170 °C, 30 min • Non-isothermal: – 40 °C: 30 min – 80 °C: 30 min – 100 °C: 30 min – 120 °C: 30 min
2	• Frequency: 1.57 Hz, Strain: 3.58 • Initialization: 100 °C	• Isothermal: 170 °C, 30 min • Non-isothermal: – 80 °C: 60 min

2.5 Differential Scanning Calorimetric (DSC) Studies

The measurements of heat flow on the samples were carried out from 30 to 300 °C with a heating rate of 20 °C/min in a nitrogen atmosphere using the DSC 204F1 from Netzch, Germany. The **ΔH exothermic** values were calculated from the area under the curve to ascertain the degree of cure.

2.6 Curing

The cure time and temperature were determined using rheological studies. The molding of test sheets (300 mm × 150 mm × 2 mm) was carried out in a hydraulic press (Make: PlusOne Machinefabrik, Belgaum, India) with a pressure of 50 kg/ cm^2 and temperature of 170 °C. These sheets were left at room temperature for 24 h followed by a post-curing cycle of 80 °C, 60 min in air circulating hot air oven.

2.7 Determination of Physical and Mechanical Properties

The tensile stress–strain properties were tested on dumbbell specimens cut from sheets with a DAK Systems Universal Testing Machine (Model No. T-72102 Ex Hi-Tech., India) at 25 °C with a crosshead speed of 500 mm/ min according to BS ISO- 37(2017) standard.

3 Results and Discussions

3.1 Rheological Measurements

Figure 4a and 4b shows the torque (S ′) versus time plot for HNBR rubber with DCP at different concentrations and fixed concentrations of coagent, respectively.

It can be seen from the figures that with the increase in concentration of peroxide elastic component of torque (S′) increased, indicating crosslinking. Adding a fixed concentration of coagent also increased S′ compared to a compound with no coagent, which means that a reaction is occurring between the unsaturated species in the coagent and peroxide, further reacting with HNBR.

Cure time analysis: Fig. 5a and 5b show a plot of cure time and scorch time versus peroxide concentration without and with coagent addition in HNBR compounds.

Scheme 1 shows the probable mechanism of crosslinking HNBR with DCP. Due meager degree of unsaturation, the cumyl radical abstracts the hydrogen predominantly from the polyethylene chain of HNBR. It can be seen that the cure time

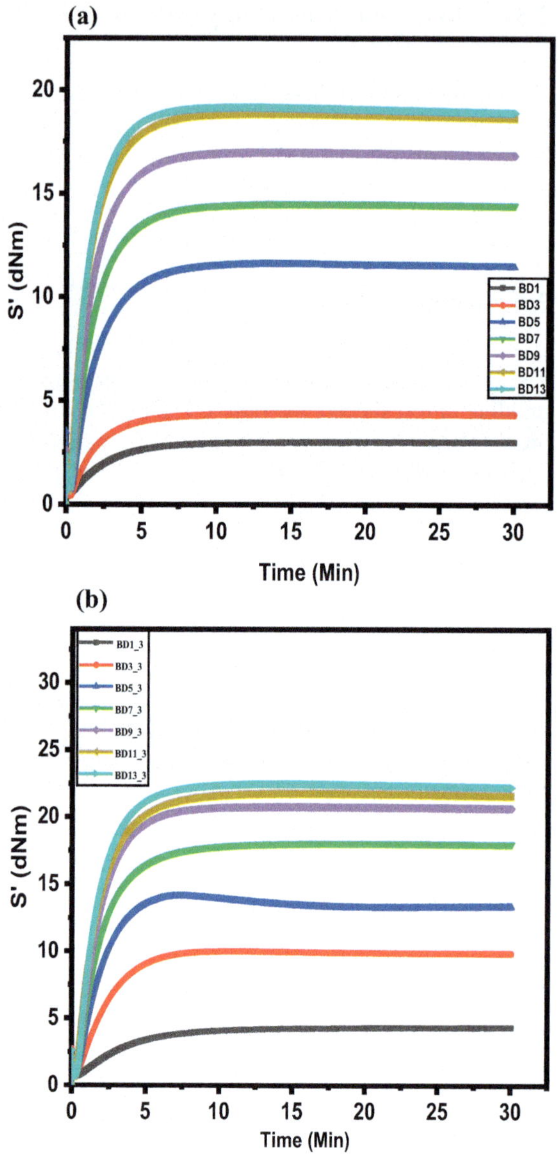

Fig. 4 **a** Torque versus time curve without coagent, **b** torque versus time curve with coagent

of HNBR compounds decreased with increasing concentration of peroxide. This is due to the increase in the reaction rate. The addition of fixed concentration of TAIC resulted in increase in cure time. This can be ascribed to the reaction of double bonds in structure (Fig. 3) with HNBR rubber and peroxide through the formation of bridge network structure [9].

Fig. 5 **a** Cure time versus peroxide concentration, **b** scorch time versus peroxide concentration

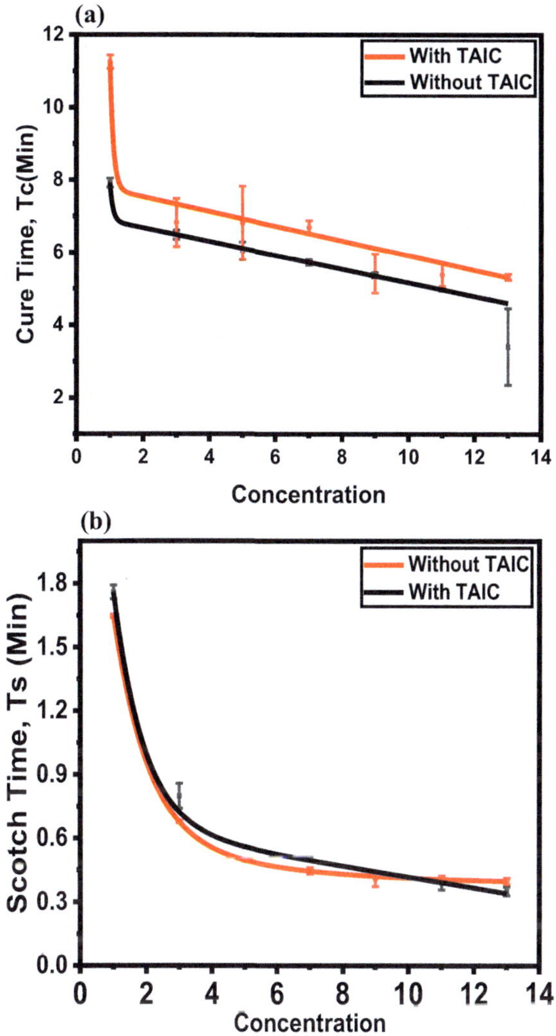

Figure 5b shows that the scorch time decreases with peroxide concentration for both cases of with and without coagent indicating loss of flow properties. Figure 6a, b shows DSC plots of HNBR rubber compounds without and with coagent.

It can be seen from the figures that there is a sizeable exothermic peak in the temperature region between 160 and 210 °C, with the peak at around 190 °C for all HNBR compounds. It can be seen that coagents have not altered the peak cure temperature of the DSC curve, indicating that the rate of the cure reaction has accelerated with the coagent. This suggests a significant increase in crosslinking due to coagent in the presence of peroxide compared to peroxide alone (Fig. 7).

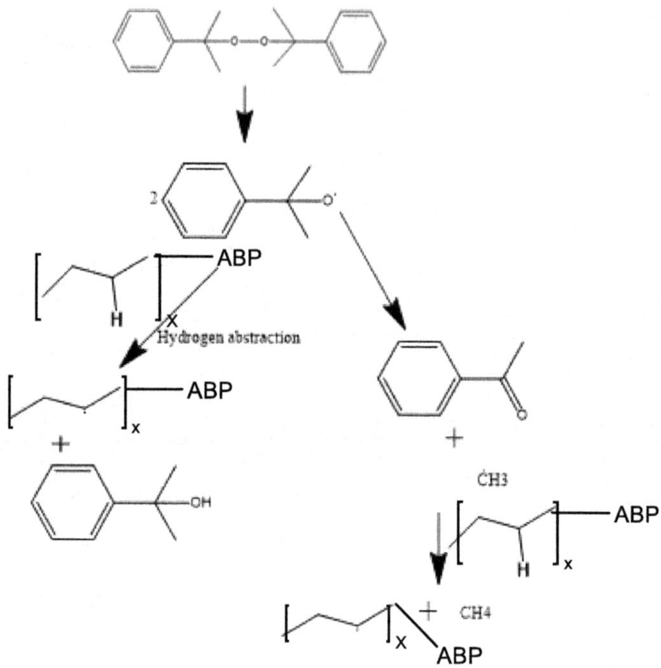

Scheme 1 Mechanism of crosslinking HNBR with DCP [35]

The (ΔH) values increase continuously without coagent and with coagent, indicating a high degree of crosslinking in the presence of a nitrogen atmosphere. DSC data, which shows a positive effect of coagent in increasing the degree of crosslinking, is supported by the trend observed from the torque difference plots ($S'_{max}-S'_{min}$), which gives crosslink density [33, 34] shown in Fig. 8.

The crosslinking process stabilizes after 9 phr for coagent addition as the radicals generated from peroxide are terminated in an air atmosphere. The probable mechanism of crosslinking of TAIC addition in peroxide curing is similar to one reported in the literature [34] and is depicted in Scheme 2.

The post-curing cycle was established by conducting non-isothermal studies on uncured HNBR samples. It can be seen from Fig. 9a that the G′ value increases from 40 to 80 °C after the initial isothermal rheometer cycle suggesting additional crosslinking formation at 80 °C due to unreacted peroxide present in the HNBR rubber matrix. The same is reconfirmed in Fig. 9b, where the isothermal cycle at 80 °C for 60-min duration indicates an increase in G′ values.

Figure 10a, b show the variation of ultimate tensile strength (UTS) and % elongation at break vs. peroxide concentration without and with coagent, respectively.

The trends clearly show that tensile strength decreases drastically with peroxide concentration and reaches a plateau from 7 phr. The trend is similar with coagent addition, where the plateau reached 3 phr of peroxide concentration. The same can

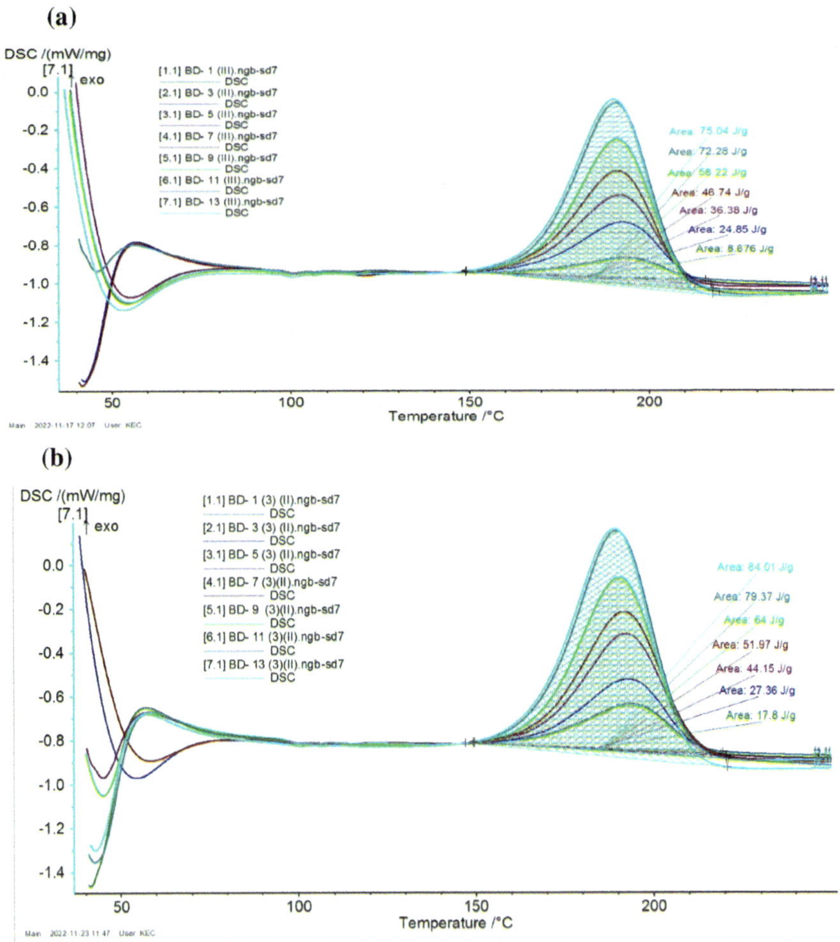

Fig. 6 **a** DSC plots with without coagent, **b** DSC plots with coagent

be observed for elongation at break variation with peroxide concentration shown in Fig. 10b. The rubber property is lost at 11 phr for peroxide, and 7 phr with coagent since elongation at break falls below 100%. The abnormal behavior can be explained by stress–strain plots shown in Fig. 11.

From plots, it can be seen that HNBR (with 39% ACN) at lower concentrations of peroxide has a tendency to undergo strain-induced crystallization (SIC) under tensile loads. HNBR rubber with 39% ACN content has alternative acrylonitrile butadiene and ethylene sequence in the micro structure, which undergoes crystallization [35–37]. This contributes excellent mechanical properties to HNBR rubber compounds up to 5 phr concentration of DCP where the stereoregular structure is not disturbed. With an increase in peroxide concentration at 7 phr and more, the significant ability to crystallize under tensile loads is lost due to the formation of an irregular crosslink

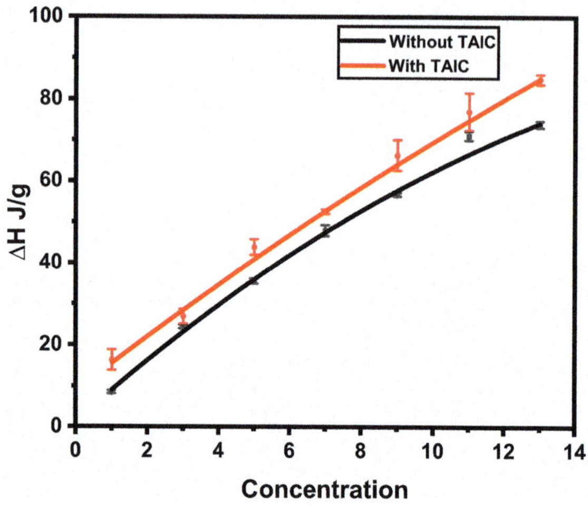

Fig. 7 Plot of ΔH versus peroxide concentration

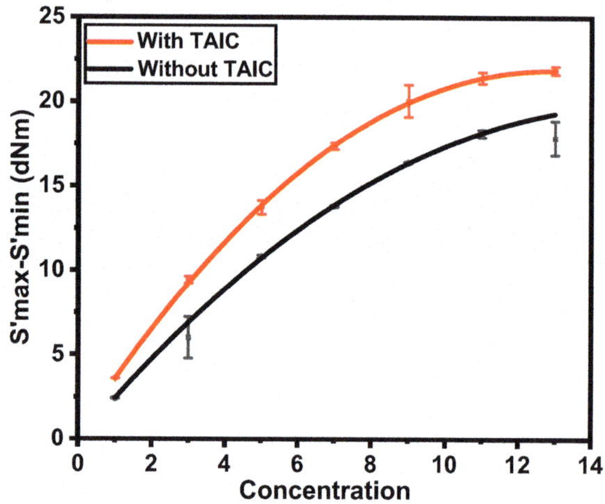

Fig. 8 Torque difference plots vs. peroxide concentration

structure. The trend is more prominent with coagents where the mechanical properties drop from 5 phr DCP concentration. The probable mechanism of peroxide crosslinking of HNBR can be described in Scheme 1, mainly by the abstraction of an H atom from the elastomer chain.

With the increase in peroxide concentration, more such crosslinks are formed, leading to chain flexibility restriction and brittle structure as seen from stress–strain plots when DCP s added beyond 7 phr.

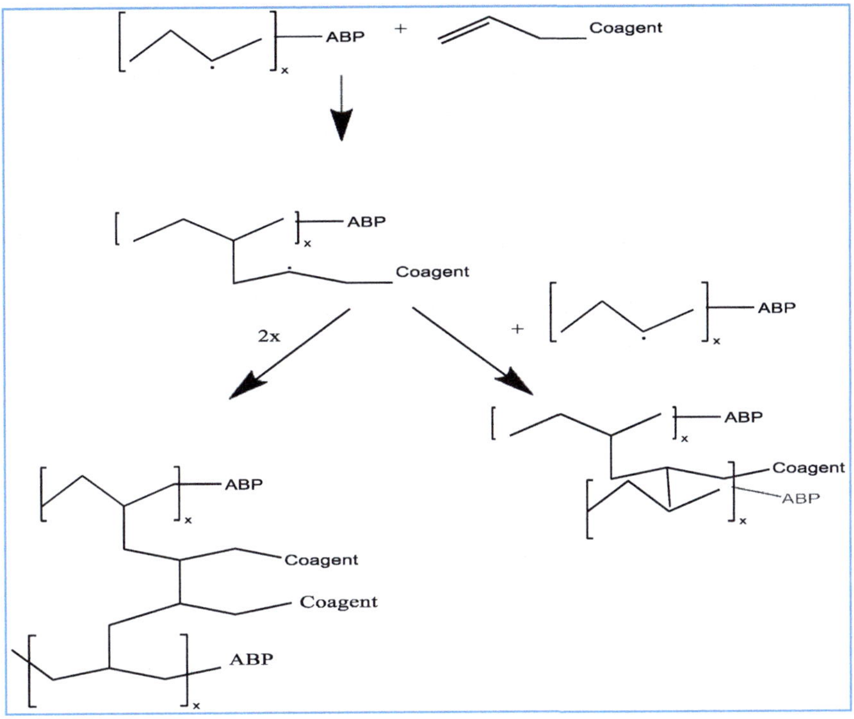

Scheme 2 Mechanism of crosslinking of TAIC addition in peroxide curing [12]

4 Conclusions

HNBR (39% ACN) with a very low degree of unsaturation was cured using peroxide DCP at various concentrations. The effect of a fixed concentration of coagent TAIC was also studied on DCP concentrations. It was observed that on the addition of coagent in peroxide-cured HNBR, the degree of crosslinking increases significantly, as observed from torque difference (S′ max−S′ min) values obtained from MDR curves and ΔH exothermic obtained from DSC curves. Further, it can be seen that the mechanical properties of DCP cured HNBR are found to be inferior by addition of TAIC coagent to that when cured by DCP alone.

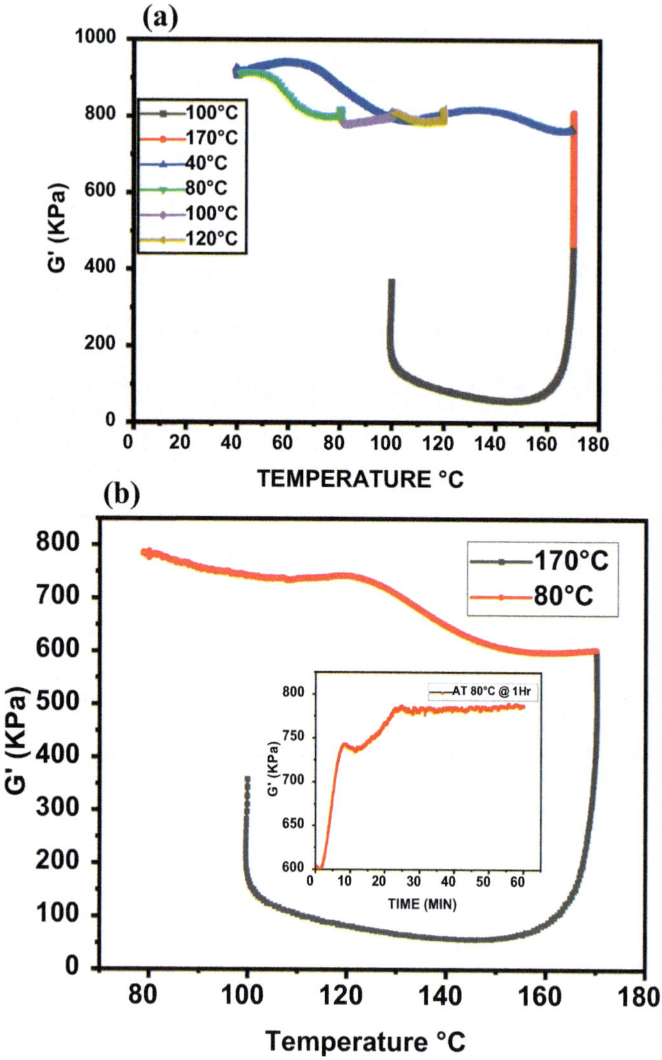

Fig. 9 **a** G′ versus temperature (°C): post-curing studies, **b** G′ versus temperature (°C), isothermal

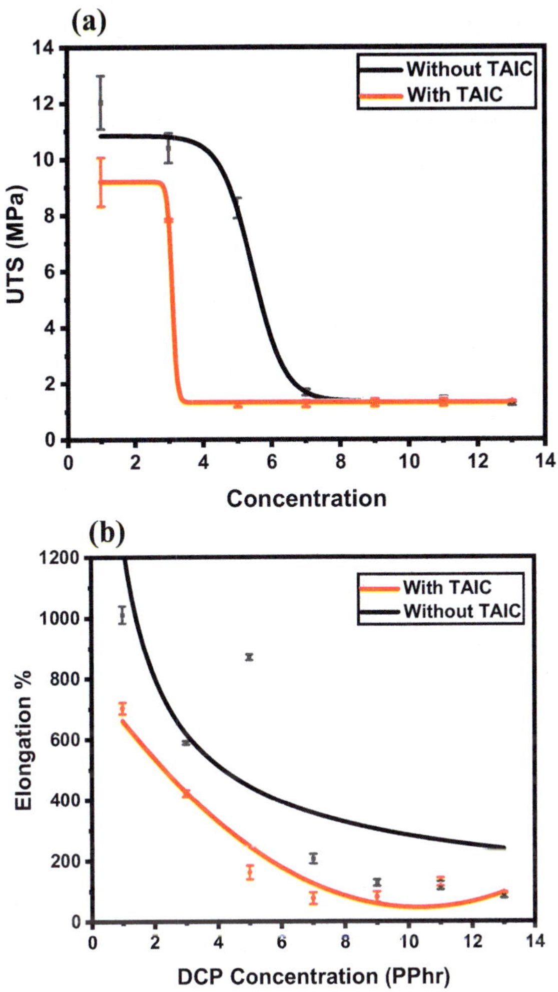

Fig. 10 **a** UTS versus DCP concentration, **b** % elongation at break versus DCP concentration

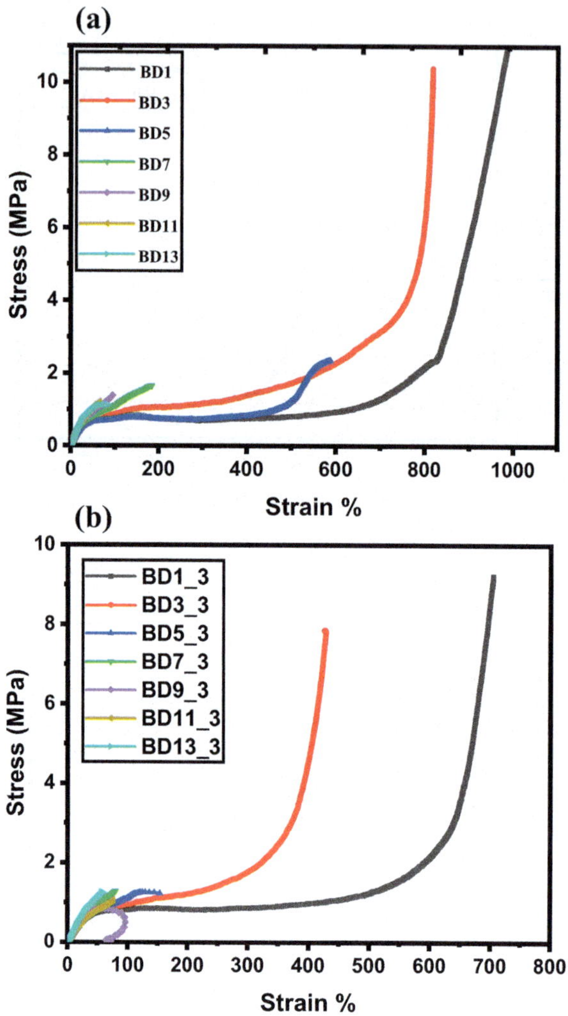

Fig. 11 **a** Stress–strain plots of HNBR cured by DCP, **b** stress–strain plots of HNBR cured by DCP and coagent

Acknowledgements The authors acknowledge Shri Abhirup Dubey, for testing rubber specimens using differential scanning calorimetry (DSC) at Defence Research Metallurgical Laboratory (DMRL). We also acknowledge the services rendered by Smt. Challa Sneha Reddy and Shri P. Rohit of Elastomers Division, Advanced Systems Laboratory (ASL), for assisting testing of rubber specimens.

References

1. Nagdi K (1993) Rubber as an engineering material: guideline for users. Hanser Publications, NY, USA
2. Claus W, Klaus R (2001) Winkelbach Hans R: Therban – the high-performance elastomer for the new millennium. Macromol Mater Eng 286:657–662
3. Mirza Jean, Leibbrandt Friedrich, Thoermer Joachim (1987) Hydrogenated nitrile rubber for improved durability of automotive rubber parts. In: International congress and exposition, Detroit, Michigan, 23–27 Feb
4. Wiseman WA, Ridland JJ (1989) Hydrogenated nitrile rubber-long term aging in automotive fluids. In: International congress and exposition, Detroit, Michigan, Feb 27–March 3
5. Wang Hui, Yang Lijuan (2013) Homogenous hydrogenation art of nitrile butadiene rubber. Polym Rev 53(2):192–239
6. Robert Keller.: Practical guide to Hydrogenated Nitrile Butadiene Rubber Technology. Smithers RAPRA Technology Ltd. Shropshire, UK (2012).
7. Hayashi Sachio (2001) Handbook of elastomers, 2nd edn. Marcel Dekker, New York, pp 785–813
8. Datta RN (2002) Rubber curing systems. RAPRA Rev Rep 12(12):28
9. Coran. Y. Aubert (2001) The science and technology of rubber, 3rd edn. Elsevier Academic Press, New York, Chap 7
10. Kruzelak Jan, Sykora Richard, Ivan Hudec.: Sulphur and peroxide vulcanization of rubber compounds. Chem. Papers. 70(12), 1533–1595 (2016).
11. Likozar Blaz, Krajnc Matjaz.: Influence of morphology on the dynamic mechanical properties of hydrogenated nitrile butadiene elastomer/coagent nanodispersions. J App. Polym. Sci. 110, 183–195(2008).
12. Maria Montserrat Alvarez Grima.: Novel Coagents for improved properties in peroxide cure of saturated elastomers." Ph.D. Thesis, University of Twente (2007).
13. Young Seok Lee., KIRyong Ha.: Effects of co-agent type and content on curing characteristics and mechanical properties of HNBR composites. Elast. And Compos. 55(2), 95–102(2020).
14. Steven K. Henning.: The Use of Coagents in the peroxide radical cure of elastomers," Proceeding of the 56th International Wire and Cable Symposium Lake Buena Vista FL, November 11–14 (2007).
15. Magdalena, Maciejewska, Marian Zaborski, Anna Masek.: Co-agents for peroxide curing hydrogenated nitrile butadiene elastomer. Polimery -Warsaw 54(11), 814–819. (2018).
16. Victor Nasreddine, Pete Spanos.: Mechanical and Dynamic Properties of Peroxide Cured HNBR Compounds. Conference: 200th Fall Technical Meeting of the Rubber Division, American Chemical Society (2021)
17. Wongwitthayakool P, Sirisihna C, Saeoui,P.: Cure and viscoelastic properties of HNBR. Elastomer and Plastics KGK. Nov 506–512 (2010).
18. Guo Jianhua, Zhicai Li.: Effect of curing systems on vulcanization mechanical properties and stress softening of hydrogenated nitrile butadiene rubber vulcanizates. J Elasto. And Plast., 1–15 (2016).
19. K. Narynbek Ulu, B. Huneau, E.verron.: Efeects of acrylonitrile content and hydrogenation on fatigue behavior of hydrogenated nitrile butadiene rubber. Fatigue Fract Eng Mater Struct. 1–17 (2019).
20. Likozar Blaz, Krajnc Matjaz.: Temperature-Dependent Dynamic Mechanical Properties of Hydrogenated Nitrile Butadiene Rubber and effect of Peroxide crosslinkers. e-Polymers No.131 (2007).
21. Thavamani P, Bhowmick AK (1992) Dynamic mechanical properties of hydrogenated butadiene rubber: Effect of crosslink density, curing system filler and resin. J Mat. Sci. 27:1243–1253
22. Yu. V. Korovina, E.I. Shcherbina.: Peroxide vulcanization of hydrogenated butadiene acrylonitrile Rubber. Kauchuk I. Rezina. No.1, 4–7 (2007).

23. Xinyan Shi, Zhu Zhu, Lingyan Jia, Qi Li, Weina Bi.: HNBE/EPDM Blends: Co-vulacanization and compatibility. J Appl Polym Sci. 3054- 3060 (2013).
24. D.Felhos, J. Karger-Kocsis, D. Xu.: Tribological testing of peroxide cured HNBR with different MWCNT and silica contents under dry sliding and rolling conditions against steel. J Appl. Polym Sci.108, 2840–2851 (2008).
25. Chunrui Sheng, Zhiwei Hu, Hoch Martin, Yongxin Duan, Jianming Zhang.: Effect of small amount of sulphur on the physical and mechanical properties of peroxide-cured fully saturated HNBR Compounds. J App Polym Sci. 41612 (2015).
26. Geralda Severe.: Characterization and Development of new Hydrogenated Acrylonitrile butadiene rubber blends. Ph.D. Dissertation, University of Akron. (1999).
27. Choudhury Anasuya, Anil K. Bhowmick, Ong Christopher.: Effect of different nanoparticles on thermal, mechanical and Dynamic mechanical properties of HNBR Nanocomposites. J Appl. Polym Sci. 116, 1428- 1441 (2010).
28. LI Qi, Zhao Suhe, Pan Yan.: Structure morphology and properties of HNBR filled with N550, SiO_2, ZDMA and two of three kinds of fillers. J App. Polym Sci. 117, 421–427 (2010).
29. Zhao Xingbo, Zhang Qiuyu, Gu Junwei.: Effects of carbon black on the properties of HNBR reinforced by in- situ prepared ZDMA. Polym -Plast Tech. and Eng.50, 1507–1510 (2011).
30. Vinicius Demetrio Da Silva, Italo Ribeiro de Barros, Aramid pulp reinforced HNBR composites with ionic liquid compatabilizers. J. Appl Poym. Sci. (2019).
31. Syed Mohammed Reffai Syed Ismail, Tuhin Chatterjee, Kinsuk Naskar.: Superior heat resistant and oil resistant blends based on dynamically vulcanized HNBR and polyamide. Polym. Adv. Technol. (2016).
32. Alan N.Gent.: Engineering with Rubber. How to Design rubber components. 3rd edn. HANSER publications. Cincinnati, USA. (2000).
33. P. Phewphong, P. Saeoui, C. Sirisinha.: The use of dynamic mechanical thermal analysis technique for determining an uneven distribution of precipitated silica in CPE/NR blends. Polymer testing. 27 No. 7, 873–880 (2008).
34. Hanafi Ismail, B. T. Poh.: Cure and tear properties of ENR 25/SMR L and ENR 50/SMR L blends. Eur Polym Jour. 2403–2408 (2000).
35. Jan Kruzelak, Richard Sykora, Ivan Hudec.: Vulcanization of rubber compounds with peroxide curing systems. Rubber Chem. Tech. 90(1), 60–68 (2017).
36. Braun D, Haufe A, Leiß D, Hellmann GP (1992) Strain induced crystallization and miscibility behavior of Hydrogenated nitrile rubber. Die Angewandhe Makromolekulare Chemie. 202/ 203:143–158(3673)
37. Akhtar S. Khan, Muneer Baig, Syed Humid, Huo Zhang (2010 Thermo-mechanical Large deformation responses of hydrogenated nitrile butadiene rubber (HNBR): experimental result. Int Jour Solid Struct 47:2653–2659)
38. Xi Zhang, Jing wu, Zhiqieng Xu (2021) comparative study on the molecular chain orientation and strain induced crystallization behaviors of HNBR with different Acrylonitrile content under uniaxial stretching. J Polym 219:123520

Hot–Wet Environmental Effects on Glass Transition Temperature of IMA/M21E Aircraft Grade CFRP Composites

Kishora Shetty, C. M. Manjunatha, and Shylaja Srihari

1 Introduction

Use of carbon fiber-reinforced plastic (CFRP) composites in aerospace is mainly due to their lower weight with high strength and stiffness [1–4]. This unique combination of properties has enabled the rapid development and use of lightweight, high performance structures for a range of applications like aircraft structural parts. In their service life, composite aircraft structures are usually come in contact with a range of environmental conditions like hot–wet (hygrothermal), chemical environments, UV radiation, biological conditions, etc. This causes degradation in material properties of these composite parts [5–7]. In case of hot–wet effects, water or moisture which gets infused into matrix or interface region between matrix and fiber of composite can act as plasticizer. This results in increasing space between the polymer chains. This significantly deteriorates the glass transition temperature (T_g) and mechanical properties. It is important to know how they are affected by differing environmental conditions [8–13]. This is an important study related to the property—retention capability of polymer composites under application environments. This study is more important and crucial for use of composites in aero-structures due to the associated stringent safety norms and airworthiness certification requirements.

In this present research work, a comprehensive approach has been adopted to study both the diffusion and degradation effects in a new carbon epoxy composite, subjected to different hot–wet conditions (temperature and humidity in terms of % water), immersion in aviation fuels like aviation turbine fuel and piston engine fuel. All these conditions have great relevance to aerospace and allied applications. The

K. Shetty (✉) · C. M. Manjunatha · S. Srihari
Academy of Scientific and Innovative Research (AcSIR), Ghaziabad 201002, India
e-mail: kishora_shetty@yahoo.com

C. M. Manjunatha · S. Srihari
CSIR-National Aerospace Laboratoires, Bangalore 560017, India

M. R. Manohara Babu et al. (eds.), *Recent Advances in Composites Engineering and Research*, Springer Proceedings in Materials 39,
https://doi.org/10.1007/978-981-99-8807-5_33

CFRP Composite HexPly® M21E/34%/UD/194/IMA material was studied. M21E thermoset resin used in this composite is a new development from Hexcel. This is a high temperature (180 °C) cure epoxy resin system with high post-cure glass transition temperature, which is well above the aircraft skin temperature. This composite was exposed to different hot–wet environments up to moisture saturation limits. The effect of moisture absorption on physicochemical properties like glass transition temperature (T_g) of the CFRP was evaluated using differential scanning calorimeter (DSC) for the specimens after conditioning in all the environments of study. T_g got deteriorated by about 8% due to hygrothermal aging.

2 Experimental

HexPly® M21E/34%/UD/194/IMA CFRP composite was subjected to following various hot–wet environmental conditions till the saturation is reached in terms of moisture absorption.

Hot–wet (hygrothermal):

- at 45 °C/85%RH (relative humidity)
- at 70 °C/85%RH
- at 55 °C/100%RH.

Fuel environments at room temperature:

- Aviation gas turbine fuel
- Aviation gasoline (piston engine) fuel.

The effect of moisture absorption on physicochemical properties like glass transition temperature (T_g) of the CFRP was evaluated using differential scanning calorimeter (DSC) for the specimens after conditioning in all the environments of study. T_g was measured on unconditioned and conditioned samples using DSC 2910 model (TA Instruments, USA), shown in Fig. 1, was calibrated with high purity indium for base line, temperature, cell constant, and specific heat.

After saturation, a specimen from each media was selected. A small sample of this saturated specimen was cut and weighed on a Sartorius balance. The hermetically sealed cups were used to hold the samples in DSC cell. The samples were heated at 10 °C/mm. Dynamic scan of the samples was recorded and T_g values were obtained.

3 Results and Discussions

Results of moisture absorption behavior of HexPly® M21E/34%/UD/194/IMA composite for various hot–wet conditions and fuel immersion conditions with maximum moisture absorption and duration for the same are given in Table 1.

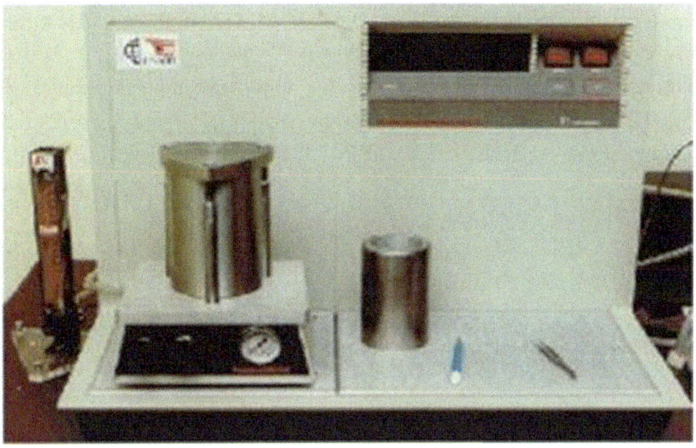

Fig. 1 Differential scanning calorimeter

Table 1 Moisture absorption and duration for saturation for HexPly® M21E/34%/UD/194/IMA CFRP composite

Aging condition	Maximum moisture absorption (Wt.)	Duration for saturation (months)
45 °C/85% RH	0.76	13
70 °C/85%RH	0.82	11
55 °C/100%RH	1.24	9
Aviation gas turbine fuel at RT	0.60	23
Aviation gasoline (piston engine) fuel at RT	0.81	23

To understand the moisture absorption effects on physicochemical property of HexPly® M21E/34%/UD/194/IMA CFRP composite, T_g was measured on samples after conditioning at three different hygrothermal environments and fuel immersion conditions. T_g using DSC was measured on moisture saturated CFRP specimens as well as unconditioned specimens. Results are presented in Table 2. Figure 2 shows the dynamic scans (thermograms) of the carbon fiber-reinforced plastic (CFRP) composite unconditioned and after conditioning in hygrothermal and fuel environments. Comparison of degradation in T_g for HexPly® M21E/34%/UD/194/IMA CFRP composite aged at various hygrothermal conditions and fuel environments is shown in Fig. 3.

It is observed that T_g of the composite degraded due to moisture absorption. The maximum deterioration was observed for both conditions 45 °C/85%RH and 70 °C/85%RH with 8% deterioration followed by 6% for 55 °C/100%RH. Fuel immersion at RT resulted in about 2% deterioration in T_g of HexPly® M21E/34%/UD/194/IMA CFRP composite. This indicated that the quicker diffusion and absorption of moisture to composite matrix as in the case of environmental conditioning at different

Table 2 Glass transition temperature (T_g) of HexPly® M21E/34%/UD/194/IMA composite after conditioning in different hygrothermal and fuel environments

Aging condition (nomenclature give is as per Fig. 2, DSC graph)	Glass transition temperature T_g (°C)
No conditioning (E1W5-AS RECEIVED CFRP LAM)	180.6
45 °C/85% RH (B1W3- CFRP LAM CONDITIONED AT 45C-85RH)	166.3
70 °C/85%RH (C1W3- CFRP LAM CONDITIONED AT 70C-85RH)	166
55 ^{0}C/ 100%RH (D1W4- CFRP Lam conditioned 55C-100RH)	169.6
Aviation gas turbine fuel at RT (A1W2-CFRP LAM-Fuel A immersion)	176.2
Aviation gasoline (piston engine) fuel at RT (E2W3-CFRP LAM-FUEL B IMMERSION)	177.3

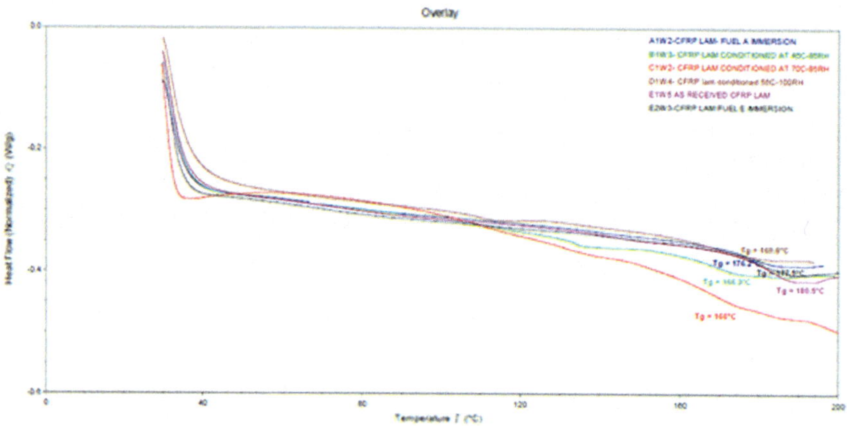

Fig. 2 DSC thermograms of moisture saturated CFRP specimen aged at different conditions and unaged specimen superimposed for comparison

temperatures and humidity levels, more is the degradation in glass transition temperature. Lowest degradation seen from samples conditioned at fuels at RT. The moisture saturation took very long time here.

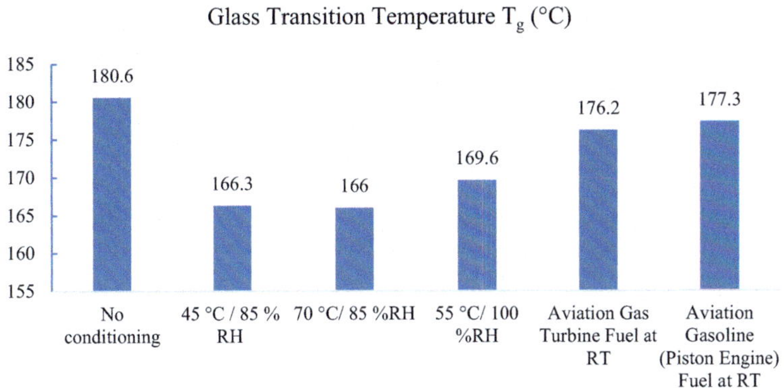

Fig. 3 Comparison of degradation in T_g of HexPly® M21E/34%/UD/194/IMA CFRP composite for not conditioned and conditioned

4 Conclusions

The T_g degradation of the HexPly® M21E/34%/UD/194/IMA CFRP composite is in accordance with the quicker moisture absorption from hygrothermal conditioning. It is evident that moisture absorption due to hygrothermal or fuel environments effects T_g of CFRP composites and this will be an important aspect to consider for application of these composite materials.

The polymeric matrix constituent of the composite materials may collect moisture and change its properties over time is one of the main concern associated with use of these materials in structural members of aircraft. It is very important to study moisture absorption behavior and its effects on properties of CFRP composites.

References

1. Shetty K, Manjunatha CM, Gururaja S, Srihari S (2022) Hot–wet environmental effects on in-plane shear strength of IMA/M21E Aircraft Grade CFRP composites. In: Jonnalagadda K, Alankar A, Balila NJ, Bhandakkar T (eds) Advances in structural integrity. Springer, Singapore. https://doi.org/10.1007/978-981-16-8724-2_26
2. Shetty K, Bojja R, Srihari S (2020) Effect of hygrothermal aging on the mechanical properties of IMA/M21E aircraft-grade CFRP composite. Adv Compos Lett
3. Shetty K, Srihari S, Manjunatha CM (2016) A survey on the environmental effects on Aerospace Fibre Reinforced Plastic Composites. In: Proceedings of 14th ISAMPE national conference on composites (INCCOM-14), 2016
4. John KM, Thirumalai Kumaran S, Kurniawan R, Park KM, Byeon JH (2018) Review on the methodologies adopted to minimize the aterial damages in drilling of carbon fibre reinforced plastic composites. J Reinf Plast Compos 1–18
5. Dogan A, Arman Y, Ozdemir O (2018) Experimental investigation of transverse loading on composite panels coated with different gelcoat colors subjected to UV radiation and hygrothermal aging. Mater Res Express 6(2):025301

6. Dogan A, Arman Y (2019) The effect of hygrothermal aging and UV radiation on the low-velocity impact behavior of the glass fiber-reinforced epoxy composites. Iran Polym J 28(3):193–201
7. Shetty K, Srihari S, Manjunatha CM (2019) Effect of hygrothermal aging on the interlaminar shear strength of a carbon fibre composite. Procedia Struct Integrity 14:849–854
8. Ogi K (2003) Influence of thermal history on transverse cracking in a carbon fiber reinforced epoxy composite. Adv Compos Mater 11(3):265–275
9. Sala G (2000) Composite degradation due to fluid absorption. Compos B 31(5):357–373
10. Tsenoglou CJ, Pavlidou S, Papaspyrides CD (2006) Evaluation of interfacial relaxation due to water absorption in fiber–polymer composites. Compos Sci Technol 66(15):2855–2864
11. Behera A, Thawre MM, Ballal A (2020) Hygrothermal aging effect on physical and mechanical properties of carbon fiber/epoxy cross-ply composite laminate. Mater Today: Proc
12. Shylaja Srihari, Rao RMVGK (1999) Hygrothermal characterisation and diffusion studies on carbon/epoxy composites. J Reinf Plast Compos 18(10):921–930
13. Ryan JM, Adams R, Brown SGR (2009) Moisture Ingress effect on properties of CFRP. In: Proceedings of 17th international conference on composite materials, Edinburgh, UK, 2009

Effect of Resin Matrices on Thermal Stability, Mechanical, and Ablation Properties of Carbon Fiber-Reinforced Resin Composites

Golla Rama Rao, Ivautri Srikanth, and K. Laxma Reddy

1 Introduction

By combining thermomechanical, thermochemical, and thermophysical forces brought about by a high-temperature, high-pressure, and high-velocity combustion process, ablation is an erosion process that removes fragments of a material [1, 2]. Ablation resistance is one of the important qualities that determine how well a heat-resistant composite material performs and how long it lasts under ablative conditions. Nozzles and reentry vehicle structures for aerospace applications are frequently subjected to such harsh environments. Carbon fiber-reinforced polymer composite materials have long been recognized as advantageous for their high-temperature ablation resistance. The ablation properties of a carbon fiber-reinforced composite are greatly influenced by the kind of matrix used, such as resin, carbon, or mixed matrix. Since carbon fibers have a high carbon content, strong thermal stability, low thermal conductivity, and limited thermal expansion, they are commonly used as reinforcement for ablation. Because phenolic resin forms char efficiently during pyrolysis, it has long been used as a typical thermosetting resin matrix for ablative composite materials [3]. The ability of the char to absorb heat generated during the ablation process makes it essential to the behavior of the ablation process. When there is a strong shear force, pressure, and velocity, the brittle, burned charred matrix may be mechanically removed.

Resole type phenolic resins are high-temperature resins with minimal toxicity and excellent smoke and fire resistance [4]. The short shelf life of phenolic resin is caused by the ease with which methylol groups included in the aromatic structure

G. R. Rao (✉) · I. Srikanth
Advanced Systems Laboratory, DRDO, Hyderabad, India
e-mail: ramaraogolla.asl@gov.in

K. Laxma Reddy
Department of Chemistry, National Institute of Technology, Warangal, India

M. R. Manohara Babu et al. (eds.), *Recent Advances in Composites Engineering and Research*, Springer Proceedings in Materials 39,
https://doi.org/10.1007/978-981-99-8807-5_34

can cross-link, resulting in variations in the components' characteristics as a function of storage time. Furthermore, phenolic resin produces volatile by-products during curing, resulting in significant volumetric shrinkage and void formation of the cured structures. The ultimate product will have porosity if pressure is not applied at the gel point of the resin. Furthermore, during the curing process, phenol and formaldehyde are released into the air, posing health risks. The development of novel ablative materials resistant to increased heat fluxes and temperatures is crucial. Two novel types of resins that can compete with phenolic resin for thermal protection system are cyanate ester and benzoxazine resins. These newer resins cure with addition polymerization, resulting in no by-products and minimal volumetric shrinkage. In addition to having attractive thermal stability and char yield, these resins have decreased moisture absorption and cure shrinkage [5–8].

K. N. Shiva Kumar et al. reported that carbon fiber–cyanate ester resin composite undergoes thermal cycling leading to reduced tensile strength and fracture strain and increased glass transition temperature [9]. Felix Abali et al. combined carbon fiber T 300 with Primaset PT-30 cyanate ester resin to create a carbon–carbon composite and then assessed its mechanical properties[10]. When S. P. Doherty et al. prepared carbon fiber–cyanate ester resin composites with high-temperature organo-modified nanoclay, they found that the addition of nanoclay improved the resin's thermal stability but only marginally increased the composites' short beam shear strength and decreased their open hole compression strength [11]. S. Rajesh Kumar et al. prepared glass fabric-reinforced bisphenol F benzoxazine (BZ) resin by mixing polyvinyl butyral (CA) and ethyl silicate (ES) in different weight ratios. It was reported that glass fabric-reinforced polybenzoxazine silicate composites (GBCE) had improved stiffness, cross-link density, service temperature, and network branching properties [12]. S. Rajesh Kumar et al. prepared ternary mixes of bisphenol F benzoxazine in different ratios to act as a hardener for polyethylene glycol diglycidal ether (EL) and bisphenol F novolac epoxy (ER) resins. Reported that, when compared to polybenzoxazine composites, these composites had better mechanical characteristics, hardness, and hydrolytic stability [13].

J. H. Koo et al. reported that comparing to silica fiber-reinforced phenolic resin composite, silica fiber-reinforced polysiloxane resin composite has a lower rate of erosion [14]. Hui Zhao et al. reported that silicon-containing polyborazine improved the mechanical properties, ablation, and thermal stabilities of carbon fiber/phenolic resin laminated composites. [15]. Donghwan Cho et al. investigated the ablation and thermal stability of four different types of carbon fiber-reinforced composites: phenolic, carbonaceous with impregnated resin, carbonaceous with pyrocarbon, and carbonaceous with phenolic matrix. The results of the ablation tests show that the composites with the highest ablation resistance are those that just have a carbonized matrix (1C/C) and those that have a carbonaceous matrix (1C/C + CVI) with pyrocarbon infiltrated using a CVI procedure [16].

The impact of different resin matrices, specifically benzoxazine, cyanate ester, and phenolic resin, on the mechanical, thermal stability, and ablation properties of carbon fiber-reinforced resin composites has not been studied before. This paper aims to close this gap. Therefore, the present study's objectives are as follows:

(i) Thermogravimetric analysis is utilized to characterize the thermal stability and char yield of benzoxazine, cyanate ester, and phenolic resins.
(ii) Fabrication of C-Ph composite, C-CE composite, and C-BZ composite and testing for mechanical and mass ablation properties.
(iii) To investigate how the mechanical characteristics, thermal stability, and ablation rate of carbon fiber-reinforced resin composites are affected by resin matrices.
(iv) The development of an alternative resin system to phenolic resin for carbon fiber-reinforced composite for thermal protection systems.

2 Materials

The benzoxazine (BZ) resin, which has a solid resin content of around 60 wt%, was purchased from M/s Anabond Limited, India. Resole type phenolic resin (Ph), which has a solid resin content of around 60 wt%, was purchased from M/s IVP Limited, India. Cyanate ester resin PT 30 S (CE) was acquired from M/s Lonza Pvt. Ltd., Switzerland, and contains approximately 70% solid resin. For reinforcement in composites, PAN-based carbon fabric (Torayca Carbon Fiber T 300, 3 K) was used because of its superior mechanical properties and the addition of an epoxy sizing agent.

3 Experimental Work

3.1 Preparation of Cured Neat Resin Systems

Neat liquid benzoxazine resin was used to cast a petri dish. The volatiles were then allowed to evaporate by heating the mixture to 60 °C for one hour. Benzoxazine resin was then allowed to cure in a hot air oven for one hour at 60 °C, one hour at 80 °C, one hour at 140 °C, three hours at 170 °C, two hours at 180 °C, and four hours at 200 °C. Clean liquid phenolic resin was put into a petri dish, and the mixture was heated to 60 °C for an hour to cause the volatiles to evaporate. The phenolic resin was then cured in a hot air oven at 90 °C for one hour, 120 °C for 1.5 h, 150 °C for one hour, and 170 °C for two hours. Neat liquid cyanate ester resin was poured onto a petri dish, and the volatiles were heated to 60 °C for an hour to cause them to evaporate. The cyanate ester resin was then heated to 120 °C for one hour, 150 °C for one hour, and 177 °C for three hours in a hot air oven before being cooled to room temperature. To achieve full polymerization, the cyanate ester was post-cured at 100 °C for 30 min, 150 °C for 30 min, 200 °C for 30 min, and 250 °C for three hours.

3.2 Preparation of Composites Using Carbon Fabric and Resin Systems

Carbon fiber–phenolic resin composite (C-Ph), carbon fiber–cyanate resin composite (C-CE), and carbon fiber–benzoxazine resin composite (C-BZ) were prepared for investigating how resin matrices affect properties of composites. Fabric layers that were 300 × 300 mm in size and 0.4 mm thick were cut. Fabric layers were impregnated with resin. By using a hand layup procedure, twelve of these layers were layered on a flat metal plate. Vacuum bagging and autoclave curing were used to these stacked layers. Under a pressure of five bars and a vacuum of one bars, curing was completed according to the same cure cycle as that described for neat resin, and then, it was cooled to room temperature. In order to achieve this, three different forms of carbon fabric-reinforced resin composites were created, with a resin content of around 35 wt%.

4 Characterization and Testing

4.1 Char Yield and Thermal Stability of Resins and Composites

To determine the char yield of neat resin systems and carbon fiber-reinforced resin composites, thermogravimetric analysis (TGA) was carried out. A sample mass of 10 mg was extracted from the composites and resin. During TGA experiments, these samples were heated at a rate of 10 °C/min to 900 °C while being continuously supplied with nitrogen at a flow rate of 60 ml/min. The temperature (T10%/°C) at which a 10% weight loss occurs and carbonaceous char (CR) were observed. Equation (1) was used to calculate oxidation index (OI) of resins and composites from the char yield (CR) [17–19].

$$\mathrm{OI} \times 100 = 1.74 \times 0.4\ \mathrm{CR} \tag{1}$$

4.2 Physical Properties of Composites

The physical characteristics of the composites, specifically their densities and fiber volume fractions, were determined using ASTM D 3171 and ASTM D 792, respectively.

4.3 Interlaminar Shear Strength (ILSS) and Flexural Strength of Composites

Samples from carbon fiber-reinforced resin composites measuring 100 mm by 10 mm by 4 mm were used for the flexural strength test, while 40 mm by 10 mm by 4 mm test samples were used for the ILSS test. For the ILSS test, ASTM D 2344 was used, while ASTM D 790 was used for the flexural strength test. The M/s ADMET Model 2505 UTM was used to conduct the tests. There were six samples examined for every kind of composite. The standard deviation and average value were calculated for every composite.

4.4 Oxyacetylene Torch Test of Composites

Oxyacetylene torch test was carried out on composite samples for evaluation for mass ablation rate as per ASTM E 285. For testing, samples measuring 100 mm by 100 mm by 4 mm were utilized. By adjusting the flame source to sample distance, heat flux was adjusted to the required level. A 500 W cm-2 heat flux was applied for 60 s to one set of samples, whereas a 125 W cm-2 heat flux was applied for the same amount of time to the other group. For ablated composites, the mass ablation rate was determined using mass loss per second. Two samples at each flow were tested in order to estimate the average ablation rate.

4.5 Microstructure Characterization of Ablated Composites

A scanning electron microscope (Table Top SEM, Make: SEC, Korea, Model: SN 4500 Plus) was used to characterize the microstructure alterations in composite samples that were subjected to an oxyacetylene torch test.

5 Results and Discussion

5.1 Char Yield and Thermal Stability of the Resins and Composites

Figure 1 shows TGA scans for neat cyanate ester resin, benzoxazine resin, and phenolic resin. As determined by TGA of resins, characteristic temperatures, namely T10 percent/°C, and char yield are given in Table 1.

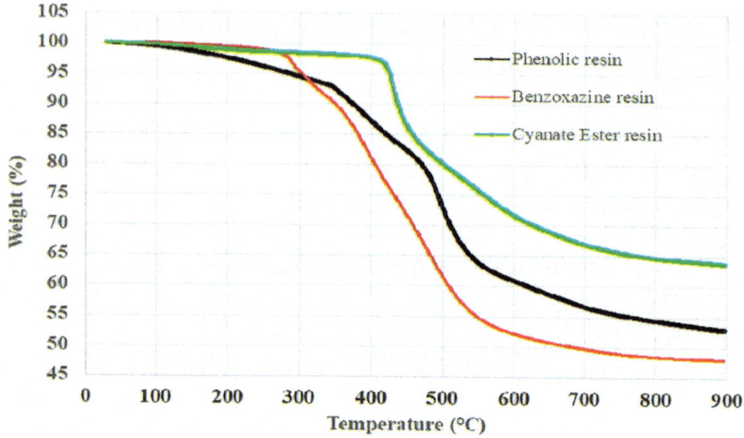

Fig. 1 TGA scans for CE resin, Ph resin, and BZ resin

Table 1 Characteristic temperatures and char yield of resins

Resin description	$T_{10\%}$/°C	Char yield at 900 °C/%	Oxidation index (OI)
Cyanate ester resin (CE)	436	64	44.54
Phenolic resin (Ph)	370	53	36.88
Benzoxazine resin (BZ)	350	48	33.41

Figure 1 shows that cyanate ester resin degrades in a single step at 430 °C, indicating a homogeneous decomposition of the resin. Char yield of cyanate ester is about 64 wt% at 900 °C. Phenolic resin shows three distinct decomposition reactions that occur at 280, 380, and 495 °C, and at 900 °C, the phenolic char production is roughly 53 weight percent. Benzoxazine resin exhibits three different breakdown reactions at temperatures of 295, 400, and 480 °C. At 900 °C, the benzoxazine char yield is approximately 48 weight percent. Ascending order for char yield of three resin systems is benzoxazine resin, phenolic resin, and cyanate ester resin. Table 1 displays the values of the oxidation index (OI) for three resins. The material under research has greater thermal stability when the oxidation index is higher [17]. Cyanate ester resin is more thermal stable than phenolic resin and benzoxazine resin.

Figure 2 shows TGA scans for composites. Characteristic temperatures, namely T10 percent/°C, and char yield of composites are presented in Table 2. Composites' 10% weight loss and char yield values have risen relative to their neat resin. This is because of composites contain 60–70 wt% of carbon fiber. C-CE composite has higher thermal stability than C-Ph and C-BZ composites.

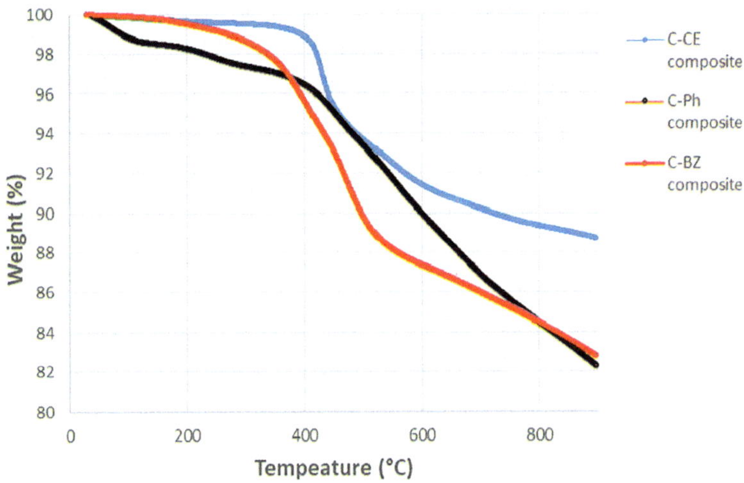

Fig. 2 TGA scans for C-CE, C-Ph, and C-BZ composites

Table 2 Thermal stability of composites

Composite	$T_{10\%}$/°C	Char yield at 900 °C/%
C-CE	721	88.72
C-Ph	580	82.27
C-BZ	480	82.81

5.2 *Density, Fiber Volume Fraction of Composites*

In comparison with the C-CE and C-BZ composites, Table 3 demonstrates that the C-Ph composite has a somewhat higher density and fiber volume fraction. When compared to cyanate resin and benzoxazine resin, phenolic resin has low viscosity. Resin was squeeze out in the fabrication of C-Ph composite and resulting in high fiber volume fraction [20].

Table 3 Physical properties of composites

Composite	Density (g/cc)	Fiber volume fraction (%)
C-CE	1.51	60
C-Ph	1.52	61
C-BZ	1.50	59

Table 4 ILSS and flexural strength (FS) of composites

Composite	ILSS (MPa)	FS (MPa)
C-CE	30 (1.1)	535 (11.1)
C-Ph	21 (0.81)	485 (15.58)
C-BZ	20	480 (3.4)

[a] Bracket values indicate standard deviation of tested values

5.3 ILSS and Flexural Strength of Composites

In Table 4, the values for the composite's ILSS and flexural strength (FS) are shown. ILSS of C-CE composite is about 50% higher than C-Ph and C-BZ composites. This is due to cyanate group (–OCN) present in the cyanate ester resin might have reacted with epoxy resin (sizing) present on the carbon fiber and may have formed covalent bond with it. This shows that cyanate ester is more compatible with PAN-based carbon fabric than phenolic resin and benzoxazine resin. Flexural strength of C-CE composite is about 10% higher than C-Ph and C-BZ composites. This is because the CE resin system and the carbon fiber's sizing portion have strong chemical bonds that allow the carbon fiber to withstand flexural loads. Therefore, more chemical linkages increase the composites' overall strength [21–23].

5.4 Ablative Performance of Composites

Table 5 shows the ablation rate of composites.

The mass ablation rates of C-CE composite are showing lower than C-Ph and C-BZ composites at both heat fluxes 125 W/cm^2 and 500 W/cm^2. The oxyacetylene torch ablation involves numerous simultaneous degrading steps of composites. Equations (2)–(4) state that endothermic pyrolysis of the resin matrix into char and other gaseous products is one kind of chemical composite breakdown [8, 24, 25]. The removal of the char and fibers is known as mechanical erosion.

$$\text{Phenolic resin} \rightarrow \text{C(Char)} + CO_2 + H_2O + CO + CH_4 - \text{Heat} \quad (2)$$

Table 5 Ablation rate of composites

Composite description	Mass ablation rate (g/sec) at heat flux: 125 Wcm^{-2}	Mass ablation rate (g/sec) at heat flux: 500 Wcm^{-2}
C-CE	0.084	0.123
C-Ph	0.100 (19%)	0.138 (12%)
C-BZ	0.115 (37%)	0.147 (19%)

[a] Bracket values indicate % of increase in mass ablation rate with respect to C-CE composite

$$\text{Cyanate ester resin} \rightarrow \text{C(Char)} + CO + CO_2 + HOCN + NH_3 - \text{Heat} \quad (3)$$

$$\text{Benzoxazine resin} \rightarrow \text{C(Char)} + CO + CO_2 + NH_3 - \text{Heat} \quad (4)$$

The microstructure of the surface that has been ablated provides information about the mechanisms used during the ablation. The ablated surface's microstructure and ablation mechanisms are explored in the following sections.

5.5 *Microstructural Behavior of Ablated Surfaces of Composites*

125 W/cm^2 and 500 W/cm^2 heat flux tests on the ablated surface's microstructures in Fig. 3 shed light on the methods employed during the ablation.

As can be seen in Fig. 3a, ablation happens in the case of C-Ph composite at low heat fluxes of 125 W/cm^2, whereas fiber breaking is clearly visible at high heat fluxes of 500 W/cm^2 (as indicated with arrows in Fig. 3d). As a result, when the C-Ph composite was evaluated at a heat flux of 500 W/cm^2, the ablation rate increased slightly. There is no discernible fiber degradation in the C-CE composite when assessed at a heat flux of 125 W/cm^2 (Fig. 3b). The areas that are indicated in Fig. 3b demonstrate that there are still some matrix traces on the surface of the ablated composite. The ablative patterns of C-CE at the greater flux of 500 W/cm^2 are distinct from those at the low flux of 125 W/cm^2, as can be observed in Fig. 3b, e. At a high flux of 500 W/cm^2, it is evident that the matrix and fibers have been severely

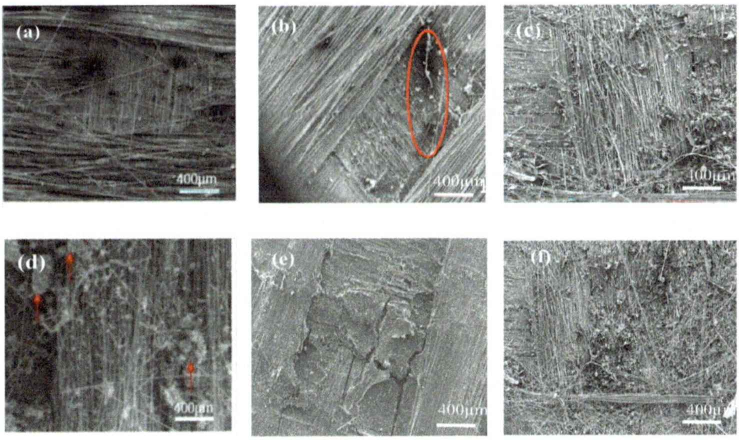

Fig. 3 Microstructure of samples tested at heat flux 125Wcm^{-2} **a** C-Ph, **b** C-CE, **c** C-BZ and samples tested at heat flux 500 Wcm^{-2} **d** C-Ph, **e** C-CE, **f** C-BZ

damaged (Fig. 3e). There is no visible fiber degradation in the C-BZ composite assessed at a heat flux of 125 W/cm^2, and Fig. 3c illustrates remnants of the burned matrix holding the fibers together. Large chunks of tows were damaged for the C-BZ composite in Fig. 3f at high heat flux of 500 W/cm^2.

6 Conclusions

Three distinct resin matrices were used to investigate the mechanical, thermal, and ablative capabilities of carbon fiber-reinforced resin composites, namely phenolic resin, cyanate ester resin, and benzoxazine resin. Summary of the research findings are given below:

- Of the three resin systems examined, cyanate ester resin and carbon fiber–cyanate resin composites have higher thermal stability and char yield than phenolic resin, benzoxazine resin, and their carbon fiber composites.
- Compared to carbon fiber–phenolic resin and carbon fiber–benzoxazine resin composites, the interlaminar shear strength of the cyanate ester composite is almost 50% greater. Compared to C-Ph and C-BZ composites, the flexural strength of C-CE composite is roughly 10% higher.
- Compared to C-Ph and C-BZ composites, the mass ablation rates of C-CE composites are lower when tested at both heat fluxes of 125 and 500 W/cm^2.
- Carbon fiber-reinforced cyanate ester resin composite is the better ablative composite when compared to traditional carbon fiber-reinforced phenolic resin composite.

References

1. D'Aleio GF, Parker JA (1971) Ablative plastics. Marcel Dekker, New York
2. Cho D, Lee JY, Yoon BI (1993) Microscopic observations of the ablation behaviours of carbon fibre/phenolic composites. J Mater Sci Lett 12:1894–1896
3. Schmidt DL (1971) Ablative polymers in aerospace technology. In: D'Alelio GF, Parker JA (eds) Ablative plastics. Marcel Dekker, New York, pp 1–39
4. Pulci G, Tirillò J, Marra F, Fossati F, Bartuli C, Valente T (2010) Carbon-phenolic ablative materials for re-entry space vehicles: manufacturing and properties. Compos: Part A 41(10):1483–1490
5. Hamerton I, Hay JN (1998) Recent developments in the chemistry of cyanate esters. Polym Int 47:465–473
6. Reghunadhan Nair CP, Mathew D, Ninan KN (2000) Cyanate ester resins, recent developments. Adv Polym Sci 155:1–80
7. Mangalgiri PD (2005) Polymer-matrix composites for high-temperature applications. Def Sci J 55(2):175–193
8. Ramirez ML, Walters R, Lyon RE, Savitski EP (2002) Thermal decomposition of cyanate ester resins. Polym Degrad Stab 78:73–82

9. Shivakumar K, Chen H (2009) Effect of thermal fatigue on tensile and flexural properties of carbon/cyanate ester pultruded composite. J Reinf Plast Compos 28:675–689
10. Abali F, Shivakumar K, Hamidi N, Sadler R (2003) An RTM densification method of manufacturing carbon-carbon composites using Primaset PT-30 resin. Carbon 41:893–901
11. Doherty SP, Takimori M, Deitzel JM, Heider D, Gillespie Jr JW, Shah A, Giaya A (2006) Thermal degradation of carbon fiber/cyanate ester resin composites filled with clay silicate nanoparticles. Presented at a meeting of the Thermoset Resin Formulators Association at the HyattRegency Montreal in Montreal, Quebec, Canada, September 11 through 12, 2006
12. Rajesh Kumar S, Krishna Mohanb S, Dhanasekaran J (2018) Novel glass fabric reinforced polybenzoxazine-silicate composites along with polyvinyl butyral for high service temperature applications. New J Chem 19:15499–16386
13. Rajesh Kumar S, Krishna Mohanb S, Dhanasekaran J (2015) Epoxy benzoxazine based ternary systems of improved thermo-mechanical behavior for structural composite applications. RSC Adv 5:3709–3719
14. Schellhase KJ, Koo JH, Buffy J, Brushaber R (2017) Development of new thermal protection systems based on silica/polysiloxane composites. In: 58th AIAA/ASCE/AHS/ASC structures, structural dynamics, and materials conference. https://doi.org/10.2514/6.2017-1367
15. Zhao H, Chen L, Yun J, Tang L, Wen Z, Zhang X, Gu J (2018) Improved thermal stabilities, ablation and mechanical properties for carbon fibers/phenolic resins laminated composites modified by silicon-containing polyborazine. Eng Sci 2:57–66
16. Cho D, Yoon BI (2001) Microstructural interpretation of the effect of various matrices on the ablationproperties of carbon-fibre-reinforced composites. Compos Sci Technol 61:271–280
17. Pashaei S, Siddaramaiah, Syed AA (2010) Investigation on thermal, mechanical and morphological behaviours of organo nanoclay incorporated epoxy nanocomposites. ARPN J Eng Appl Sci 5:76–86
18. Jeevananda T (2001) Siddaramaiah, Thermal and morphological studies on ethylene-vinyl acetate copolymer–polyaniline blends. Thermochemica Acta 376:51–61
19. Kumar H, Anil Kumar A, Siddaramaiah (2006) Physico-mechanical, thermal and morphological behaviour of polyurethane/poly (methyl methacrylate) semi-interpenetrating polymer networks. Polym Degrad Stab 91:1097–1104
20. Asaro L, Rivero G, Manfredi LB, Alvarez VA, Rodriguez ES (2016) Development of carbon fiber/phenolic resin prepregs modified with nanoclays. J Compos Mater 50(10):1287–1300
21. Zhao Y, Wang K, Zhu F, Xue P, Jia M (2006) Properties of poly(vinyl chloride)/wood flour/ montmorillonite: effect of coupling agents and layered silicate. Polym Degrad Stab 91:2874–2883
22. Wu Q, Lei Y, Yao F, Xu Y, Lian K (2007) Properties of HDPE/clay/wood nanocomposites. J Plast Technol 27:108–124
23. Lei Y, Wu Q, Clemons CM, Yao F, Xu Y (2007) Influence of nanoclay on properties of HDPE/ wood composites. J Appl Polym Sci 106:3958–3966
24. Bessire BK, Lahankar SA, Minton TK (2014) Pyrolysis of phenolic impregnated carbon ablator (PICA). Appl Mater Interfaces 7:1383–1395
25. Hemvichian K, Kim HD, Ishida H (2005) Identification of volatile products and determination of thermal degradation mechanisms of polybenzoxazine model oligomers by GC-MS. Polym Degrad Stab 87(2):213–224

SPRINGER NATURE

GPSR Compliance

The European Union's (EU) General Product Safety Regulation (GPSR) is a set of rules that requires consumer products to be safe and our obligations to ensure this.

If you have any concerns about our products, you can contact us on ProductSafety@springernature.com

In case Publisher is established outside the EU, the EU authorized representative is:

Springer Nature Customer Service Center GmbH
Europaplatz 3
69115 Heidelberg, Germany

The manufacturer's authorised representative in the EU is Springer Nature Customer Service Centre GmbH, Europaplatz 3, 69115 Heidelberg, Germany. If you have any concerns regarding our products, please contact ProductSafety@springernature.com

Printed and bound by CPI Group (UK) Ltd, Croydon, CR0 4YY
01/10/2026
02225902-0013